Software Defined Chips

Shaojun Wei · Leibo Liu · Jianfeng Zhu ·
Chenchen Deng

Software Defined Chips

Volume I

Science Press
Beijing

Springer

Shaojun Wei
School of Integrated Circuits
Tsinghua University
Beijing, China

Leibo Liu
School of Integrated Circuits
Tsinghua University
Beijing, China

Jianfeng Zhu
School of Integrated Circuits
Tsinghua University
Beijing, China

Chenchen Deng
Beijing National Research Center for
Information Science and Technology
Tsinghua University
Beijing, China

ISBN 978-981-19-6993-5 ISBN 978-981-19-6994-2 (eBook)
https://doi.org/10.1007/978-981-19-6994-2

Jointly published with Science Press
The print edition is not for sale in China mainland. Customers from China mainland please order the print book from: Science Press.

This Springer imprint is published by the registered company Springer Nature Singapore Pte Ltd.
The registered company address is: 152 Beach Road, #21-01/04 Gateway East, Singapore 189721, Singapore

Contents

Chapter 1
Introduction

We are firmly convinced that when a special purpose configuration may be accomplished using available facilities, a new level of inventiveness will be exercisable.
—Gerald Estrin, Western Joint Computer Conference, 1960.

In recent years, with the rapid social and technical development, the demand for high performance, energy efficiency and flexibility of computing chips has been growing. A large number of emerging applications have far greater demand for computing power than before. In the past few decades, the technological advancement of integrated circuits is an important measure to improve the capability of computing architectures. However, this measure is gradually ineffective as the Moore's Law and Dennard scaling law slow down or even come to an end. The occurrence of the well-known power wall makes the power constraints on integrated circuits tighter in many applications. The performance gains brought by the technological advancement of integrated circuits are getting smaller and smaller, which severely limits the computing power of the hardware architecture. Therefore, computer architects have to divert their attention from performance to energy efficiency. Meanwhile, the flexibility has also become a design consideration that cannot be ignored. With the springing up of emerging applications, increasing requirements, rapid advancement of technological capabilities, and increasingly rapid software upgrades, hardware implementations that cannot adapt to software changes will face short life cycles and high non-recurring engineering (NRE) costs. Overall, energy efficiency and flexibility have become the most important evaluation criteria for computing architectures.

However, for mainstream computing architectures, meeting all these new requirements is extremely challenging. Specifically, application-specific integrated circuits (ASICs) can achieve high energy efficiency but have insufficient flexibility; von Neumann processors, such as general purpose processors (GPPs), graphics processing units (GPUs), and digital signal processors (DSPs) are flexible enough but have low energy efficiency. Field programmable gate arrays (FPGAs) have been widely used in communications, networking, aerospace, national defense and other fields due to their ability to customize large-scale digital logic and to quickly develop

S. Wei et al., *Software Defined Chips*,
https://doi.org/10.1007/978-981-19-6994-2_1

prototypes. However, their intrinsic properties such as single-bit programming granularity and static configuration have caused problems such as low energy efficiency, limited capacity, and low usability, which cannot meet the ever-increasing application requirements. In recent years, FPGAs have received continuous technological upgrades by increasing the hardware scale, adopting heterogeneous computing and high-level language programming. However, due to their intrinsic properties, the above-mentioned problems have not been addressed radically. Without fundamental changes in the infrastructure, the future of FPGAs has several limitations. Adopting the mixed-grained programming dominated by coarse-grained programming and dynamic configuration, the software defined chips (SDCs) can fundamentally solve the above-mentioned technical problems that restrict the development of FPGAs, and meet the requirements for energy efficiency and flexibility simultaneously. The mixed-grained programming can greatly reduce resource redundancy and improve energy efficiency of chips. The dynamic configuration allows lifting restrictions on the hardware capacity by time-division multiplexing, improving programmability and usability by using high-level languages. The Defense Advanced Research Projects Agency (DARPA) of the United States invested 71 million dollars in the Electronics Resurgence Initiative (ERI) in 2018, forming the top teams in the United States to carry out joint research on the Software-Defined Hardware (SDH). The Horizon 2020 programme of the European Union (EU) has also attached great importance to this direction and provided sustained R&D support. The SDC has become an indispensable research direction of strategic importance for world powers.

This chapter first introduces the conceptual evolution of SDCs in the context of the development of integrated circuits and computing architectures, then discusses the innovation of SDCs from the development, principles and problems faced by traditional programmable devices, and finally introduces the current research of traditional programmable devices and SDCs.

1.1 Concept and Background

The SDC is a novel chip design paradigm (including the hardware and software). Figure 1.1 shows a summary of the comparison between chip architecture design paradigms. Specifically, the design paradigm of GPPs such as central processing units (CPUs), GPUs and DSPs is software programming. Application algorithms are converted to impersive instructions or long instructions, single-instruction multiple-data (SIMD) instructions etc. Instructions and corresponding pipelining, scheduling, and control mechanisms are executed to physically implement application functions. Since the design mode cannot well reflect hardware features and therefore the hardware advantages cannot be fully exploited which means this computing pattern is inefficient. The design paradigm of traditional programmable logic devices is hardware programming. Hardware resources are abstracted to several spatially distributed

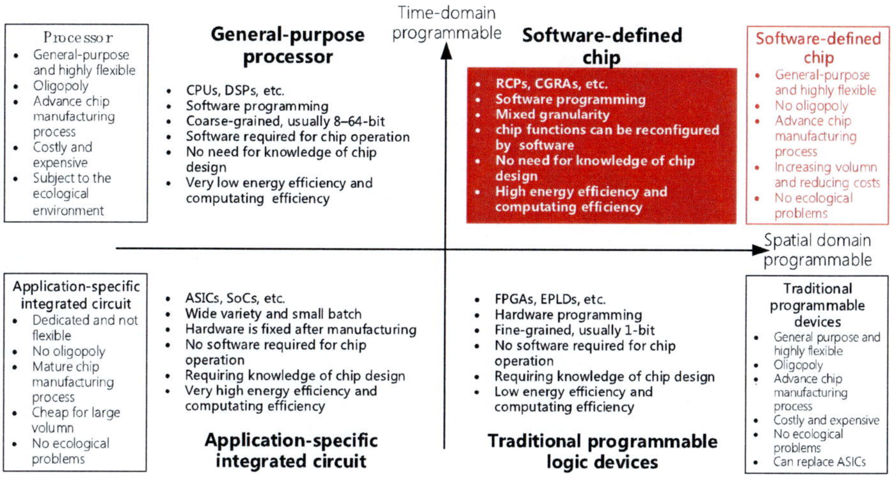

Fig. 1.1 Comparison of chip architecture design paradigms (CGRA refers to coarse-grained reconfigurable architecture; SoC refers to system on a chip; EPLD refers to erasable programmable logic device.)

functional clusters, and the combination of these clusters can fulfill different functional requirements. Since no abstraction of hardware features is included in software, the usability in this pattern is poor. Figure 1.2 summarizes the problems behind the chip architecture design paradigm. Software implementations are sequential and allow temporal switching with high flexibility. Hardware implementations are parallel and can be spatially extended with high performance and efficiency. The software implementations represented by GPPs ignore the spatial parallelism of hardware, while the hardware implementations represented by traditional programmable devices do not exploit the TDM to reuse hardware resources. In short, GPPs can be regarded as "software implemented by hardware", while traditional programmable devices can be seen as "user-configurable hardware".

The SDC, as a new chip architecture design paradigm, is aimed at removing the barrier between software and hardware, and directly defining the hardware's functions and rules using software at runtime, so that hardware functions can be changed in real time as software changes (hardware allows not only switching functions continuously in the time domain but also programming circuit functions in the spatial domain). Moreover, it permits real-time functional optimization, with full consideration given to the high efficiency of hardware and the high flexibility of software [1].

Similar to the concepts of software-defined radio (SDR) and software-defined network (SDN), the SDC reflects the concept of software-defined everything (SDE). The underlying motivation behind these concepts is that the development of the information society is much faster than the updates of many underlying facilities, and the infrastructure is expected to be flexible so that they can be updated less frequently, thus avoiding the huge design and cost overhead associated with their updates. For

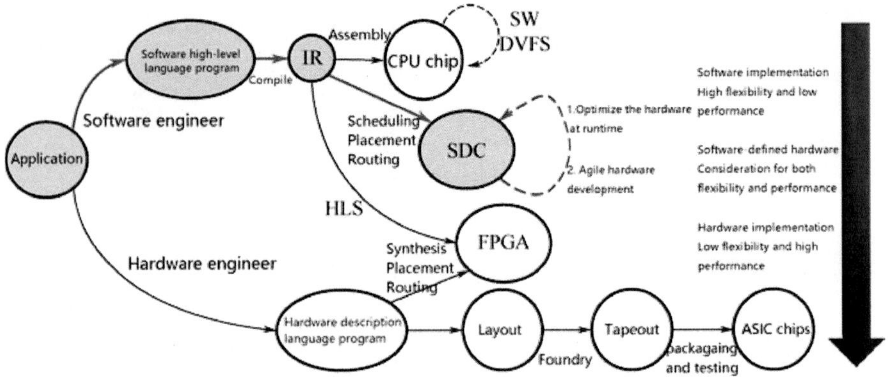

Fig. 1.2 Problems behind the chip architecture design paradigm

example, the SDR has been popular since the 1990s. That is a time of rapid development of mobile communications, when multiple digital wireless communication standards such as the Global System for Mobile Communications (GSM) and the code-division multiple access (CDMA) coexisted but were not compatible with one another. It is desired to convert the signals in different frequency bands into digital signals that can be processed by software in a unified manner, thus keeping up with the development of wireless technologies. The OpenRAN concept that has become popular in recent years, is essentially a software-defined radio access network (RAN), an extension of the SDR and SDN. With the rapid development of 5G technology, there are fewer manufacturers in the market who can adapt to new technologies and standards and make sustained investments in R&D, and OpenRAN is a way to cope with this situation and has received support from many countries (including the United States) and organizations (such as O-RAN Alliance).

The emergence of SDCs is a reflection of the fact that the development of integrated circuits and computing architectures fails to suit social development needs. The following sections describe the background, reasons and inevitability of the emergence of SDCs from these two perspectives.

1.1.1 Development Background of Semiconductor Integrated Circuits

The development of integrated circuit fabrication processes has increased the integration capabilities of silicon chips. Since the late 1970s, the integrated circuit process based on metal-oxide semiconductor (MOS) transistors has become a popular choice, beginning with the design of N-channel metal oxide semiconductor (NMOS) circuits and then that of complementary metal-oxide-semiconductor (CMOS) circuits. The amazingly rapid development of the integrated circuit technology has become a key factor in improving chip performance. In 1965, Gordon Moore predicted that the number of transistors in an integrated circuit would double every year. In 1975, he revised the forecast to roughly doubling every two years, still an astonishing prediction that has thus far proved accurate [2]. However, the improvement of integration began to slow down around 2000, which differed from Gordon Moore's prediction by more than 15 times in 2018. The difference has a tendency to increase, as shown in Fig. 1.3a [3]. Moore's Law is gradually failing and the prediction is getting closer to the physical limit [4]. Meanwhile, another prediction made by Robert Dennard (Dennard scaling), which states that as transistors get smaller their power density stays constant [5], has completely broken down around 2012, as shown in Fig. 1.3b [3]. This makes the power wall a critical problem for GPPs and energy efficiency a key metric for designing integrated circuits.

In addition, the development of integrated circuit fabrication processes has brought about the increase in design and manufacturing costs. As shown in Fig. 1.3c, the increasing cost of fabrication and design has resulted in fewer and fewer companies with the state-of-the-art chip R&D capabilities in the industry. The number of companies that now are able to follow the advance 5 and 7 nm processes is only a few such as Apple, Intel, Qualcomm, and NVIDIA (note that they all design general-purpose chips). At the same time, the one-time investment in chips increases and requires mass production to compensate the cost. Therefore, the chip design needs to balance the requirements for increasingly high performance and sufficient flexibility wherever possible.

1.1.2 Development Background of Computing Architectures

In the modern information society, a large number of emerging applications, such as neural networks, cloud computing, and bioinformatics, have an increasing demand for computing power. As described in Sect. 1.1.1, the development of integrated circuit processes has brought steady improvements in performance, efficiency, and capacity over the last few decades. In contrast, the development of computer architectures has brought much smaller advances. Nowadays, the design of mainstream architectures of computing chips such as CPUs, GPUs, and DSPs is still based on von Neumann structures with generic and Turing-complete computational processes controlled by instruction streams. However, to ensure flexibility, the GPP is designed to contain a large number of non-arithmetic logic units such as instruction fetch and instruction decode, so it has considerable additional energy overhead during computation and low energy efficiency for computation. For example, a typical CPU fabricated with 14 nm process, the Xeon E5-2620, out-of-order execution (OoOE) 4-way multiple-issue processor, needs to consume 81 nJ energy to complete a single iteration of the Smith-Waterman algorithm (for genetic analysis). In contrast, the domain-specific processor, Darwin fabricated with 40 nm process requires only 3.1 pJ energy to complete an iteration, of which only 0.3 pJ is consumed by computation and 2.8 pJ is consumed by memory access and control [6]. It can be seen that for a modern microprocessor, the actual power consumption of computating resources is

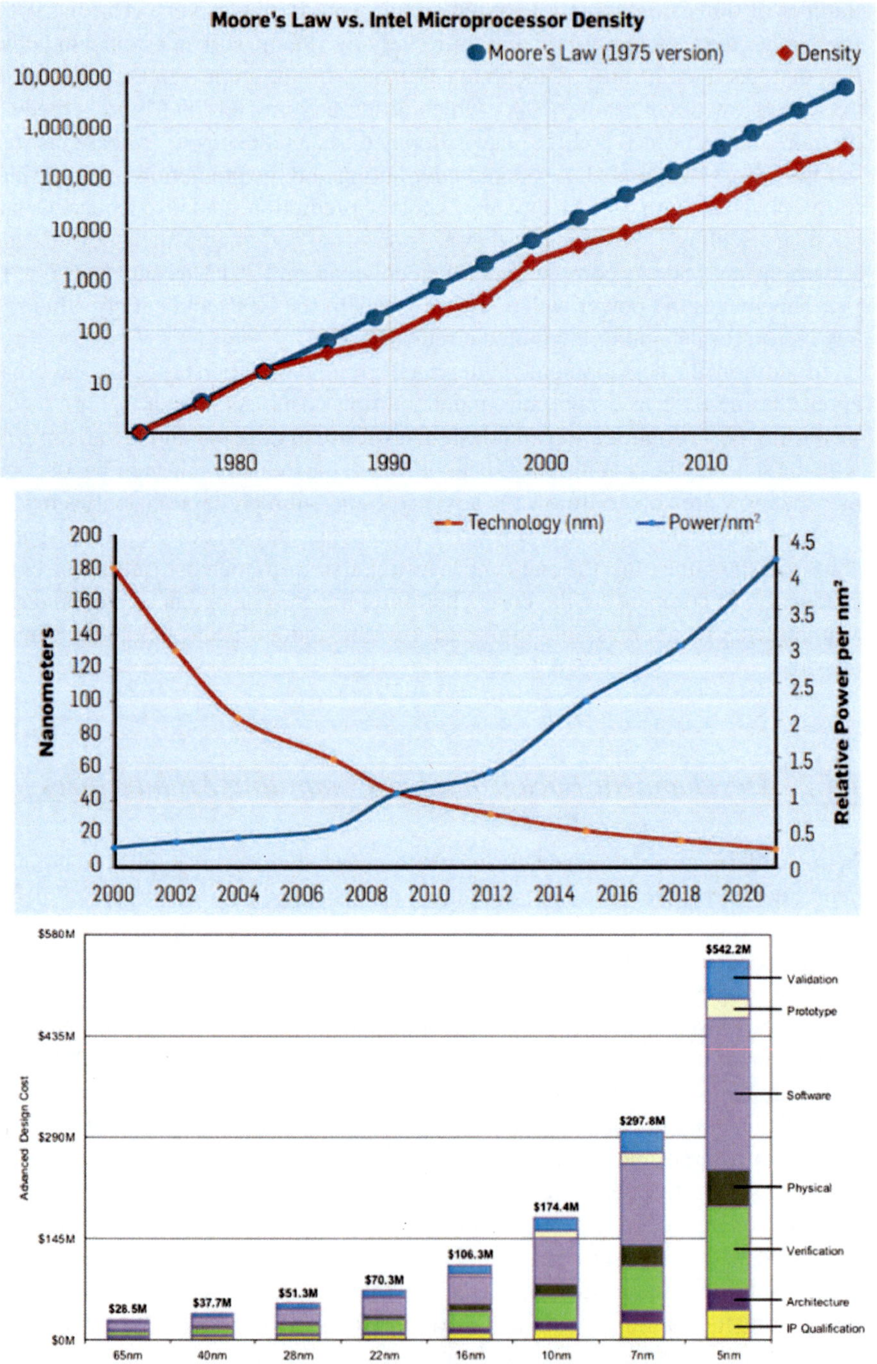

Fig. 1.3 Development of integrated circuits (see the color illustration)

very low, even less than 1% of the total, and a large amount of power is consumed by instruction fetch, decode, and reordering. The additional energy overhead exacerbates the power wall problem of today's GPPs. Specifically, the thermal management of integrated circuits determines the upper limit of power consumption; the end of Dennard scaling means that the highest operating frequency can hardly be increased and the era of continuously improving the chip performance using new processes is coming to an end; considerable additional computating overhead means that the efficiency is low. The advance of process can no longer improve the computing power as before, and therefore it is quite urgent to design new computing chip architectures.

The Makimoto's Wave is a summary and prediction of the development of computing chips. In 1987, Tsugio Makimoto, previously chief engineer of Hitachi, proposed that semiconductor products would cyclically alternate between standardization and customization every ten years, meaning that the chip design paradigm would swing between flexibility (generality) and efficiency (specificity). In 1991, he officially published this idea in the *Electronics Weekly* as the Makimoto's Wave. It has been proven correct by the rapid development of programmable chips in recent years, thus receiving positive responses from numerous programmable device companies and exerting a wide impact [7]. Professor Hartenstein of the University of Kaiserslautern, Germany, further referred to it as Makimoto's Law. He revised the fifth wave (1997–2007) as an inefficient field-programmable general-purpose computing architecture and the sixth wave (2007 onwards) as a long-wave architecture featuring coarse-grained reconfigurable computing [8]. Professor Hartenstein believed that Makimoto's Law could become an industrial law like Moore's Law, guiding the continued rapid development of integrated circuit technology as the feature size of integrated circuits is approaching its physical limit. On the basis of the Makimoto's Wave, Xu Juyan, academician of the Chinese Academy of Engineering (CAE) proposed the semiconductor cycle [9], or Xu's Cycle, in 2000, as shown in Fig. 1.4. He pointed out that the mainstream computing architecture from 2008 to 2018 was SoCs and that from 2018 to 2028 should be highly efficient programmable SoCs, that is, user reconfigurable SoCs (U-rSoCs).

The concept of SDC follows the trend from 2018 to 2028 predicted by the Makimoto's Wave and Xu's Cycle. It is the further development of the concept of general-purpose programmable computing, from software programming of CPUs to hardware programming FPGAs, and finally to SDCs that will improve the programming efficiency and effectiveness through the innovation of the chip architecture design. SDC will be the mainstream trend of the future chip design.

1.1.3 SDCs Versus Programmable Devices

SDCs mean that the function of hardware are defined directly by software at runtime, while programmable devices refer to the chips that hardware functions can be reconfigured after fabrication. The concept of programmable devices is broader and was proposed much earlier. There are two major application sceanarios for traditional

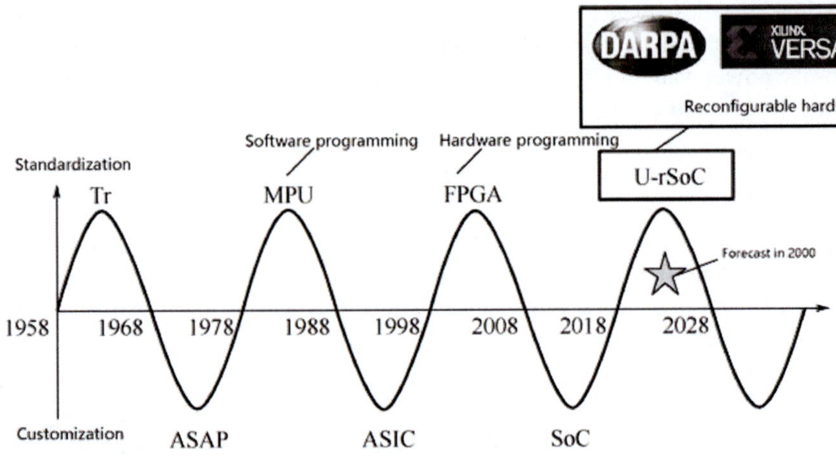

Fig. 1.4 Xu's cycle

programmable devices. One is for verification of digital circuits. The other one is for application accelerations by replacing ASICs, with programmable algorithms and relatively low overall cost. SDCs are mainly used for the acceleration for computate- and data-intensive applications. Therefore, SDCs can be considered as a novel type of programmable devices. It takes much less time to reconfigure hardware after fabrication. The runtime reconfiguration means thatthe computing capacity is expanded significant At the same time, SDCs sacrifice some flexibility which means that they can hardly be used for verification. However, the high efficiency in performing compute- and data-intensive algorithms is achieved in return.

1.1.4 SDCs Versus Dynamically Reconfigurable Computing

The concept of reconfigurable computing was proposed before the concept of programmable devices. In the 1960s, Professor Gerald Estrin from the University of California, Los Angeles proposed a special reconfigurable computing hardware architecture capable of receiving external control signal and forming hardware to accelerate specific computational tasks by means of tailoring and reorganization, which was the earliest design concept of reconfigurable computing [10]. However, limited by manufacturing and design level at the time, the concept of reconfigurable computing had not gained much attention from academia and the industry until the 1990s. In 1999, Professor André DeHon and John Wawrzynek from the research group of Reconfigurable Architectures, Systems, and Software (BRASS), University of California, Berkeley, explicitly defined reconfigurable computing as a computating structure with the following characteristics [11]: (1) Chips can be customized to solve various problems after fabrication; (2) Computations can be performed by spatially mapping the tasks to the chips. Based on the first characteristic, reconfigurable computing is a type of programmable devices. So what is the difference between reconfigurable computing and SDCs?

Programmable devices such as FPGAs are sometimes considered to be static reconfigurable computing. However, the reconfigurable computing mentioned in this book mainly refers to the dynamically reconfigurable architecture represented by CGRAs. Dynamically reconfigurable computing can change hardware functions at runtime, while static reconfigurable computing cannot do that. Thus, there is an intersection between dynamically reconfigurable computing and SDCs. Dynamically reconfigurable computing focuses on function reconfiguration, while SDCs go a step further by requiring hardware functions to be defined by software and to be optimized at runtime, thus achieving greater flexibility and efficiency. Dynamically reconfigurable computing is an inevitable choice for the underlying hardware structure of SDC, but not all dynamically reconfigurable computing can be regarded as SDC.

Figure 1.5 shows the evolution of the SDC and briefly describes the relationship among the SDC, programmable devices and dynamically reconfigurable computing. The programmability has increased significantly, from the oldest programmable logic array (PLA), to generic array logic (GAL), and to FPGAs. The high level synthesis (HLS) tool further increases the software programmability of FPGAs, enabling programming in high-level languages, but the programming efficiency is still low. Run-time reconfiguration (RTR) accelerates the hardware reconfiguration of FPGAs and changes the completely static reconfiguration to partially dynamic reconfiguration, but the reconfiguration still takes a long time. The CGRA has become one of the most important carriers for implementing SDCs, with significant increases in software programmability and hardware reconfiguration speed.

Research on dynamically reconfigurable architectures represented by CGRAs in the past 30 years can be roughly divided into three phases.

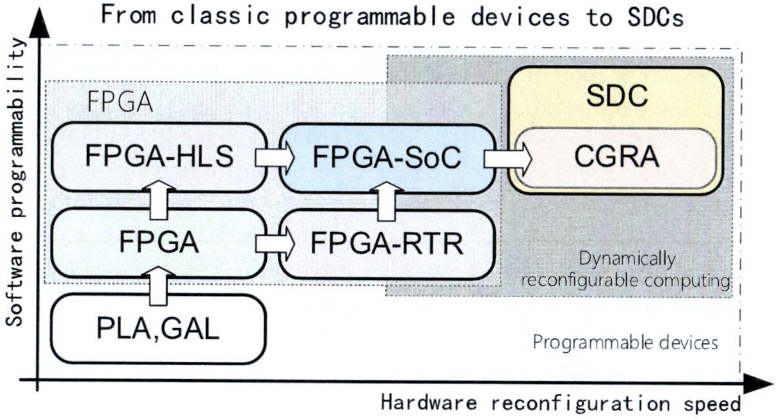

Fig. 1.5 Evolution of the SDC concept

1. Starting stage(1990–2000)

In the early 1990s, fine-grained reconfigurable devices like FPGAs gradually started to develop into the communication and even general-purpose computing domain; however, the single-bit programming caused problems such as long compilation time and considerable configuration switching overhead. Researchers started to conduct research on coarse-grained reconfigurable architectures and focused on the design of hardware architectures, exploring the architecture-level design space in multiple dimensions like processing elements (PEs), interconnect structures, and on-chip memory.

PEs of early dynamically reconfigurable architectures differed significantly in reconfigurable PE functions and datapath granularity depending on target applications. Dynamically reconfigurable architectures like MATRIX [12], REMARC [13], and PipeRench [14] use static scheduling. That is, the sequence of instruction execution and data transfer of PEs is compiled statically by the compiler. The hardware architecture is designed to be simple and straightforward, with high computational performance and energy efficiency for regular applications. On-chip interconnects are essentially different from CPUs. Specifically, interconnects between CPU instructions exchange data through memory, while computing instructions in CGRAs exchange data via on-chip interconnection after being mapped to different PE in the array. This design has a significant impact on the performance of the computating architecture. Early designs included point-to-point interconnects, buses, and crossbar switches, but all suffered from poor scalability. Network-on-chip (NoC) based on topologies such as the mesh topology can achieve a better balance between scalability, bandwidth, area and latency. The memory access is designed to meet the data requirements of the PE array (PEA) by using the scratchpad and data controller that is customized depending on the target application domain.

2. Initial development period (2000–2010)

With the diversification of applications, designs of PEs based on multiple instructions started to proliferate in order to efficiently support irregular applications. The PE was designed to be more flexible through the addition of internal scheduling logic, including the use of the predicate mechanism within the unit to support control flow and tagged token to support various forms of multi-thread execution. At the same time, the PE was able to support larger scale applications with fewer computing resources through time-domain computing, and the total number of operations that the PEA can support could be up to the product of the number of PEs and the maximum number of instructions within a single PE. Typical multi-instruction PEs like TRIPS [15], WaveScalar [16], Mophosys [17], and ADRES [18] all appeared in this period. Function configuration was an important part at runtime for a dynamically reconfigurable array, and partial reconfiguration allowed separation of configuration from computation and improved hardware utilization of the array. XPP-III [19] used the fine-grained configuration, that is, a single configuration is done for a single PE, and Chimaera [20] used the row configuration, that is, the array is configured sequentially in rows. In this way, when a particular PE or PE row is being configured, the PEs that have been already configured in the datapath can start working.

As the increasing demand for the computing power of chips, computing resources of dynamically reconfigurable arrays were growing rapidly. In order to use these hardware resources efficiently and conveniently, the development and research of compilation systems became the key in this period. At that time, static compilation was the dominant compilation technology, that is, the high-level programming language code describing application functions is translated into the machine code that can be recognized by the underlying hardware before the execution of applications. By predicting hardware behaviors at runtime, the compiler adopted technologies such as loop unrolling, software pipelining, and polyhedral model to optimize the scheduling strategy [21, 22] and improve the throughput and efficiency of the computation.

3. Rapid improvement period (2010–present)

In this period, the design space of PEs has been explored thoroughly, and the proposal of new PE structures is no longer a research focus. Although multi-instruction PEs can improve flexibility and performance, the poor utilization leads to a waste of resources on non-critical paths and consumption of static power that reduces the overall energy efficiency of the architecture. In recent years, dynamically reconfigurable computing chips have been used more often as domain-specific accelerators. Efficient acceleration can be achieved in specific domain with similar computational characteristics and hardware is tailored to support these specific computing patterns. Therefore, for energy efficiency purposes, the architectures of PEs with single instructions and static scheduling are generally used, such as Softbrain [23], DySER [24], and Plasticine [25]. The PE contains only single instructions with no additional instruction cache and scheduling logic. Only the core computing functions that meet the application requirements are retained to ensure a simplified structure, thus maximizing performance and energy efficiency. At the same time, the simple array structure requires fewer configurations. To further reduce the configuration overhead, the fast configuration mechanism is applied in DySER. The asynchronous execution of tasks in different arrays is triggered by stream in Softbrain to hide the configuration time by parallelizing configuration and computation.

With increasingly abundant computing resources integrated on the chip, the scale of the mapping problem in static compilation of SDCs grows exponentially. Numerous studies have attempted to adopt dynamic compilation due to its dynamic reconfiguration characteristics to improve the utilization of hardware resources and implement multi-thread processing, thus further enhancing the energy efficiency of the chip. Based on hardware virtualization technology, the dynamic compilation uses the configuration flow or instruction flow that has been compiled offline to convert the static compilation result into a dynamic configuration flow that conforms to the dynamic constraints at runtime. Dynamic compilation is usually classified into dynamic generation of configurations based on instruction flow, such as DORA [26] and dynamic transformation of configurations based on configuration flow. However, current compilation systems still require a lot of manual assistance to ensure compilation quality, and their automated implementation is still today's research focus.

1.2 Development of Programmable Devices

SDCs are a new type of programmable devices that have significant advantages over traditional programmable devices in terms of software programmability and hardware reconfiguration speed. This section will take the classic programmable devices represented by FPGAs as an example to introduce the emergency, different stages of development and technical principles of FPGAs, as well as problems faced by their development, and analyze how SDCs will solve these problems.

1.2.1 Historical Analysis

As a kind of classic programmable devices, FPGAs have increased their capacity from less than 100,000 transistors to over 40 billion transistors since their invention in 1984, increasing performance by two orders of magnitude and dramatically reducing cost and power consumption per unit of function. Combined with the historical review of a Xilinx Fellow Dr. Trimberger on the occasion of the 30th anniversary of FPGAs [27], the development of FPGAs is roughly divided into four phases: invention, expansion, accumulation, and systematization. The following details the development of programmable devices by describing the four development phases of FPGAs.

1. Invention phase (1984–1991)

The first FPGA, XC2064, was invented by Ross Freeman, who was a co-founder of Xilinx. At that time, transistors are enormously expensive, and circuit designers always aimed to make the most of every transistor in their designs. Freeman pioneered the design of a chip containing 64 logical modules with programmable functions and interconnects, which meant that some of the transistors were idle many times. However, Ross Freeman was convinced that the development of Moore's Law would eventually lead to a significant drop in transistor costs and FPGAs would shine. History has also confirmed his bold and prescient idea. The Xilinx FPGA chip was taped out on the 2.5 μm process, and the programmability of the logical modules was achieved by a 3-input LUT based on static random access memory (SRAM). They could be programmed repeatedly by changing the data in the memory. However, the SRAM covered most of the area on the chip. In this phase, there was another type, antifuse-based FPGAs, which were more area-saving than SRAM-based FPGAs, but could only be programmed once. They had a large market share at the time by sacrificing reprogrammability for a reduction in area.

2. Expansion phase (1992–1999)

In the 1990s, Moore's Law continued to advance rapidly, with transistor integration doubling every two years. Each doubling of the number of available transistors on a silicon chip allowed the largest size of the FPGA to be doubled as well, while reducing the cost per unit of function by half. More importantly, the successful application of chemical mechanical polishing (CMP) allowed more metal layers to be stacked and the cost of interconnects to fall faster than that of transistors, thus enabling the addition of programmable interconnects to adapt to larger capacities. The increase in area resulted in higher performance, better functionality and ease of use. The rapid growth in capacity made it impossible to complete synthesis and placement & routing (P&R) manually, and design automation became essential. Meanwhile, SRAM replaced anti-fuse as the mainstream technology adopted by FPGAs. That's because anti-fuse could only be programmed once, which limited the applications of FPGAs, and anti-fuse took longer to be implemented on new processes compared with SRAM, resulting in relatively slow performance improvement. In this phase, LUTs gradually became the dominant logical architecture adopted by FPGAs and continue to be so today.

3. Accumulation phase (2000–2007)

With the increasing cost and complexity of silicon chip fabrication, the risk of customizing made many ASIC users gradually turn to FPGAs. The advance of the Moore's Law made FPGAs bigger, but customers were not willing to pay high fees for mere area increase. The pressure on reductions in cost and power consumption led to a shift in architectural strategies from programmable logic to specialized logical modules. These modules included large-capacity memory, microprocessors, multipliers, and flexible input/output (I/O) interfaces. Meanwhile, FPGAs were widely used in the communication system as they incorporated specialized high-speed I/O transceivers and a large number of high-performance multipliers, thus enabling mass data forwarding without affecting the throughput. In this phase, the addition of specialized modules was the main feature, and FPGAs were gradually no longer used only as a generic alternative to ASICs, playing an increasingly important role in digital communication applications.

4. Systematization phase (2008–present)

In this phase, FPGAs have evolved into reconfigurable SoCs that, in addition to programmable logic modules, also have microprocessors, memory, analog interfaces, and NoCs to meet the system requirements of SoCs. To meet requirements of emerging applications, FPGAs also incorporated specialized acceleration engines. For example, Xilinx's adaptive compute accelerate platform (ACAP) introduced in 2018 heterogeneously integrated the scalar, vector and programmable logic blocks to meet the computing needs of different algorithms. Complex systems require efficient design tools to ensure ease of use and risk control, usually at the cost of a certain level of performance, flexibility and energy efficiency. Modern FPGA EDA software have enabled modeling of systems using C, CUDA, and OpenCL to simplify the design. However, there are still many difficult problems for HLS.

1.2.2 Technical Principles of FPGAs

The hardware programming of FPGAs is different from the software programming of GPPs. In particular, the logical functions of FPGAs are implemented by configuring logical blocks. Figure 1.6 shows a typical FPGA architecture. It mainly consists of (1) configurable logic blocks (CLBs) that are fundamental building blocks for computing and memory, (2) programmable interconnects that use configurable routing switches to achieve flexible interconnections between different logic blocks and between logic blocks and I/Os, and (3) programmable I/Os that are interface provided by the chip for connecting peripheral circuits to satisfy different matching and drive requirements.

The hardware programmability of FPGAs is mainly implemented by CLBs based on LUTs or based on multiplexers (MUXs). The following illustrates the implementation principles: Fig. 1.7 shows an LUT-based programmable logic block, implemented using SRAM technology. The LUT consists of a memory cell SRAM and a MUX to choose the memory bits. The 2-input LUT in the figure allows implementing arbitrary logical functions with two variables. When the value stored in SRAM is the output of the truth table of a logical function, the LUT implements this

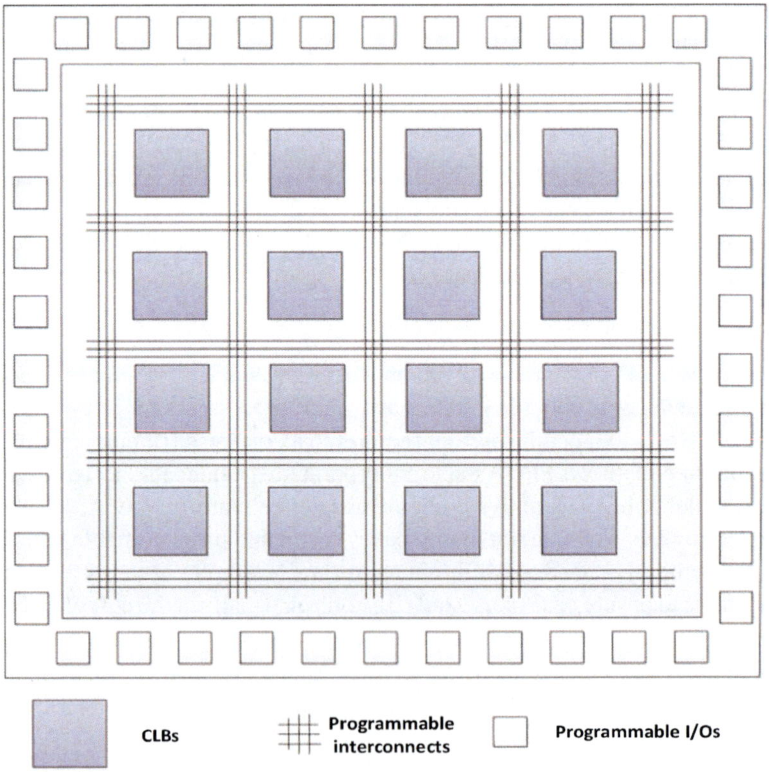

Fig. 1.6 Typical FPGA architecture

logical function. In this way, different logical functions can be implemented on the same circuit by configuring the contents of the LUT. For example, the exclusive OR (XOR) logic can be implemented by filling the output of the truth table in Fig. 1.7b into the memory cell in Fig. 1.7a. Commercial FPGA logic blocks commonly used are usually based on 4-input or 6-input LUTs. A 4-input LUT can be considered as a 16 × 1-bit SRAM with 4-bit address pins.

Figure 1.8 shows a programmable logic unit based on a 2-input MUX. Changing the configuration of the input and MUX allows implementing different logic gate functions. For example, when the inputs A and B are 1 and 0 respectively, and the selector of MUX is M, logic NOT gate can be implemented, i.e. $\overline{M}$. Multiple 2-input MUXs can be combined into more complex digital logic.

Programmable interconnects, connecting programmable logic blocks and I/O interfaces in FPGAs, are another important part of implementing hardware programming. There are various ways to implement flexible interconnects.

Figure 1.9 shows an example of a mesh-based programmable interconnect network. The SRAM-based connect box used between programmable logic tile blocks enables routing by configuring the SRAM to secure interconnect between

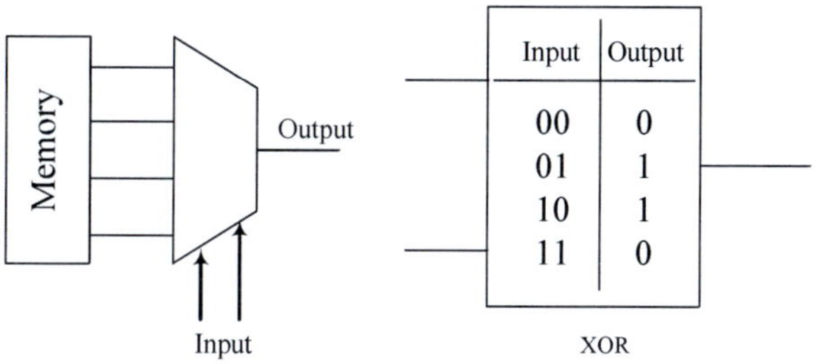

Fig. 1.7 Principle of the SRAM-based LUT [28]

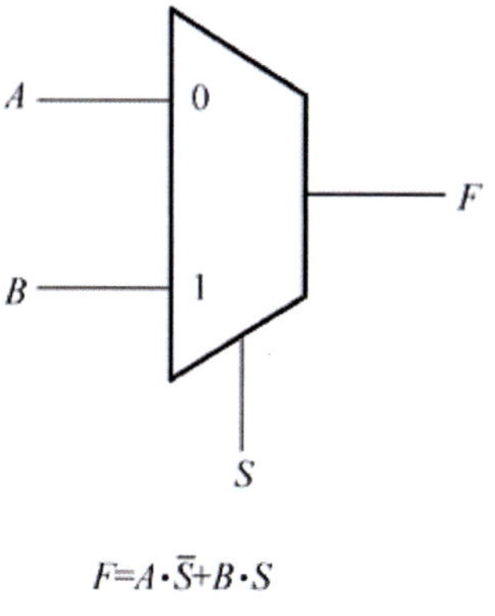

A	B	S	$F=$
0	0	0	0
0	M	1	M
0	N	1	N
0	N	M	MN
M	0	N	$M\overline{N}$
N	0	M	$\overline{M}N$
N	1	M	$M+N$
1	0	M	$\overline{M}$
1	0	N	$\overline{N}$
1	1	1	1

(a) (b)

Fig. 1.8 Programmable logic unit based on a 2-input MUX and logical functions that can be implemented [28]

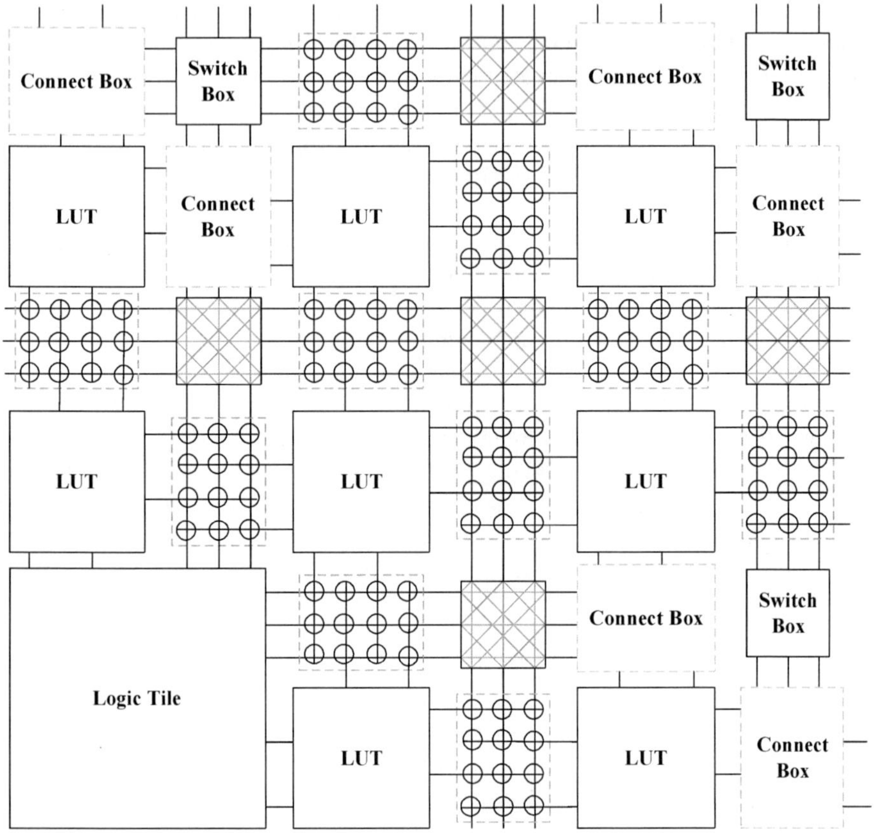

Fig. 1.9 Example of a programmable interconnect network [28]

programmable tile logic blocks with the upper, lower, left and right adjacent channels. At the intersection of the vertical and horizontal channels, an SRAM-based switch box is used. Similarly, the SRAM configuration can enable connections between channels and interconnections between two logic tile blocks, thus implementing target logic functions.

1.2.3 Challenges to FPGA Technical Paradigm

With the continuous integration of various processor cores, and customization of intellectual property (IP) and various standard interfaces, FPGAs are gradually evolving into an integrated platform for programmable heterogeneous computing with collaborative evolution of software and hardware. However, the single-bit fine-grained programming and static configuration of FPGA have not changed essentially, and are becoming the bottleneck limiting the development of FPGAs from the following aspects: (1) Low energy efficiency. Compared to ASICs, FPGAs have low performance, high power consumption, and high static power consumption. (2) Limited capacity. Most of the hardware resources are used for interconnection, and only a small part is for computation. (3) High use threshold. Poor programmability makes development difficult, and software programmers without deep knowledge of circuit designs cannot implement efficient design. In recent years, FPGAs have received continuous technological upgrades by increasing the hardware scale and adopting heterogeneous computing and high-level language programming. However, without changes to their intrinsic properties, the above-mentioned problems have not been addressed radically.

1. Energy efficiency

The bit-level reconfiguration granularity causes the low energy efficiency of FPGAs. SDCs can be simply classified into fine-grained and coarse-grained, based on the granularity of the reconfigurable hardware modules. Normally, the configuration granularity less than 4 bits is considered to be fine grained while that greater than 4 bits is coarse-grained. Due to the single-bit configuration granularity, FPGAs are typical fine-grained SDCs. The advantage is that computing with any precision can be supported. However, due to fine-grained programming, FPGAs have considerable area and power overhead, which reduces the energy efficiency and hardware utilization. Meanwhile, they have huge routing overhead and large volume of configuration contexts that directly lead to long configuration time. According to related documents [29–31], more than 60% of the area and power in FPGA implementations is consumed in the configuration of interconnects, and about 14% of the dynamic power is consumed by on-chip interconnects [32].

2. Resource capacity

The static configuration property of FPGAs exacerbates the problem of limited capacity due to the occupation of considerable resources by interconnects. In spite of multiple configuration patterns of FPGAs, the common way is to load configurations from off-chip memory at power-up. Depending on circuit functions, the configuration of FPGAs usually takes hundreds of milliseconds, or even seconds. The circuit function cannot be changed at runtime after the completion of configuration. A new function can be switched to and loaded only after the computational task being performed by the FPGA is interrupted. With increasingly complex circuit functions to be mapped, FPGA commercial companies can only continue to increase the hardware resource capacity of FPGAs by using advanced manufacturing processes. Figure 1.10 shows the statistics of LUTs, DSP modules and memory resources of Xilinx's Virtex series FPGA devices on different processes. It can be seen that with advances in process technology, there is an overall trend of rapid growth in the number of resources. At the same time, the number of LUTs,

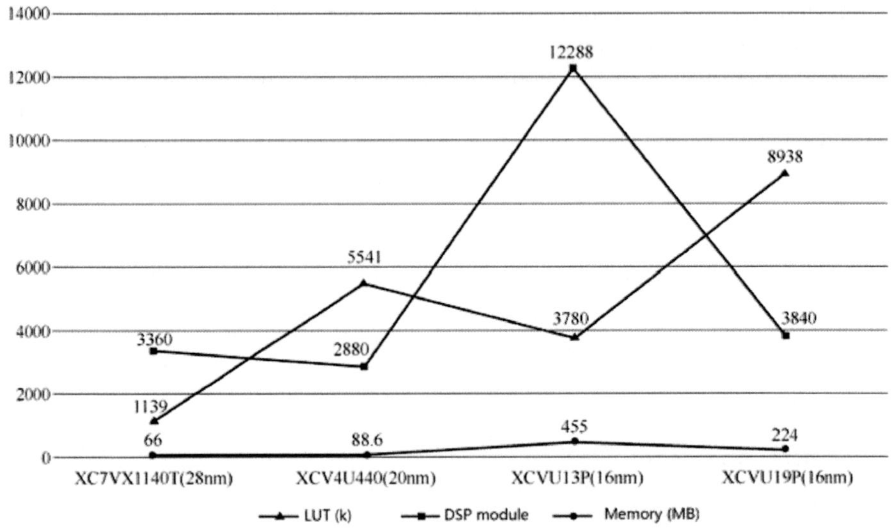

Fig. 1.10 Resource statistics on Xilinx's virtex series FPGAs

DSP modules and memory resources will be increased accordingly depending on the different application requirements.

3. Programming difficulty

The user-friendliness of the programming tools and development environment directly affects the user experience and application of FPGAs. With the increase of integrated resources and system complexity of the FPGA, the traditional development based on hardware programming has become a limiting factor that constrains the further promotion of FPGA applications. The evolution of Xilinx's development tools is used here to illustrate this issue. Efficient programming tools and user-friendly development environments are required to provide convenient application interfaces, highly scalable programming models, efficient HLS tools, and programming languages capable of integrating hardware and software features, and high-performance hardware code generation. Figure 1.11 shows three FPGA development platforms developed by Xilinx, i.e. integrated software environment (ISE), Vivado and Vitis. The ISE design kit mainly supports devices prior to the 7 series such as Spartan-6 and Virtex-6 series. FPGA developers are required to acquire sound knowledge of hardware and to be able to perform hierarchical customization from circuit function definition, register transfer level (RTL) design to synthesis constraints depending on the priorities of design goals, thus achieving near optimal implementation results on FPGAs.

Vivado is an HLS development suite introduced by Xilinx in 2012 to support high-end devices of the 7 series and above, including Zynq, Ultrascale, Ultrascale+, MPSoC, and RFSoC. The platform further extended the range of FPGA users, allowed algorithm developers to use C/C++ language to achieve fast mapping between algorithms and FPGA configurations directly, and greatly improved FPGA

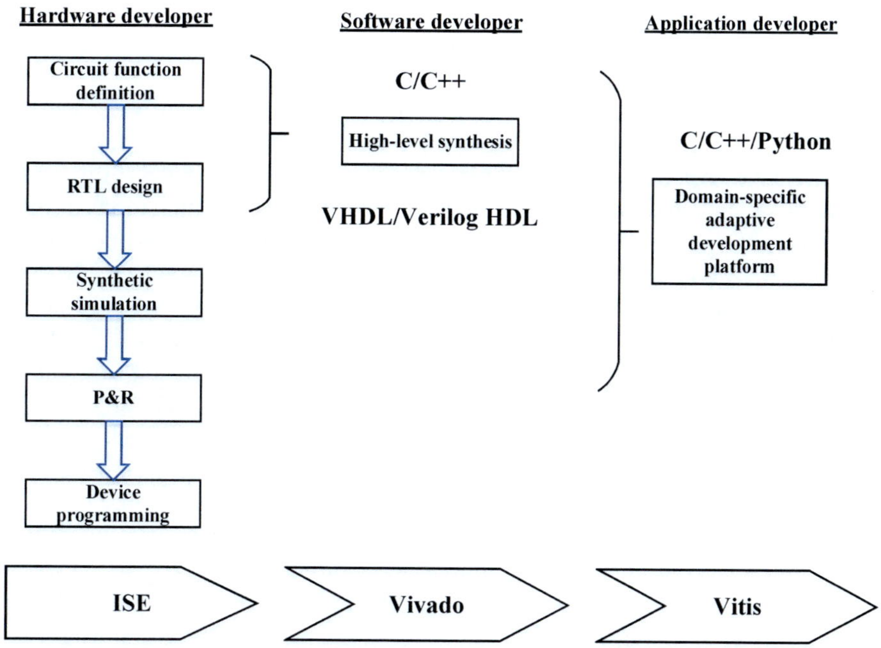

Fig. 1.11 Evolution of Xilinx's FPGA development tools

development efficiency, in spite of the poor implementation results compared to the traditional development process. As FPGA chip architectures evolved by generations and application requirements continued to grow, Vivado's new functions were increasingly developed, such as SDSoC for embedded system developers, SDAccel for data center applications, and AI toolkits for AI applications. However, developers are required to have strong FPGA hardware development capabilities, as it was intended for hardware and it involved hardware design and simulation. But subject to a high threshold, hardware developers are relatively few, compared to millions of software developers. Obviously, if the development threshold can be lowered so that more software developers can be engaged, this will undoubtedly greatly enrich Xilinx's application ecosystem.

In recent years, with the rapid development of deep learning-based machine vision, autonomous driving, data center and IoT applications, hardware circuit designs have gradually shown the technical trend of application-driven architecture innovation. At the same time, traditional FPGA devices are gradually growing into heterogeneous integrated platforms with enormouse resources. Following this trend, Xilinx launched its unified software development platform Vitis in 2019. Developers can automatically adapt the appropriate hardware architecture to the software or algorithm code without extensive hardware expertise. The coverage of developers can be further expanded, and data scientists and application developers can focus only on the optimization and development of algorithm models.

1.2.4 Innovation of SDCs

SDCs enable dynamic and real-time definition of chips through software, and reconfiguration of circuit functions is achieved in nanoseconds according to algorithmic requirements. Applications in multiple domains can be implemented agilily and efficiently. The two reasons that they can break through the three limitations of FPGAs are: (1) SDCs use a programmable architecture with mixed-grained programming dominated by coarse-grained programming instead of the fine-grained LUT logic of FPGAs, which significantly reduces resource redundancy, enables 10–100× better than the energy efficiency of FPGAs, and reaches the same level as ASICs; (2) SDCs support dynamic configuration (configuration time can be shortened from milliseconds of FPGAs to nanoseconds), and the capacity can be expanded through fast TDM in hardware, no longer limited by the chip's physical size. In other words, similar to a CPU that can run software code of any size, an SDC can implement digital logic of any size. Moreover, dynamic configuration fits the sequentialization feature of software programs better than static configuration, and is more efficient when programming in high-level languages. Software developers without circuit designs background can program SDCs efficiently. The improvement of the usability will enable agile development of the chip, speed up the iteration and deployment of applications, and expand the scope of use. In summary, the mixed-grained programming dominated by coarse-grained programming and dynamic configuration are the fundamental reasons for SDCs to outperform FPGAs by orders of magnitude in overall performance, as shown in Fig. 1.12. Therefore, the SDC seems to be a more appropriate direction of technical breakthroughs. A detailed explanation will be given in Chap. 2.

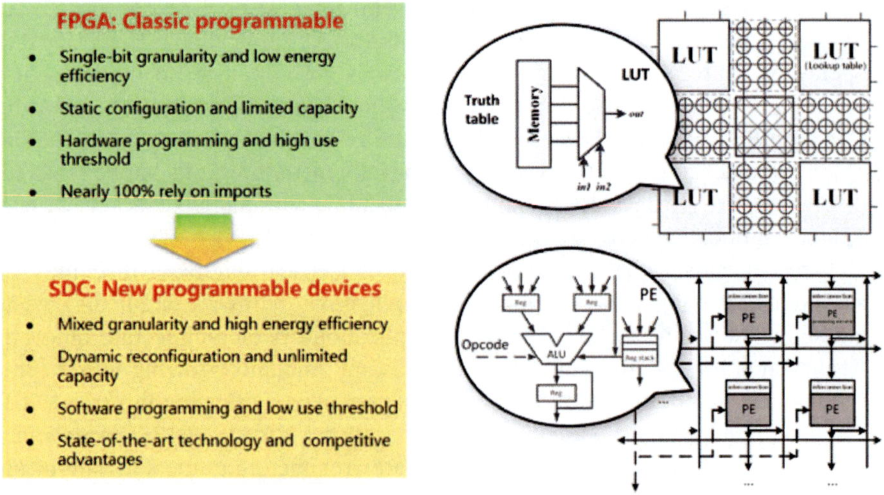

Fig. 1.12 Key advantages of SDCs over classic programmable devices

1.3 State-of-the-Art Programmable Devices

1.3.1 Development of FPGAs

Classic programmable devices represented by FPGAs have been widely used and had an important position in communications, networking, aerospace, national defense and other domains due to its ability to customize massive digital logic and to quickly complete prototypes. Table 1.1 compares the FPGAs made in China and Xilinx's FPGAs [33]. The international development trend of FPGAs is mainly as follows:

(1) Utilizing the advanced integrated circuit manufacturing technology is the simplest and most direct practice to improve the overall capacity, performance, and energy efficiency of FPGAs. Xilinx's latest model uses the 7 nm fin field-effect transistor (FinFET) process of Taiwan Semiconductor Manufacturing Company Limited (TSMC), while Altera uses Intel's 14 nm Tri-gate process.
(2) Integrating heterogeneous computing structures, such as on-chip memory resources and high-speed interfaces is a trend. CPUs, GPUs, DSPs, DDR3/DDR4 memories,PCIe interfaces, USBs, and Ethernet interfaces are integrated to achieve fully programmable SoCs and even a flexible and adaptive acceleration platform, ACAP [34].
(3) The use threshold is lowered gradually. At the very beginning, very high speed integrated circuit hardware description language (VHDL) is used for programming. Then in the Vivado, C-based programming can be performed. Now in the Vitis, thehigh-level frameworks can be used in combination with Tensorflow, Caffe, C, C++, or Python to develop acceleration applications. Although

Table 1.1 Comparison of FPGA's application status

Classification	An FPGA company in China	Xilinx
Fabrication process	28 nm	7 nm
Capacity	Maximum 0.35 millionLUTs	Maximum 5 millionLUTs
Hardware architecture	FPGA + MCU	ACAP multi-core heterogeneous processing platform (CPU, FPGA, RF, NoC, and AI processor)
Software	Some having the full process technology of commercial software	Software programming suites Vivado and Vitis: AI, HLS, synthesis, P&R, IP, DSP, and SoC
Product	3 series, over 10 types of chips	10 generations (latest UltraScale + and Versal), 30 series, hundreds of types of chips
Application domains	Some domains of communication equipment, industrial control, and consumer electronics	Communication equipment, industrial control, data center, automotive electronics, consumer electronics, and military and aerospace

manual optimization may be inevitable during the programming process, the development efficiency has been greatly improved.

It should be noted that the above-mentioned technical development is actually an extension and improvement of the existing technical route of FPGAs, and will not eventually solve the problems of low energy efficiency, limited programmable logic capacity, and high programming difficulty mentioned in previous sections.

1.3.2 Research Status of SDCs

The SDC has emerged in recent years, with few specific studies. However, related topics such as the spatial architecture, dynamically reconfigurable architecture, and HLS have been the focus in areas such as computer architectures, solid-state circuits, and electronic design automation. The exploration of new reconfigurable programmable devices and related technologies featuring dynamic reconfiguration, coarse-grained computing, and software-defined hardware (SDH) has been carried out widely.

Figure 1.13 shows that the DARPA identified SDH technology as a supporting technology for the electronic development in the next decade [1] in 2017, hoping to build "runtime-reconfigurable hardware and software that enables near ASIC performance without sacrificing programmability for data-intensive algorithms." SDH will enable (1) the dynamic optimization of software code and hardware structures when input data changes, and (2) the reuse of hardware for new problems and new algorithms. Therefore, DARPA believes that the key to SDH implementations is fast hardware reconfiguration and dynamic compilation. According to DARPA's program plan, SDH should enable at least 100 $\times$ better than the energy efficiency of GPPs in the next five years, and achieve a reconfiguration speed up to 300–1000 ns. This program is the first to explicitly propose the research goal of SDH, guiding the future direction of the domain. SDCs are not essentially different from SDH. According to Fig. 1.2, the agile development of chips is an additional feature of SDCs. Meanwhile, the EU's Horizon 2020 programme also has similar plans in SDH, but with a greater focus on specific applications such as communications [35].

Extensive research related to SDCs has been carried out. Table 1.2 lists some representative products. The European Space Agency (ESA) used the IP of PACT's CGRA devices on Astrium's satellite payloads in 2010 [36]. The dynamically reconfigurable structure ADRES, proposed by IMEC around 2004, has been used in Samsung's biomedical [37] and HDTV [38] series products. Renesas Technology of Japan, has used its DRP structure proposed in 2004 [39]. With the addition of architectures such as coarse-grained programmable computing arrays, Xilinx's new product Versal could represent the evolution of software-defined SoCs [34]. In academia, research teams from Stanford University, UCLA, Massachusetts Institute of Technology (MIT) and others have also conducted long-term research in this direction,

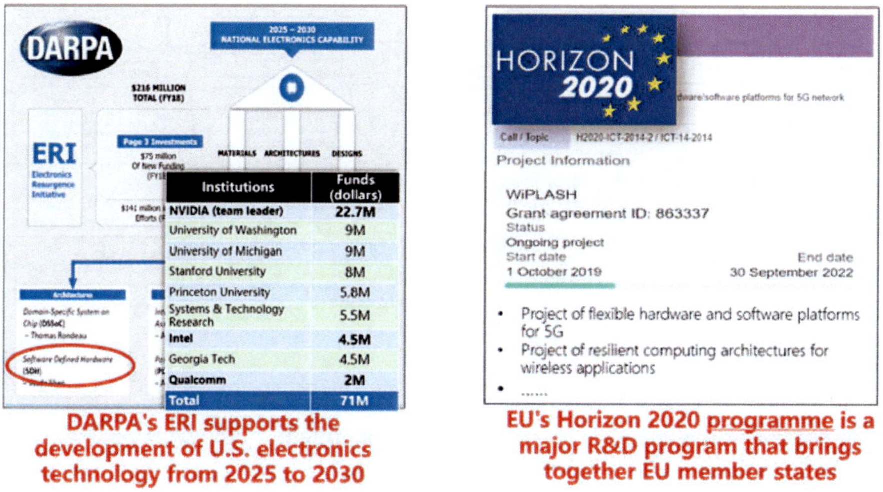

Fig. 1.13 DARPA's ERI and EU's Horizon 2020 program

and research results have been continuously published in top conferences in related domains.

In China, the research on SDCs has been supported by ministries at all levels for nearly 20 years as shown in Fig. 1.14. The National Natural Science Foundation of China (NSFC) launched the "Fundamental Research on Semiconductor Integrated Chip System" major research plan in 2002, to make preparations for the R&D of the basic theory of reconfigurable computing chips. In the last 10 years, the NSFC has supported projects related to reconfigurable computing almost every year. The Ministry of Science and Technology (MOST) has supported the R&D of reconfigurable computing chip technology by setting up the 863 program "Embedded Reconfigurable Mobile Media Processing Core Technology" in the 11th Five-Year Plan and the 863 program "R&D of Key Technology of Reconfigurable Processors for General-purpose Computing" in the 12th Five-Year Plan. In the meanwhile,

Table 1.2 Industrialization of SDC-related research

Company	Product Series	Application	Time
PACT	XPP	Artificial satellite	2003
Samsung	ULP-SRP	Biomedical	2012
IPFLEX	DAPDNA-2	Image processing	2012
Wave Computing	DPU	AI	2017
Renesas Technology	Stream transpose	Digital audio and video	2018
Tsing Micro	Thinker	AI	2018
Xilinx's FPGAs	Versal	AI	2019
Wuxi Micro Innovation	S10/N10	Information security and network	2019

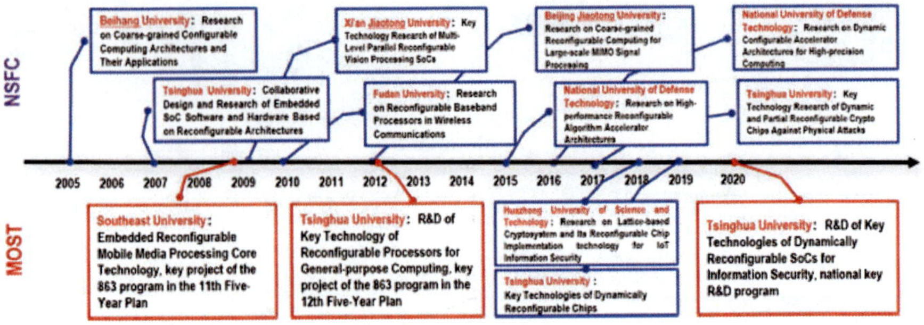

Fig. 1.14 Domestic research support for SDCs

multiple startups based on reconfigurable computing technology incubated, such as Tsing Micro and Wuxi Micro Innovation.

References

1. DARPAp[EB/OL]. https://www.darpa.mil. Accessed 25 Nov 2020.
2. Moore GE. Progress in digital integrated electronics. In: IEEE international electronic devices meeting. 1975. p. 11–13.
3. Hennessy JL, Patterson DA. A new golden age for computer architecture. Commun ACM. 2019;62(2):48–60.
4. Moore GE. No exponential is forever: but "forever" can be delayed! In: IEEE international solid-state circuits conference. 2003. p. 20–23.
5. Dennard RH, Gaensslen FH, Yu HN, et al. Design of ion-implanted MOSFET's with very small physical dimensions. IEEE J Solid-State Circ. 1974;9(5):256–68.
6. Dally WJ, Turakhia Y, Han S. Domain-specific hardware accelerators. Commun ACM. 2020;63(7):48–57.
7. Makimoto T. Towards the second digital wave. Sony CX-News. 2003;33:1–6.
8. Hartenstein R. Trends in reconfigurable logic and reconfigurable computing. In: The 9th International conference on electronics, circuits and systems. 2002. p. 801–08.
9. Juyan X, Yongsheng Y. Semiconductor cycle and reconfigurable chips. Embedded Syst Appl. 2005;2:2–5.
10. Estrin G. Organization of computer systems: the fixed plus variable structure computer. In: IRE-AIEE-ACM'60 (Western). 1960. p. 33–40.
11. Dehon AE, Wawrzynek J. Reconfigurable computing: what, why, and implications for design automation. In: Proceedings 1999 design automation conference. 1999. p. 610–15.
12. Mirsky E, Dehon A. MATRIX: a reconfigurable computing architecture with configurable instruction distribution and deployable resources. In: FPGAs for custom computing machines. 1996. p. 1–8.
13. Miyamori T, Olukotun K. REMARC: reconfigurable multimedia array coprocessor (abstract). IEICE Trans Inf Syst. 1998;E82D(2):261.
14. Goldstein SC, Schmit H, Moe M, et al. PipeRench: a coprocessor for streaming multimedia acceleration. In: Proceedings of the 26th international symposium on computer architecture. 1999. p. 28–39.
15. Burger D, Keckler SW, McKinley KS, et al. Scaling to the end of silicon with EDGE architectures. Computer. 2004;37(7):44–55.

16. Swanson S, Michelson K, Schwerin A, et al. WaveScalar. In: International symposium on microarchitecture. 2003. p. 291–02.
17. Singh H, Lee MH, Lu GM, et al. MorphoSys: an integrated reconfigurable system for data-parallel and computation-intensive applications. IEEE Trans Comput. 2000;49(5):465–81.
18. Mei BF, Vernalde S, Verkest D, et al. ADRES: an architecture with tightly coupled VLIW processor and coarse-grained reconfigurable matrix. In: Field programmable logic and application. 2003. p. 61–70.
19. Schüler E, Weinhardt M. Dynamic system reconfiguration in heterogeneous platforms. Dordrecht: Springer, Netherlands; 2009.
20. Hauck S, Fry TW, Hosler MM, et al. The chimaera reconfigurable functional unit. In: Proceedings the 5th annual IEEE symposium on field-programmable custom computing. 1997. p. 87–96.
21. Ebeling C. Compiling for coarse-grained adaptable architectures. Technical report UW-CSE-02-06-01. Washington: University of Washington; 2002.
22. Park H, Fan K, Kudlur M, et al. Modulo graph embedding: mapping applications onto coarse-grained reconfigurable architectures. In: International conference on compilers, architecture and synthesis for embedded systems. 2006. p. 136–46.
23. Nowatzki T, Gangadhar V, Ardalani N, et al. Stream-dataflow acceleration. In: International symposium on computer architecture. 2017. p. 416–29.
24. Govindaraju V, Ho C, Nowatzki T, et al. DySER: Unifying functionality and parallelism specialization for energy-efficient computing. IEEE Micro. 2012;32(5):38–51.
25. Prabhakar R, Zhang Y, Koeplinger D, et al. Plasticine: a reconfigurable architecture for parallel patterns. In: International symposium on computer architecture. 2017. p. 389–02.
26. Watkins MA, Nowatzki T, Carno A. Software transparent dynamic binary translation for coarse-grain reconfigurable architectures. In: IEEE international symposium on high performance computer architecture. 2016. p. 138–50.
27. Trimberger SM. Three ages of FPGAs: a retrospective on the first thirty years of FPGA technology. Proceedings IEEE. 2015;103(3):318–31.
28. Rabaey J, Chandrakasan A, Nikolic B. Digital integrated circuits a design perspective. 2nd ed. New York: Pearson; 2002.
29. Calhoun BH, Ryan JF, Khanna S, et al. Flexible circuits and architectures for ultralow power. Proc IEEE. 2010;98(2):267–82.
30. Poon KK, Wilton SJ, Yan A. A detailed power model for field-programmable gate arrays. ACM Trans Des Autom Electron Syst. 2005;10(2):279–302.
31. Kuon I, Tessier R, Rose J. FPGA architecture: survey and challenges. Delft: Now Publishers Inc.; 2008.
32. Adhinarayanan V, Paul I, Greathouse JL, et al. Measuring and modeling on-chip interconnect power on real hardware. In: IEEE international symposium on workload characterization. 2016. p. 1–11.
33. Anlogic Infotech. http://www.anlogic.com. Accessed 26 Nov 2020
34. Voogel M, Frans Y, Ouellette M, et al. Xilinx versal™ premium. IEEE Computer Society; 2020. p. 1–46.
35. European Commission[EB/OL]. https://cordis.europa.eu/project/id/863337. Accessed 26 Nov 2020
36. PACT[EB/OL]. http://www.pactxpp.com. Accessed 26 Nov 2020
37. Kim C, Chung M, Cho Y, et al. ULP-SRP: ultra low power Samsung reconfigurable processor for biomedical applications. In: 2012 International conference on field-programmable technology. 2012. p. 329–34.
38. Kim S, Park YH, Kim J, et al. Flexible video processing platform for 8K UHD TV. 2015 IEEE hot chips 27 symposium. 2015. p. 1.
39. Fujii T, Toi T, Tanaka T, et al. New generation dynamically reconfigurable processor technology for accelerating embedded AI applications. In: 2018 IEEE symposium on VLSI circuits. 2018. p. 41–42.

Chapter 2
Overview of SDC

Gain efficiency from specialization and performance from parallelism.
—William J. Dally, Yatish Turakhia, and Song Han, Communications of the ACM, 2020.

As modern society develops in the direction of digitalization, intelligence and automation, the demand for computing is increasing. Traditional infrastructure providers usually improve computing services by introducing more GPPs, but the high energy consumption and costs are quickly becoming a bottleneck limiting this approach. Many infrastructure providers, such as Microsoft and Baidu, are introducing hardware accelerators and dedicated computing architectures to improve computing services significantly. Meanwhile, companies represented by Amazon, Microsoft and Alibaba have started to provide cloud computing services based on reconfigurable hardware, and flexible hardware infrastructure as a service is gradually becoming a popular model. Therefore, the efficiency, flexibility and ease of use have become key metrics in the design of new hardware architectures.

This chapter focuses on the technical principles, characteristic analysis and key research problems of SDCs. First, it describes the basic principles and makes detailed comparisons. Then, it analyzes the characteristics of SDCs in terms of performance, energy efficiency, reconfiguration and programmability, and describes their key advantages and development potential. Finally, the key research problems of SDCs are described.

2.1 Basic Principles

The SDC, a new chip architecture design paradigm, allows directly defining the hardware functions and rules at runtime using software, so that hardware functions can be changed and optimized in real time as software changes. There are already many mature computing chip architectures (such as GPPs, GPUs, ASICs and FPGAs) that have well-developed tool chains and ecosystems, so why is it necessary to explore

S. Wei et al., *Software Defined Chips*,
https://doi.org/10.1007/978-981-19-6994-2_2

a new chip architecture design paradigm? Is it worth constructing software, hardware and ecosystems at such a huge cost for the advantages of SDCs?

2.1.1 Necessity Analysis

The development of SDCs is driven by a combination of factors:

(1) In terms of the upper-layer application requirements, the social and technical development has brought massive amounts of data, including network data, sensor data, logistics data, measurement and control data, and biological data. Emerging areas such as big data, cloud computing, artificial intelligence, bioinformatics have become popular. Then, processing and analyzing massive amounts of data has become the core capability to meet the needs of the development of the information society, and providing efficient computing has become a major driver of the development of new architectures. For example, the number of layers, size, and number of parameters of deep neural networks (DNNs) is increasing as a high accuracy is required, and the amount of data and computational load has reached megabyte (MB) per round and million operations per second (MOPS) [1].

(2) In terms of the lower-layer integrated circuit fabrication process, the development of manufacturing technology is slowing down significantly. The integration of transistors predicted by Moore's Law has gradually slowed down and will soon reach its physical limit. The energy efficiency gains predicted by Dennard scaling law have completely failed. Moreover, subject to the heat dissipation technology, the continuous improvement in performance has become increasingly difficult, and the reduction in process feature sizes can no longer enable substantial performance and energy efficiency gains [2]. For example, the total computational load for training the largest DNN at the time has doubled every 3–4 months since 2012. In contrast, the number of transistors per integrated circuit predicted by Moore's Law doubles every 2 years, which differed from the actual value by already 15 times with the slowdown of the development of integrated circuits in recent years. As a result, there is a widening gap between the growth rate of application requirements and that of integrated circuit technology. Today, it is difficult to imagine how slow it is to train neural networks with GPPs, and we must be prepared for future demands that far exceed today's computing power. The SDC discussed in this book is one of the important countermeasures at the level of architecture designs.

(3) In terms of computing chip architecture designs, ASICs have done it to the extreme in specialization and parallelization to meet the application demand for computing power and the circuit demand for computing efficiency. Compared to general-purpose architectures, such as CPUs and FPGAs, ASICs can minimize redundant logic by specialization and maximize the peak performance by parallelization. Specifically, the instruction set architecture, such as instruction fetch,

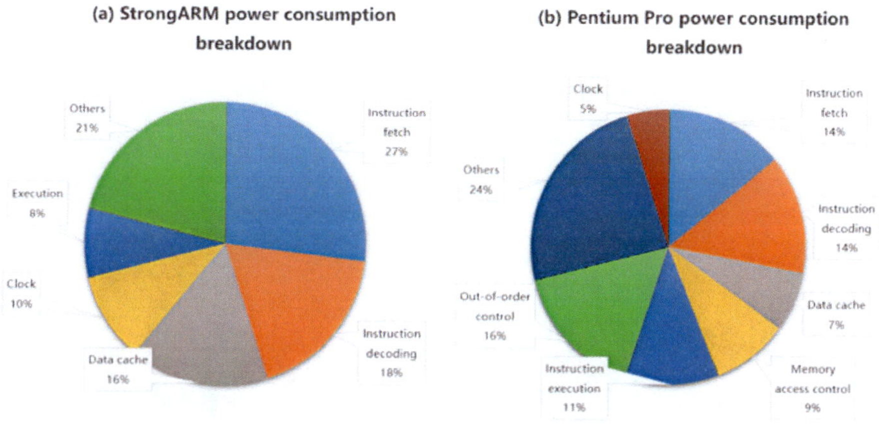

Fig. 2.1 GPP power consumption breakdown (see the color illustration)

instruction decode and register file access are converted to circuits. Some non-computing functions such as instruction fetch and instruction decode consume more than 80% of the CPU's power, as shown in Fig. 2.1 [3]. Configurable computing and interconnect of programmable devices are converted into specific functions. Redundant arithmetic and interconnect usually consume about 95% of the area/power of FPGAs [4]. The peak performance of ASICs is improved by maximizing the parallelism. The instruction-level parallelism, data-level parallelism and loop-level parallelism are extracted statically, and the paralleled execution of the entire chip is driven by data flow. Overall, the energy efficiency of ASICs can be 1000 or even 10,000 times better than that of GPPs. Therefore, ASICs have become an important option in many emerging domains to meet increasing demand. For example, in the domain of DNNs, Google designed the tensor processing unit (TPU) series to accelerate the training and inference of networks [5]; in the domain of bioinformatics, Stanford University proposed the Darwin processor to accelerate genome alignment applications [6].

However, ASICs face very serious challenges from the perspective of practicality, due to their high chip design cost, NRE costs, and time cost. Since circuit functions have been consolidated into ASICs, the chip can only support a specific application. Figure 2.2 shows the total cost of hardware development for a function-specific chip, which increases exponentially with the advancement of the technology node [7]. To share this high cost, ASICs are often limited to mature and large-volume applications such as communications and networks; otherwise, the cost would be unaffordable. In fact, chip products that can use the state-of-the-art integrated circuit process are all functionally flexible, such as CPUs, GPUs and FPGAs, relying on large volume to share the cost and be able to meet the needs of the continued development of algorithms. Therefore, at the chip architecture design level, it is difficult for ASICs with fixed functions to keep up with the development of emerging applications. For example, in the domain of DNNs, due to the very fast algorithm updates and iterations,

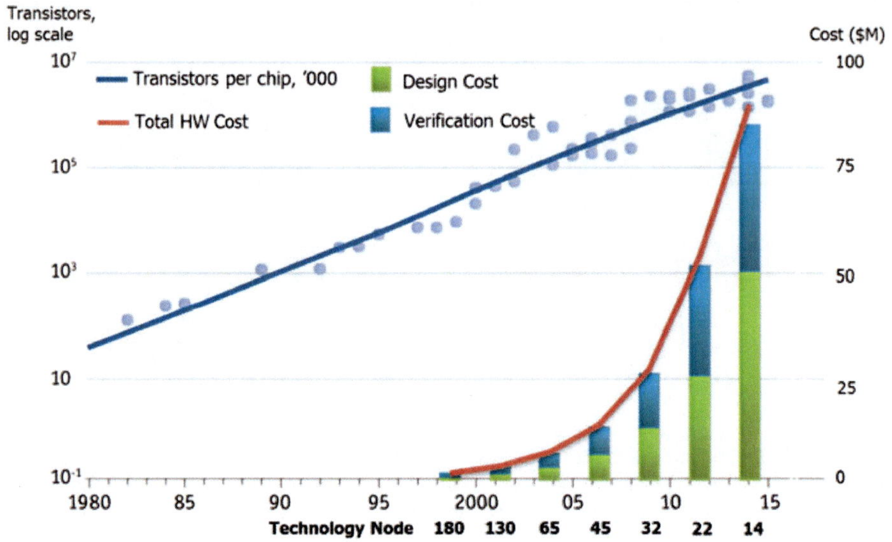

Fig. 2.2 Exponential increase in design costs of integrated circuits

and small quantities of single function versions, the optimal time to market is much less than the R&D cycle of ASICs. The increasing costs in design, production and time of integrated circuit processes are the third driving factor of the development of SDCs.

Finally, ASIC designs, customization and parallelization in particular, are not always excellent, from the view of fluctuating application scenario requirements. Fixing circuit functions before runtime means that hardware circuits can not adjust with runtime requirements, i.e., ASICs cannot be truly customized for unpredictable real-time workloads and requirements before runtime. For example, the load of cloud servers is fluctuating. If ASIC chips are used to handle such tasks, the low load will still cause low utilization in spite of the efficient data-driven computing, thus causing a waste of static power. Therefore, in a dynamic scenario, the computing power of ASIC chips will be partially wasted and the high energy efficiency will be impaired.

In summary, it is necessary to conduct research on SDCs mainly because it is increasingly difficult for the existing computing architectures to meet the following development needs:

(1) Emerging applications are increasingly demanding in terms of computing power and efficiency (customization and parallelization).
2) Advances in integrated circuit manufacturing processes are becoming less effective in improving computational efficiency (customization and parallelization).
(3) The agile design of a programmable chip architecture is an inevitable choice in the future in the context of high manufacturing costs (function flexibility).

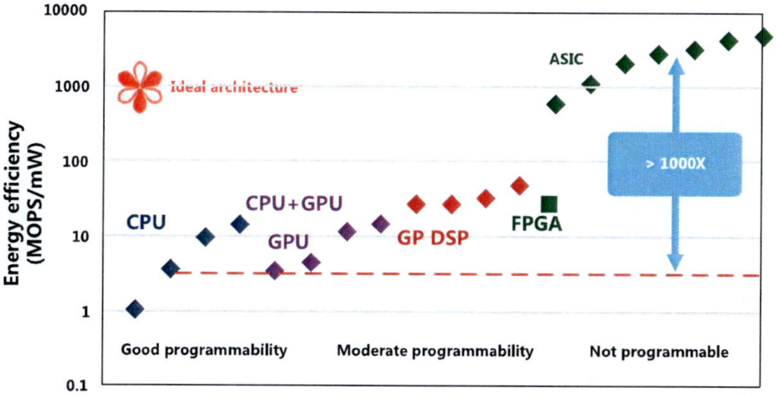

Fig. 2.3 Comparison of existing chip design paradigms (see the color illustration)

(4) Dynamic reconfiguration and dynamic optimization capabilities of the chip architecture are the key to suit the needs of more application domains (runtime flexibility).

As shown in Fig. 2.3, the current mainstream computing chips are chosen in a trade-off between flexibility and energy efficiency. ASICs sacrifice flexibility for high energy efficiency, while GPPs sacrifice energy efficiency for high flexibility. They lie roughly on a straight line, meaning that the high flexibility and high energy efficiency has never been achieved simultaneously. The Turing Award winner Prof. David Patterson predicts that the next decade is the golden age of architecture development [8]. Since the existing architecture technology can hardly meet the above development needs, the breakthrough innovation of architecture research will become crucial. So is it possible to implement the "ideal architecture" in Fig. 2.3 that can meet all the above development needs and how to implement it? This is actually the key problem in the discussion of SDCs.

2.1.2 Technical Implementations

The main body of the "Software Defined Chip" is the chip, and "software defined" describes the key features of the chip in the design and programming process. The design refers to the process of a chip from hardware functional requirements to manufacturing, while programming refers to the process of a chip from application functional requirements to executable configuration contexts. Figure 2.4 provides the explanation of the SDC meaning from these two dimensions.

In terms of programming, the SDC adopts the software programming and compilation process similar to CPUs, with two major differences: (1) Programming model. The SDC uses the domain-specific language (DSL) to bridge the gap between hardware and software and provide as many hardware implementation details as possible,

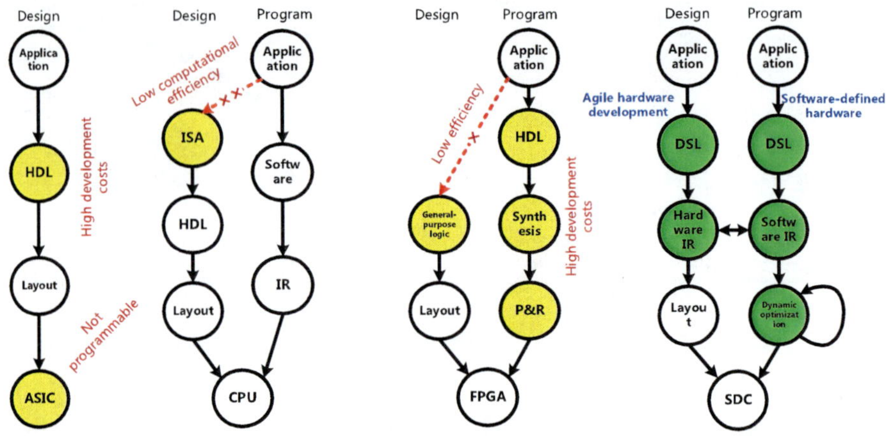

Fig. 2.4 Explanation of the SDC meaning from hardware design and software programming dimensions

thus enhancing the efficiency of chip software programming; (2) Dynamic optimization mechanism. The SDC uses dynamic compilation or dynamic scheduling to bridge the gap between hardware and software in terms of detail abstraction. By maximizing automatic optimization of hardware implementation details, the complexity of software programming is reduced. Bridging the gap between software and hardware is extremely difficult due to the contradictory properties of them. For a detailed discussion of the differences between software and hardware, see Sect. 2.3.

In terms of chip design, SDCs use the similar chip design process of ASICs but different hardware development languages. ASICs use standard HDLs (such as Verilog and VHDL), while SDCs use DSLs to design hardware functions. Note that the DSL here is not the same as the DSL mentioned above for software programming development. Its main functions are to replace the HDL, avoid designing circuits from the RTL, and provide efficient hardware design models, such as CGRAs and systolic arrays, so as to significantly improve the efficiency of hardware development and achieve agile chip development.

The whole development process of the SDC is shown in Fig. 2.5. The figure shows the software development process on the right half. (1) SDCs are mainly intended for emerging applications, such as artificial intelligence, cloud computing, multimedia and wireless networks. These applications are mainly for mobile devices limited by batteries, Internet of things (IoT) devices and warehouse-level data centers, and are data- and compute-intensive with demands for performance and energy efficiency that can hardly be met by traditional computing architectures. (2) To meet the needs of emerging applications, software developers need to describe application functions in domain-specific high-level languages such as Halide, OpenCL, and TensorFlow. Halide is a DSL in C++ for image processors, with a low programming difficulty and high development efficiency. For the local Laplace transform algorithm processing, Adobe could only get a speed-up of 10 times after 3 months of manual optimization of the C++ code, but Halide is able to speed up by more than 20 times using a few lines of code [9]. Moreover, Halide is used in the software development of coarse-grained reconfigurable computing platform of Stanford University. OpenCL is a C99-based general-purpose parallel programming model for heterogeneous systems, providing a parallel computing mechanism based on task partitioning and data partitioning. There are already FPGA vendors providing OpenCL-based development tool chains. Programming models such as TensorFlow and PyTorch are intended for machine learning, with high development efficiency, and are the programming languages of most programmable chips such as TPUs for AI applications. The DSL is aimed to provide an efficient computing paradigm. For example, Halide supports separation of computation from scheduling, and interleaving of data loading and computation; TensorFlow and PyTorch provide multi-layer network computation abstraction; OpenCL provides a task and data partitioning mechanism. DSLs are mostly functional extensions to traditional high-level programming languages, which improve efficiency without compromising the overall flexibility and ease of use. Many reconfigurable computing architectures in the research community do not yet support the above-mentioned complex programming languages, mainly due to the huge amount of engineering. However, the common efficient computing paradigms such as hardware-software partitioning, loop optimization, branch optimization, and parallel pattern have been implemented by adding compilation guidance statements, which are primary DSL implementations. (3) The next step in DSL programming is compilation. Currently, the low level virtual machine (LLVM) framework is one of the most widely used compiler frameworks [10]. LLVM was originally a research project at the University of Illinois at Urbana–Champaign with the goal of providing a compilation optimization strategy based on the static single-assignment (SSA) form, and supporting any static and dynamic programming languages. The key advantage of LLVM is its openness, which provides intermediate representation (IR) and its optimization methods independent of the programming language and the target hardware. Therefore, LLVM can be used in various domains and on different platforms. It supports compilation front ends in languages such as Halide, OpenCL, and TensorFlow, and has been applied to the compilation and development of various programmable chips. Note that the IR of LLVM is mainly intended for CPUs in the form of syntax tree. However, this form of IR is not suitable for spatial computation, so mostly reconfigurable devices further convert the LLVM IR into the IR based on task and data flow graph (TDFG) and perform optimization and back-end mapping on such IRs. The IR based data flow graph (DFG) allows exploring loop-level parallelism (LLP) in a spatial structure. For example, the goal of modulo scheduling is to find invariant kernels with the minimum initiation interval, so as to maximize loop parallelism. Finally, the DFG IR will be mapped to a specific spatial structure, which can actually be abstracted as the problem of finding isomorphic graphs. (4) The dynamic optimization usually includes dynamic compilation and dynamic scheduling. Dynamic compilation can be seen as an extension of the previous compilation step, which regenerates the compilation mapping results within the runtime constraints, thus enabling optimization for different targets. For example, when the system has a limit on the power consumption, the initial configuration can be dynamically converted to use fewer resources, sacrificing execution performance for ultra-low power consumption. When there are multiple faults in the hardware, the configuration can be dynamically converted to bypass the faulty nodes. When the requirements for hardware security is high, the idle computing resources can be used to generate highly secure configurations that can resist side channel attacks. In this way, the optimization for different needs such as energy consumption, security, and reliability can be achieved. Dynamic scheduling is a hardware optimization technique based on compilation, which implements dynamic task management and allocation for SDCs with multiple independent or loosely coupled hardware computing resources, and it requires that tasks can be dynamically compiled for heterogeneous arrays so that they can be freely scheduled by the underlying hardware. Dynamic scheduling can also implement automatic hardware optimization for software by developing scheduling strategies with different optimization objectives. (5) Due to the dynamic compilation technique, the actual configurations executed may be completely different from those statically generated by the compiler, meaning that the specific underlying hardware information does not need to be exposed to the static compilation process, similar to hardware virtualization. Meanwhile, the presence of dynamic scheduling also eliminates the need for programmers or compilers to conduct fine-grained management of the task execution process and hardware resource allocation, thus enabling convenient and efficient SDH.

The left half of Fig. 2.5 shows the hardware development process of the SDC. The process starts from application to functional description in DSL, IR optimization, and to generation of the chip structure described in RTL. Hardware development and software development have the same "Application > DSL > IR > Target" process, but there are significant differences. Both the DSL and IR used for the hardware development process are totally different. The DSL used in software development describes the algorithm execution process of the application. For example, in a DFG, nodes represent algorithmic operations and edges represent algorithmic value transfer or data and control dependences. However, the DSL used in hardware development describes hardware modules. If a graph structure is used to represent a hardware DSL,

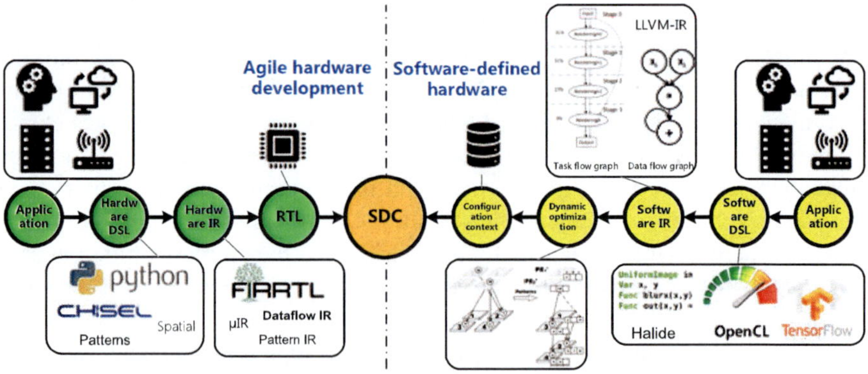

Fig. 2.5 Complete development process of the SDC

nodes may represent concurrently executing hardware modules such as PEs, memory blocks, and routing networks, while edges may represent hardware channels such as interconnects and caches. Commonly used hardware DSLs are closer to the parallel computing of hardware instead of sequential execution model of software. Some hardware DSLs even directly use functional programming languages, which are often called hardware construction languages (HCLs) to distinguish them from the more abstract HLS or the lower-level hardware description languages (HDLs). Chisel is an HCL developed based on Scala which supports hardware design using highly parameterized generators [11]. The functions in Scala are used to describe circuit components, Scala's data types to define interfaces to components, and Scala's goal-oriented programming features to build circuit library files. As a result, Chisel supports readable, reliable and type-safe circuit generation, enabling a significant increase in the robustness and productivity of RTL designs. Note that Chisel is intended for all types of circuits, and it can be used to build spatial structures like SDCs. Therefore, the productivity will be further improved if more specialized hardware DSLs are used, such as Spatial, which is specifically designed for spatial accelerator architectures. On the contrary, software languages (such as C) or software DSLs (such as OpenCL and Halide) are also used in hardware development. Typical examples are HLS tools such as Vivado, Legup and Catapault-C. Although HLS tools have been widely available for FPGA development, their efficiency in practice is unsatisfactory. HLS tools can convert the annotated software programming language directly into RTL code for accelerator hardware, but the code generated by such tools is far from optimized, and the performance and efficiency of the implemented circuits is low. HLS tools still require a lot of manual operations to ensure efficiency, and the more the programmer knows about the underlying hardware design, the better the results will be, making it difficult to achieve agile hardware development with practical value. The major constraints on HLS tools still come from the programming model, which mixes algorithmic functional correctness and microarchitecture optimization design in a single programming model and merges space exploration of software and hardware in a huge design space. In this way, each modification to the accelerator design requires code reconfiguration, which is actually a relatively simple, brutal and inefficient approach to design space exploration. The current high-level synthesizers are embedded in software compilers, converting C code to the RTL. They rely on software IRs to represent the hardware model and on compiler transformations to optimize the hardware microarchitecture design. In the classic loop unrolling optimization, HLS tools can simply translate into multiple parallel hardware instances of the loop body. However, software IRs used by HLS tools are not suitable for the research on microarchitecture optimization mainly for two reasons [12]. (1) Software IR conversions use a control flow-driven von Neumann execution model, which limits the hardware design types and C language behaviors to be optimized. Most HLS tools focus on converting loops into statically scheduled circuit structures. (2) Software IR represents execution behaviors rather than structured components of the microarchitecture, so it is difficult to understand how a single IR conversion changes the output RTL code, and it is difficult to quantify the performance and power overhead. Moreover, this difficulty is particularly evident for cases where multiple IR conversions exist. HLS developers are also well aware of the difficulties of optimizing the hardware microarchitecture based on software IR. They encourage programmers to raise the microarchitecture description to the level of behavior description in C language. This binds the behavior correctness to the structural description of the microarchitecture. However, the two have totally different programming paradigms, and relying on constant modifications to the function code to optimize the microarchitecture is very time- and labor-consuming. In summary, the hardware development process of SDCs is based on hardware DSLs and hardware IR, and programming abstraction for design and optimizing the hardware microarchitecture needs to be addressed. There are a few key points in the overall development process of the SDC:

(1) Both the software and hardware development process are driven by application requirements, but the two development processes should minimize intercoupling, which can simplify HLS and further improve the development speed and development efficiency of the SDC.
(2) The software and hardware development process can mutually improve during iterations. Specifically, software compilation needs to be based on the hardware design model, while the iterative optimization and evaluation of hardware RTLs also needs the support of software compilation.
(3) The software and hardware development process have different method of optimizing applications. Software optimization focuses on the mapping and implementation abstraction of specific algorithms, while hardware optimization focuses on the architecture design abstraction of common patterns of computing, data reuse, control in the domain. Therefore, the DSL and IR used in these two processes are also totally different.
(4) In terms of the HCL and HLS, hardware microarchitecture optimization is iterative and very time-consuming. The key to the SDC's ability to solve this problem lies in the hardware-oriented DSL and IR, as well as the hardware's dynamic optimization capabilities.
(5) The overall objective of the SDC is to achieve high flexibility, high performance and high energy efficiency. The goal of the software development process is parallelization by tailoring software to maximize the parallelism; the goal of the hardware development process is domain specialization by tailoring hardware to achieve the most appropriate degree of specialization.
(6) Based on the above discussion, the following specifies the software and hardware development methods of SDCs in detail.

1. Hardware development of SDCs

Hardware development of SDCs mainly features application-driven, software (DSL) construction and automatic implementation. The traditional development process of digital ASICs is driven by applications and implemented by manual hardware construction. It requires low-level hardware abstraction and hardware designers need to specify and adjust the vast majority of details of the hardware implementation, including registers, timing, control, interfaces, and logic. The current design process of digital ASICs is much simpler than that of analog circuits. The front-end automatic synthesis tool can automatically generate gate-level net lists based on RTL designs, while the back-end P&R and other tools can further automatically generate the layout, back-annotate circuit information, and establish a highly reliable test environment. Relying on standard design libraries and third-party IP, the design of digital ASICs has become relatively simple. In spite of this, the design of large-scale digital ASICs is still costly and risky, and has a long cycle. Figure 2.6 summarizes several hardware design development processes that can alleviate the above problems [12].

Firstly, the HLS can automatically generate RTL designs based on software languages or DSLs. The problem with this approach is that there is a huge semantic and pattern gap between the language describing the application and the hardware design. In addition, large design space, and too much missing design information are also its limitations. The HLS is an advanced method, but the current level of automatic optimization of compilation tools is far from satisfactory. Meanwhile, this approach does not provide good support for manual optimization and therefore there are a number of difficulties currently.

Secondly, HCLs such as Chisel and Delite Spatial are used for application development and then compiled to generate RTL designs. Different from the HLS, the HCL is not a high-level language, but a hardware-oriented DSL. Chisel has a low-level hardware abstraction and is mainly for GPPs. Spatial has a higher-level abstraction

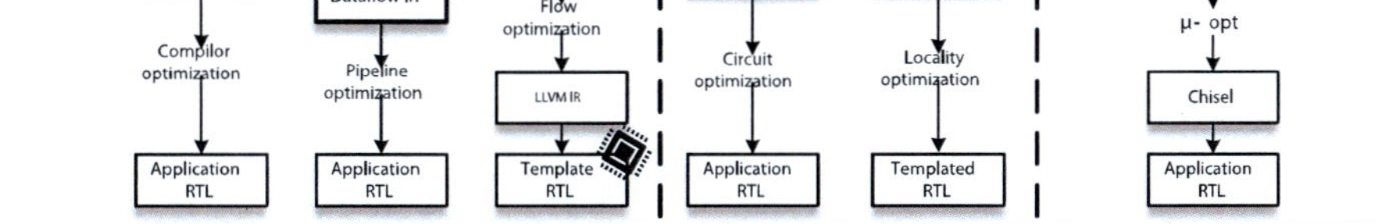

		High -level synthesis (HLS)		Hardware construction language (HCL)		HLS+HCL
		HLS tool chain Legup, Vivado, Catapault-C, AHIR, Pegasus	**Domain-specific IR** OpenCL, SPIR, Halide-HLS, Gorilla++	**Modular RTL** FIRRTL	**Pattern-based HCL** SPATIAL, Pattern	μ- IR μ- opt
Design	Input	C program	DSL	Chisel	DSL	C++/Clik/Tensorflow
	Output	Flat dataflow	pipeline of ker	Flat dataflow	Pattern-based dataflo	Hierarchical data flow
	IR nodes	Compiler operation	Kernals, streams	Logic such a mux	Concurrent mode, memory	Polymorphic operations, dataflow nodes, and memory
Target optimizatio	Higher parallelism	—	√Kernals	—	√mode	√General-purpose
	Operation pipeline	√	—	√	—	√
	Locality	—	√Streams	—	√Streams	√Partitioned and shared
	Resource	—	—	—	—	√Microarchitecture-level
	Higher operations	—	—	—	—	√Tensor operations

Fig. 2.6 HLS-based, HCL-based and mixed hardware development

and is mainly for spatial computation structures. The HCL technology allows users to explicitly perform iterative hardware optimization, with better final effects than the HLS. The prerequisite is that programmers need to have a certain understanding of hardware.

Finally, there is a mixed hardware development approach, which combines the above two techniques. Specifically, automatic compilation is adopted to generate the initial hardware description IR. A series of microarchitecture optimization techniques [12] (Fig. 2.7), such as parallelization, pipelining, data localization, and tensor computation are adopted to generate optimized hardware construction IRs or program that is finally translated into the RTL design. This approach nicely combines the advantages of compilation and manual operations, and is more efficient in optimizing iterations while avoiding excessive hardware details in manual design.

The mixed hardware development has explicitly provided platforms (such as hardware description IR) and reference methods for iteration optimization, and these optimization methods are decoupled from functional correctness. However, the design optimization process with manual effort is still very complex. Recent studies have found that specific hardware structure templates facilitate the design and exploration [13]. As shown in Fig. 2.8, the hardware description IR uses the pattern-fixed architecture description graph (ADG) instead of the general-purpose IR. This approach helps to simplify the overall optimization while improving the ease of use and productization degree of the generated hardware structures. The ADG is actually a modeled abstraction of the hardware architecture, and we will discuss the design of popular computation models, execution models, and microarchitecture models in spatial computation, respectively. The architecture description concept can be extended more widely and deeply by the systematic model discussion.

(1) Computation model

All SDCs belong to multiple instruction, multiple data (MIMD) computation according to Flynn's taxonomy [14]. Furthermore, the classification of computation models based on spatial array configurations was introduced, since instructions cannot fully reflect the computational mechanism of SDCs, as shown in Fig. 2.9 [15].

The first category is the single configuration, single data (SCSD) model. It refers to a spatial compute engine that performs a single configuration on a single dataset. SCSD is a basic implementation of spatial computation. All operations of the application or algorithm kernels are mapped to the underlying hardware, allowing instruction-level parallelism (ILP) to be fully exploited. Despite that computational is limited by the hardware size, SCSD is a powerful general-purpose compute engine that can be adapted to different programming models. For example, Pegasus is an IR of the SCSD model, and the application-specific hardware (ASH) is a microarchitecture template. Pegasus can generate the appropriate ASH configuration for the application [16]. The SCSD model mainly exploits ILP. As shown in Fig. 2.9b, configurations 1–3 must be mapped to three different timeslots in the SCSD model, which does not support thread-level parallelism in the PE array.

The second category is the single configuration, multiple data (SCMD) model. It refers to a compute engine that performs a single configuration on multiple datasets

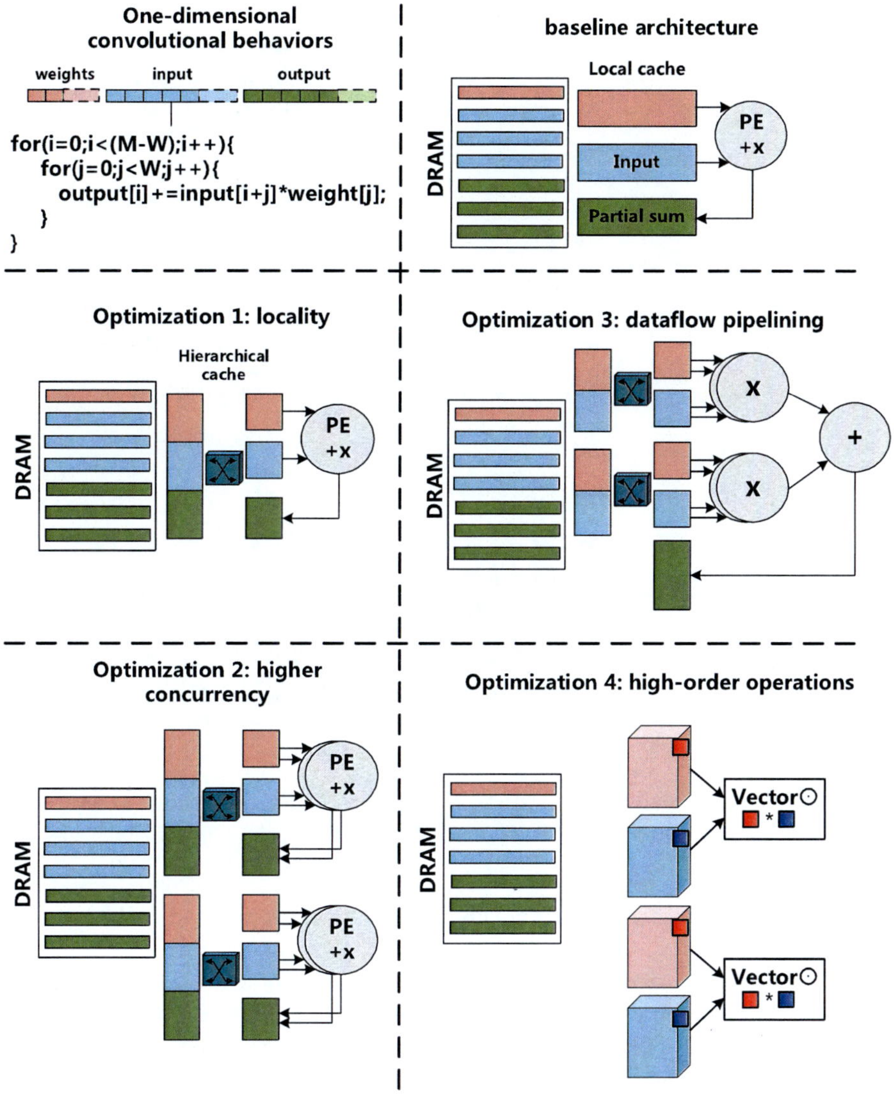

Fig. 2.7 Important optimization techniques for hardware development (with one-dimensional convolution as an example): localization, pipelining, spatial parallelism and tensor calculation (see the color illustration)

spatially distributed, and can be considered as a spatial implementation of the SIMD or single instruction, multiple threads (SIMT) model. The SCMD model is widely applicable to stream and vector computing applications, such as multimedia and digital signal processors, and is therefore adopted by many domain-specific SDCs [17]. The model mainly exploits data-level parallelism (DLP). As shown in Fig. 2.9a,

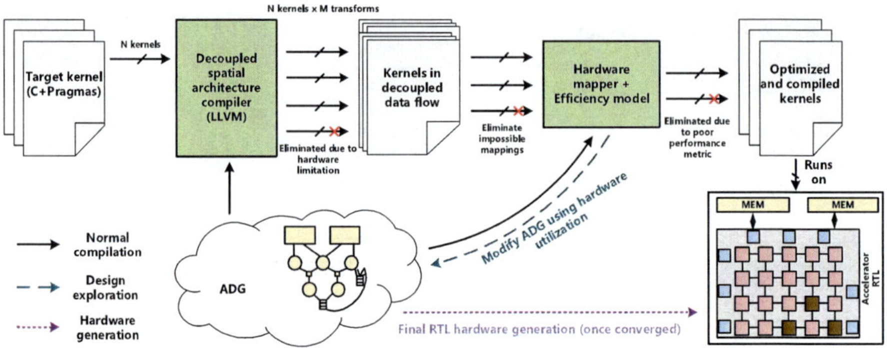

Fig. 2.8 Development of hardware accelerators based on specific templates [13]

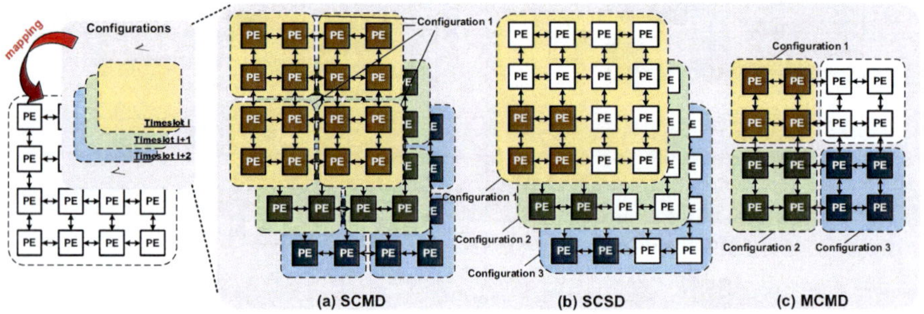

Fig. 2.9 Computation models of the SDC (Configurations 1–3 are independent and asynchronous; different colored rectangles represent different configurations; black represents idle configurations)

the configuration of multiple threads on one time slice in the SCMD model is identical.

The third category is the multiple configuration, multiple data (MCMD) model. It refers to a compute engine that performs multiple configurations (from multiple processes or multiple threads of the same process) on multiple datasets. Therefore, the model requires hardware to provide multithreading mechanisms, including simultaneous multithreading (multiple threads running on different PE subarrays) and temporal multithreading (multiple threads running on the same PE subarray). Usually, inter-thread communication can be implemented by using the message passing or shared memory technique. Due to the distributed interconnection of SDCs, message passing is used more often, rather than shared memory as in multiprocessors. On SDCs, inter-process communication models typically fall into the following categories: The communicating sequential processes (CSP) is a non-deterministic communication model for concurrent computing, where inter-process communication is implemented in a blocking manner by message passing channels without buffering, and process synchronization can also be implemented by these communication channels. Another inter-process communication model often used in SDCs is the Kahn process network (KPN), in which deterministic processes perform asynchronous and non-blocking communication (blocking when the FIFO memory is full) by using a set of first in, first out (FIFO) memories with buffering. Tartan implements asynchronous and handshake communication between PEs to improve energy efficiency [18]. The KPN model has been widely used in a variety of dataflow SDCs, such as TIA and Wavescalar, which employ FIFOs with matching capabilities as asynchronous communication channels between PEs [19, 20]. The MCMD model mainly exploits thread-level parallelism (TLP). As shown in Fig. 2.9c, the MCMD model supports simultaneous multithreading and temporal multithreading.

(2) Execution model

SDCs can be classified based on the execution model, which mainly includes the scheduling mechanism and the execution mechanism of configurations. (1) Scheduling of configurations is the process of loading configurations from memory and mapping them to hardware arrays. To simplify the hardware design, the loading order and mapping location of the configuration can be statically determined by the compiler. For example, FPGAs rely entirely on the compiler for the scheduling of configurations. Alternatively, for higher performance, the scheduling of configurations can be dynamically determined by the hardware scheduler at runtime depending on the system status (e.g., data tokens and branch conditions) and resource usage. For example, superscalar processors enable dynamic instruction scheduling through mechanisms such as predictors, scoreboards, and reservation stations. (2) Execution of configurations is the process of executing the operations contained in the configuration. If different operations are executed in the order determined by the compiler, this operating mechanism is called sequential execution. If the execution of an operation is dynamically determined at runtime by the readiness of its input data, it is called dataflow execution. The dataflow mechanism can be further divided into the static and dynamic dataflow models [21]. In the static dataflow model, the communication channel does not have buffers and the execution can be triggered when the input data is ready and the output channel is not occupied; otherwise, the output channel will be blocked and thus the execution is delayed. This model allows only one thread to be executed at a time. In the dynamic dataflow model, the communication channel has buffers to reduce the impact of blocking the output channel. Meanwhile, it uses unique tags to distinguish data of different threads, allowing multiple threads to execute simultaneously. When the operands are ready and the data tags match, the execution can be triggered. Therefore, SDCs can be classified into four major categories according to the execution model, as illustrated in Fig. 2.10: (1) static-scheduling sequential execution (SSE), (2) static-scheduling static-dataflow (SSD) execution, (3) dynamic-scheduling static-dataflow (DSD) execution, and (4) dynamic-scheduling dynamic-dataflow (DDD) execution. Note that the SSD model adopts static dataflow mechanism to schedule the operations within a configuration (at instruction level) and relies on the compiler to schedule different configurations statically (at thread level). This case is different from that of SSE whose configuration is required to contain only parallel operations or statically sequenced operations. The SDCs that adopt the former two execution models are more suitable as spatial

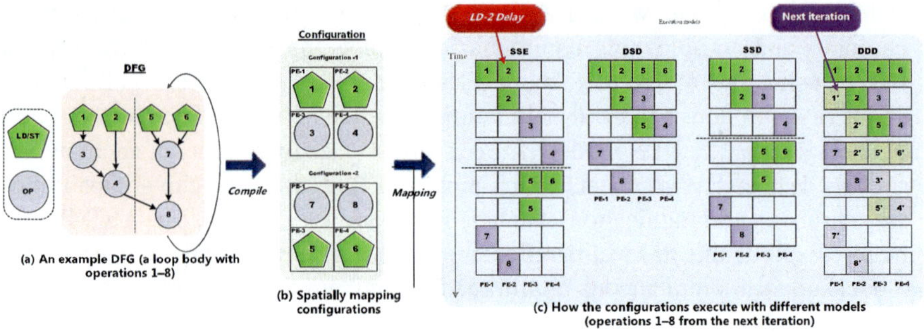

Fig. 2.10 Comparison between different execution models of SDCs Note that DSD and DDD generally require a capability of partial reconfiguration (or MCMD computation)

accelerators, such as Plasticine [22], DySER [23] and CCA [24], while the SDCs that adopt the latter two execution models are more suitable as spatial dataflow machines, such as TRIPS [25] and Wavescalar [20].

(3) Microarchitecture model

The microarchitectures of SDCs have been extensively studied in previous works. There are many different classifications based on microarchitectural characteristics, such as network/interconnect topology, data path granularity, reconfigurable logic function, memory hierarchy, operation scheduling, reconfiguration mechanism, custom operations, coupling with the host, and resource sharing with the host. At the microarchitectural level, it is easy to distinguish two SDCs, but it is difficult to propose complete classification rules with clear boundary. For example, a previous work performed an architecture exploration with different interconnect topologies on ADRES, implying that interconnect topologies are not essential to characterize ADRES [26]. The situation with most microarchitectural characteristics is similar. An individual SDC design could have a series of application-dependent variations with different granularity, reconfigurable logic functions, memory hierarchy or integration methods. Relatively speaking, it is most difficult to distinguish between different SDC architectures based on microarchitectural characteristics.

The mainstream computation model of SDCs has been SCSD for a long time because SCSD has balanced energy efficiency and flexibility. SCMD could enable high performance while maintaining great energy efficiency for data-intensive applications, but it is less flexible for general control-intensive applications. In contrast, MCMD increases control overhead and thus might decrease the computational efficiency, but it adopts the technique of multi-threaded processing to improve the throughput of large SDCs [27]. However, as the scale of SDCs increases, MCMD instances are observed more frequently. The additional area and power overhead caused by complicated control scheme is a key problem that must be solved to maintain high energy efficiency. This could be a trade-off between performance and energy efficiency in architecture designs depending on the application demands and application scenarios. For example, Google still uses classic systolic arrays (SCSD) on TPU

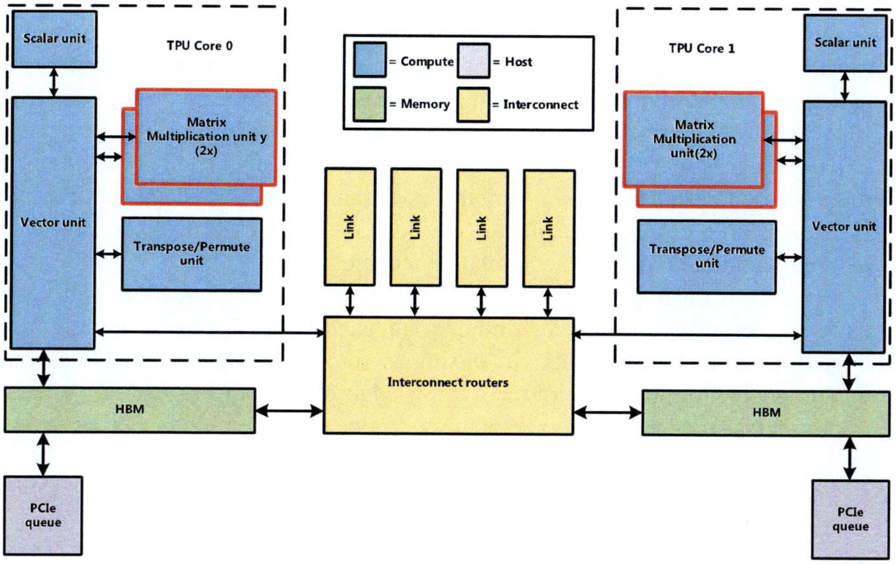

Fig. 2.11 Architecture design of TPU (see the color illustration)

for accelerating DNNs because of the special target application and power budget, as shown in Fig. 2.11 [5]. On TPU2 and TPU3 products, the parallelism is improved to support more efficient models such as SCMD. In architecture designs, these three computation models are frequently adopted according to the target domain.

The most popular execution model of SDCs is SSE because SSE provides an easy-to-use substrate for computation and control. Currently, various compilation techniques are available for static compilation optimization of SSE SDCs [28]. However, the compiler has a inherent defect in the static optimization for irregular applications because the computational tasks in irregular applications usually rely on the input dataset. The algorithmic parallelism cannot be statically determined at compilation, and working loads change greatly on-the-fly. As a result, a growing number of SDCs adopt the other execution models that employ dynamic scheduling or dataflow mechanism to exploit dynamic parallelisms. These models enable high performance for more applications and alleviate burdens on compilers [29, 30]. In recent years, the execution models of SDCs are evolving from static scheduling to dynamic scheduling and from sequential execution to dataflow execution. This trend is quite analogous to the evolution of CPU architectures from the very long instruction word (VLIW) relying on static compilation to out-of-order execution (OoO) relying on dynamic scheduling. Although the dynamic scheduling and dataflow mechanism consume additional power, they are worthwhile if greater performance improvement can be achieved.

2. Software Development of SDCs

Compared with diverse hardware development methods, software development methods of SDCs are relatively clear and mature. They usually follow the process, as shown in Fig. 2.12. For a specific application, the programmer programs with the specific DSL, writes the corresponding algorithm code, and performs certain manual optimization; the front end of the compiler converts the software code into software IRs; the back end of the compiler maps software IRs to hardware IRs to generate the initial configuration; finally, the initial configuration is converted according to the runtime resource and power consumption constraints through dynamic optimization. In the preceding process, the software development of SDCs is mainly characterized by programming models and compilation techniques.

In the software programming process, engineers convert applications with natural language descriptions into programming language code. The programming model directly determines the productivity of software programming, which includes information about the required time and knowledge of hardware for programmers to implement the target application functions. A good programming model can increase the work efficiency of programmers while reducing the work load of subsequent optimization (static and dynamic compilation). How to implement a good programming model will be discussed in details in Chap. 1 of volume 2. The following discusses the types of programming models that might be used for SDCs in terms of the design space.

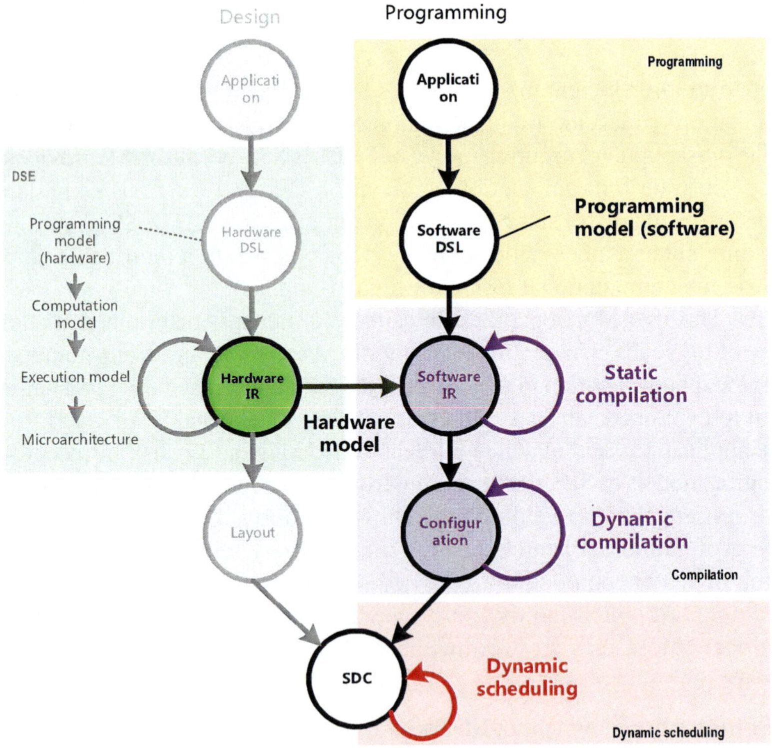

Fig. 2.12 Development process of general-purpose SDCs

(1) Programming model development

SDCs can be classified into three major categories based on execution features.

The first major category uses an imperative programming model and imperative languages, such as C/C++. An imperative model uses an ordered sequence of statements or commands or instructions to control the system states and therefore cannot express any parallelism semantically (note that some extensions of imperative languages can express explicit parallelism, such as PThreads; these extensions fall into the category of parallel programming models). This model can be used by all imperative hardware, such as most processors. An SDC's operation is controlled by its configuration sequence, so an imperative model can also be used. Because imperative languages are relatively easy for programmers and convenient to integrate with GPPs, many SDCs are programmed with imperative models.

The second major category uses a parallel programming model. For simplicity, the concept of parallel programming model used in this book refers to the programming models that can express specific parallelism. This concept comprises (1) a declarative programming model (such as functional languages, dataflow languages, and HDLs), expressing parallelism implicitly, and (2) a parallel/concurrent (imperative) programming model (such as OpenMP, PThreads, Massage Passing Interface (MPI), and CUDA), expressing parallelism explicitly or partially explicitly with directives. The declarative programming model uses declarations or expressions instead of imperative statements to describe the logic of computation. This model does not include any control flow, so computation parallelism is implicitly expressed. The concurrent programming model uses multiple concurrent imperative computations to build a program, thus expressing computation parallelism explicitly. Different from imperative programming models that rely on compilers and hardware to exploit parallelism, parallel programming models allow programmers to express parallelism explicitly with the provided interfaces, thus alleviating the burdens on compilers and hardware and consequently improving compilation and operational efficiency. Although there are fewer SDCs adopting parallel programming models [30], most SDCs are essentially suitable for these models. The reason is that SDC hardware is not imperative, but performs computations in parallel when it processes one configuration that contains multiple operations.

The third category, transparent programming, does not have any programming job or static compilation for a specific SDC architecture. This category requires dynamic compilation and relies on hardware to translate and optimize common program representations at runtime, such as instructions. Thus, its underlying computation models, execution models or microarchitectures could be totally transparent to programmers. For example, the configurations of architectures like DORA, CCA, and DynaSPAM are generated according to runtime instruction flow [31, 32]. Notably, although some SDCs (such as PPA [33]) have dynamic conversion mechanisms for configuration flows, the original configuration still needs to be generated by static compilation, so they do not belong to transparent programming. SDCs with the transparent programming model can often be ported to new computing architectures without modifications to the source program, thus greatly increasing productivity. Moreover, the hardware can perform dynamic optimizations depending on some runtime information, which are not possible with compilers. However, SDCs supporting transparent programming models are often faced with poor performance, reduced energy efficiency, and design difficulties due to the need for additional hardware modules for runtime parallelism exploitation.

In order to provide user-friendly software interfaces, the primary programming model for most SDCs is imperative programming. However, the fundamental contradiction between spatial computation architectures and imperative programming has not yet been efficiently solved. SDCs supporting high-level imperative programming languages often require complex manual optimizations to achieve higher performance, which raises the use threshold of SDCs, reduces productivity, and severely limits their application scope. Although declarative programming and parallel/concurrent programming are more challenging for programmers, these programming models are more suitable for the spatial computation model of SDCs, enabling more parallelism, and the adoption of such programming models is necessary for the development of SDC applications. Currently, some new SDCs have adopted a parallel programming model. In the future, from the programming efficiency perspective, some SDCs will continue to use imperative programming and use additional programming extensions to exploit more parallelism, both general-purpose and domain-specific parallel patterns.

(2) Development of compilation and dynamic optimization techniques

Compilation is a software technique that automatically converts one programming language into another language (assembly language for the target hardware). The compilation of SDCs takes longer because their assembly languages are complex configurations. Unlike the instruction flow of traditional processors, configurations contain a wider range of execution details, including data communication through interconnects, synchronization modes of multiple parallel PEs, usage of multiple on-chip storage resources, and data consistency maintenance. As a result, the target design space of converting configurations is extremely complex, thus becoming a technical bottleneck for software development of SDCs. The dynamic optimization technique is mainly to optimize software or hardware in real time during execution. The dynamic optimization technique is mainly intended for irregular applications, of which some key features and data are not known before operation, and the execution process of the program software and hardware needs to be adjusted accordingly at runtime in order to achieve optimal working efficiency.

SDC compilation converts programs (software DSLs) into executable configurations. As shown in Fig. 2.12, depending on the time (before operation or at runtime), SDC compilation can be roughly divided into two phases: static compilation and dynamic compilation. Static compilation is done by the compiler before runtime, while dynamic compilation is done by the hardware and software at runtime [34]. Dynamic compilation determines the mapping from software tasks to hardware resources based on real-time power overhead and requirements for performance, security and reliability, and it is one of the most important means to achieve dynamic optimization of software configurations [35]. As shown in Fig. 2.12, in addition to dynamic compilation, dynamic scheduling is another important means to achieve dynamic optimization of hardware and software, which aims to reduce the waiting cycles for the execution of configurations and improve the resource utilization by dynamically adjusting the execution order of the configuration flow after the final configuration flow is generated [27]. Dynamic compilation and dynamic scheduling are complementary. Specifically, dynamic compilation determines the spatial mapping of configurations (i.e., P&R), while dynamic scheduling determines the execution timing of configurations (this book argues that dynamic optimization of SDCs actually includes dynamic compilation and dynamic scheduling).

Broadly speaking, all techniques that convert one programming language to another (note that IRs are also programming languages) can be considered as compilation techniques, so the compilation techniques of SDCs can be broadly classified into static compilation, dynamic compilation, and dynamic scheduling (usually considered as a hardware mechanism) shown in Fig. 2.12. Figure 2.13 shows a generalized classification of the compilation techniques of SDCs. The existing techniques are classified depending on whether dynamic mapping and dynamic execution are enabled. Static compilation determines the spatial mapping of task resources and the execution order of configurations before runtime; dynamic compilation changes the spatial mapping while maintains the execution order at runtime; dynamic scheduling changes the execution order while maintains the spatial mapping at runtime; elastic scheduling changes both the execution order and the spatial mapping of configurations. These techniques are different, but they are not contradictory in use. For example, static compilation is often the basis for dynamic compilation and dynamic execution, and dynamic compilation and dynamic scheduling can be used together.

The following introduces the compilation and dynamic optimization techniques [36] of SDCs one by one, combined with Fig. 2.14. Different compilation and dynamic optimization techniques determine the application's parallelism to be exploited and utilization of spatially parallel hardware resources.

(1) Static compilation will be described in detail in Chap. 4. The technique relies on the compiler to statically determine the temporal and spatial mapping of the task. The compiler software needs to use complex optimization algorithms to eliminate dependences and extract parallelisms. For example, a compiler can obtain coarse-grained LLP through pipeline transformation and develop DLP by concurrently executing multiple groups in iteration space. Static compilation often struggles to achieve the desired results when dealing with irregular applications that contain a large number of control flows. As shown in Fig. 2.14, the execution of the statement "s + = d" depends on the condition statement "d $\geq$ 0".

In the first and fourth iterations, the statement "s + = d" does not need to be executed, but the compiler needs to reserve all potential dependences, thus generating over-sequentialized scheduling results. For example, Fig. 2.14b-1 shows a conservative static pipeline generated, in which the statement "s + = d" takes up resources but is not actually executed, resulting in wasted waiting time and compromised

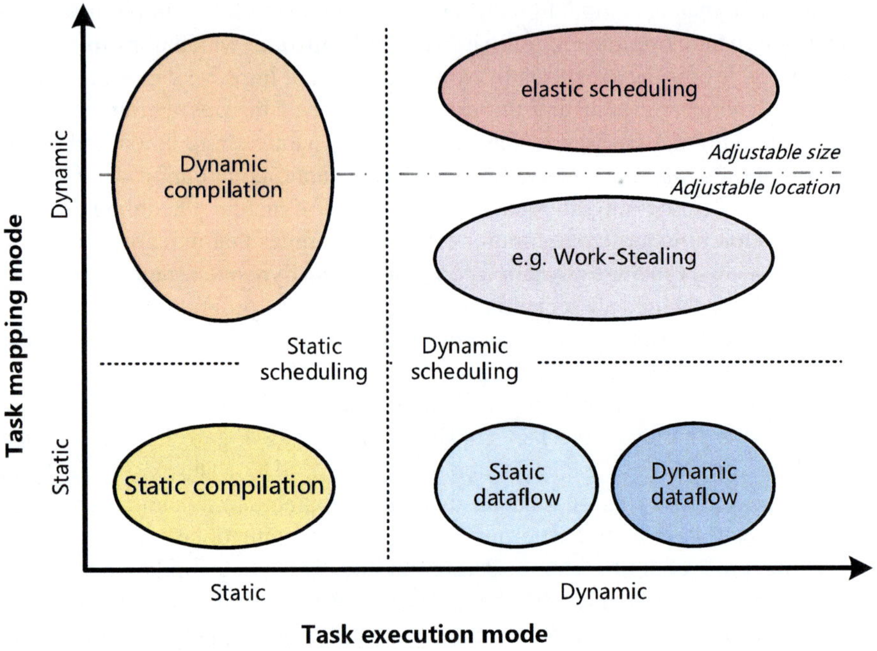

Fig. 2.13 Broad classification of compilation techniques of SDCs: dynamic mapping and dynamic execution (see the color illustration)

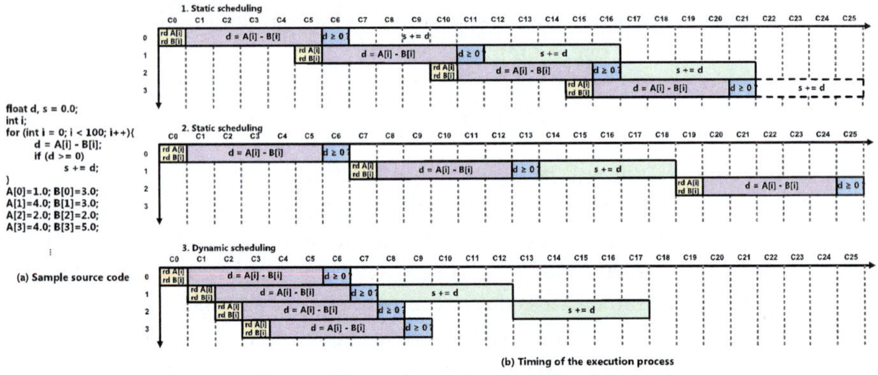

Fig. 2.14 Comparative analysis of dynamic scheduling and static scheduling (see the color illustration)

performance, while Fig. 2.14b-2 shows that a finite-state machine (FSM) is generated to sequence executions, which ensures that there is no redundant waiting time, but the entire sequential execution process degrades performance. The task-level speculation (TLS) technique of the compiler can help mitigate the impact of such problems by breaking infrequent dependences, but require the branches to possess good predictability.

(2) Dynamic scheduling (based on the preceding dynamic scheduling execution model) will be described in detail in Chap. 3. The technique is compiler-independent and relies on the dynamic triggering and execution of hardware tasks to improve task parallelism. Many newly proposed hardware architectures such as Stream-dataflow [30] and DPU [37] use a dataflow execution model so that dependences can be determined at runtime depending on data flow. As shown in Fig. 2.14b-3 , the execution order of all statements depends only on the readiness of their dependent data. Such a dynamic execution pattern drastically reduces the waiting time for the execution of each statement, thus shortening the overall execution time. For example, the statement "s + = d" of the second iteration can be significantly executed earlier, because the statement "s + = d" of the first iteration does not need to be executed, and the dependence on s can be determined when determining "d $\geq$ 0", thus making the statement "d = A[i]B[i]" of the second iteration also be executed earlier (in contrast, the statement "d = A[i]B[i]" of the third iteration is also executed earlier, but with little effect). Dynamic execution can mitigate uneven task loads and avoid idle hardware resources. However, this is followed by a higher task scheduling cost, including the storage and solving of data/control dependences for multiple tasks, which not only requires additional area and power consumption, but also causes certain delays on the execution of each task.
(3) Dynamic compilation will be described in detail in Chap. 4. This technique dynamically adjusts the spatial mapping relationship between hardware resources and software tasks. For example, a task can be mapped to more hardware resources by task duplication, or to fewer hardware resources by folding. It does not usually involve dynamic exploitation of parallelism and analysis of dependences whose complexity would be difficult to handle at runtime. Dynamic compilation is based on the decoupling of software and hardware, which allows software to be mapped to different hardware resources for execution. The advantage of dynamic compilation is the flexibility of the system, which allows the system to be optimized for multiple objectives in real time. Figure 2.14 does not show the effect of the dynamic compilation technique. In practice, for execution blocks (e.g., "s + = d") on critical path, dynamic compilation can be used to adjust the hardware resources used to change the execution time and power consumption, enabling the flexibility of dynamic optimization.
(4) Elastic scheduling is a combination of dynamic compilation and dynamic scheduling. The comparison shows that the role of dynamic scheduling is to regulate the timing of task execution (in the time domain), while the role of dynamic compilation is to regulate the spatial parallelism of task execution (in the spatial domain). Elastic scheduling allows complete and dynamic solving of potential dependences and adaptive management of computational load imbalance. Parallel XL is a task-based computation model on FPGAs which is based on the continuation passing model and the work stealing task mapping method

[29], so it is a complete elastic scheduling strategy that supports dynamic task regulation in the time and spatial domains. But work stealing is not quite suitable for SDCs, because it requires computing units to provide additional logic support and a bus to support task communication, which will reduce efficiency. Work stealing is mainly intended for multi-core architectures to frequently switch fine-grained tasks. However, this will cause an increase in the number of spatial structure reconfigurations for SDCs (the cost of reconfigurations of SDCs is more expensive), thus reducing energy efficiency [38].

In summary, the division of programming models, computation models, execution models, and compilation techniques is a top-down approach in hierarchical system design. There are specific correspondences between the models at different levels. (1) The SCSD model can implement programs with imperative languages and HDLs, and can be implemented on static-scheduling sequential or dataflow execution models. (2) The SCMD model can implement programs with imperative languages and dataflow and functional languages, and can be implemented on the dynamic-scheduling static dataflow execution model. (3) The MCMD model can implement programs with concurrent programming models, and can be implemented on the dynamic-scheduling dynamic dataflow execution model. These correspondences expose the intrinsic difference from one SDC design to another, proving the reasonableness of the preceding taxonomy. Since these correspondences are not one-to-one, it is a challenge to determine which combination generates a more efficient SDC architecture.

The conclusion is that there is no universally most efficient general-purpose architecture, and that the design of efficient SDC architectures is closely related to the characteristic of application. For example, for regular applications, static optimization is sufficient to exploit all possible parallelisms, so there is no need to use dynamic scheduling at all, and using the static-scheduling sequential execution model and static compilation is sufficient; for irregular applications, static optimization may be efficient in some cases and inefficient in others, and dynamic scheduling and elastic scheduling techniques are necessary. This problem will be discussed in detail in Chap. 3.

2.1.3 Technical Comparison

1. Abstraction Model

The SDC is a design paradigm for domain-specific computing architectures, and Fig. 2.15 compares SDCs with GPPs.

Figure 2.15 deals with the common abstraction layers of computer architectures, which are first explained as follows: (1) A programming model is an abstraction of an underlying computer system that allows for the expression of both algorithms and data structures [39]. This model bridges the gap between the underlying hardware

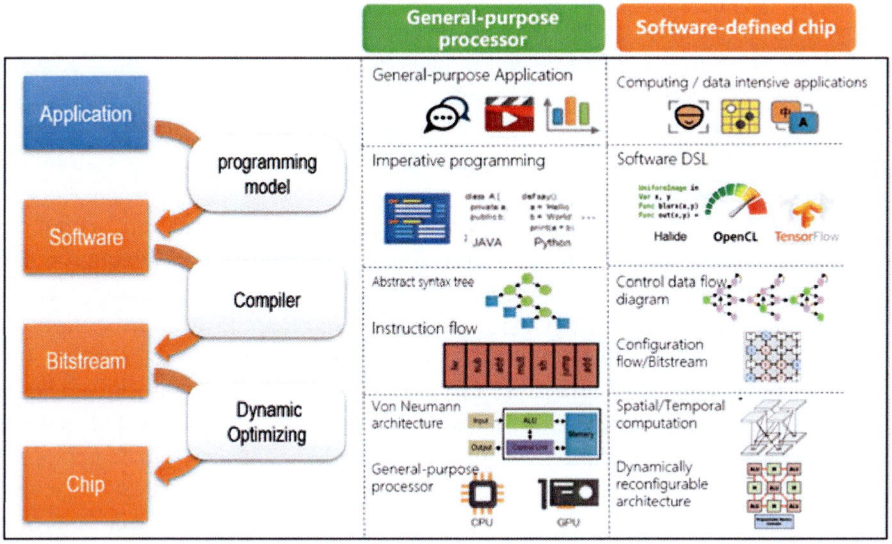

Fig. 2.15 Architecture comparison between SDCs and GPPs

and the supported software layers available to applications. In fact, all programming languages and application programming interfaces (APIs) are instantiations of programming models. The programming model can describe software applications as well as program hardware [40]. This model abstracts the hardware details so that programmers can specify parallelism without worrying how this parallelism is implemented in hardware. Meanwhile, the model determines which part of algorithmic parallelisms within applications can be explicitly expressed by the programmer, thus simplifying compiling the algorithms. For example, a typical multi-thread programming model, e.g., PThread, abstracts hardware resources as threads so that programmers can represent the parallelism of the application as coordinating threads. (2) An IR can often be seen as a special programming language or computation model [41], which is a data structure or code used internally by the compiler to represent the source code [42]. The IR aims to help with further processing, such as code optimization and conversion. A good IR must be able to accurately represent the source code without losing information.

Moreover, it is independent of any particular source or target language, and any target hardware architecture. IRs can take the forms of data structures in memory, or program-readable code for a special tuple/stack. In the latter case, it is also called an intermediate language. (3) The execution model is an abstract representation of microarchitectures, which defines the execution scheme of the hardware computation and characterizes the core working mechanisms of hardware, such as the triggering, execution and termination of instructions/configurations. Therefore, the execution model provides the basic framework for microarchitecture designs, determining the execution order of underlying hardware and the implementation of parallel patterns. The programming model layer transforms a target application into a program with specific explicit parallelisms. The IR layer transforms the programs into IRs that consist of operations, data sets and threads in parallel. The execution model layer maps these IRs onto the underlying microarchitecture, generating (offline or on-the-fly) a runtime bitstream that is directly run by the hardware.

Figure 2.15 shows the key features that distinguish SDCs from traditional computing chips at all levels.

At the application level, SDCs are often more suitable for data- and computation-intensive applications. The hot functions (occupying most of the execution time) of these applications consist of a large number of computations or data accesses, with the execution of control flow often taking a relatively shorter time. Moreover, the control flow is usually quite regular and has less impact on parallelization. Of course, it can also implement general-purpose algorithms, but the efficiency of the implementation will be somewhat reduced. Thus, the application-layer characteristics of SDCs can be summarized as: application-specific flexibility. SDCs have a high degree of post-fabrication flexibility (i.e., the chip's function can be changed after fabrication). Their hardware can be defined by software at runtime, but PEs are not as powerful as those of GPPs, and their interconnects are not as complex as those of FPGAs. This architecture is flexible enough for one or multiple specific domains, meeting requirements of the majority applications in the domain.

At the programming model level, SDCs often use DSLs. Compared with those programming languages that are general across domains, DSLs are intended for programming applications in specific domains, and often provide additional specialized functions. GPPs generally use flexible and easy-to-learn imperative programming languages, such as C++, Java, and Python. However, DSL are often more complex and more difficult to learn. They can be extensions of imperative programming languages, such as Halide, OpenCL, and PyTorch, or other types of programming languages such as Chisel and Scala.

At the IR level, SDCs prefer IRs in the graph form for code optimization and mapping, such as control/data flow graph (CDFG). In contrast, GPPs generally use an abstract syntax tree (AST) and linear IRs for optimization and assembly. Linear IRs such as SSA can directly represent the processor's pseudocode, which are actually similar to assembly languages and well suited for linear computational structures like GPPs. SDCs require optimization of circuit functions, and IRs also need to adopt a graph structure to fully exploit potential parallelisms.

At the compilation level, SDCs require the configuration flow that explicitly defines the usage of parallel operators, interconnects, on-chip memories, and external interfaces. In contrast, GPPs require the instruction flow that is extremely flexible and easy to use, and can contain far less hardware optimization information than SDCs. SDCs use a combination of static compilation and dynamic optimization, and their configurations can be dynamically generated or transformed according to runtime requirements. This is similar to dynamic compilation techniques of GPPs, such as just-in-time (JIT) compilation frameworks like Java and Google's V8.

At the execution model level, SDCs use a dynamically reconfigurable spatial architecture model, which is a 2D or 3D (i.e., dynamically reconfigurable) computing process that allows multiple nodes to communicate rapidly in parallel or through local interconnects. In contrast, GPPs are based on linear computing processes at a single computational node and are essentially sequentialized. Although SDCs are often implemented on coarse-grained dynamically reconfigurable architectures, their granularity actually varies widely from application to application, and are mixed-grained mostly. For example, SDCs used to implement encryption/decryption applications may contain a number of fine-grained components.

In summary, SDCs have distinctive features at all abstraction levels of chip architecture designs, including spatial computation model, dynamic reconfiguration capabilities, high-level language programming and domain-specific capabilities, which are fundamentally different from GPPs. Table 2.1 further compares SDCs with mainstream computing architectures, and specifically analyzes the key advantages of SDCs.

The difference between SDCs and general-purpose architectures (such as FPGAs and GPPs) is domain-specific flexibility. It tailors the hardware to target applications and keeps redundant resources minimized. As a result, for the target domain, SDCs are typically 1–2 orders of magnitude more energy-efficient than FPGAs and 2–3 orders of magnitude more energy-efficient than GPPs [44]. For general applications, the advantage of SDCs typically shrinks. Therefore, domain-specific flexibility proves to be one of the critical reasons for SDCs' balance between energy efficiency and flexibility.

In spatial computation, SDCs take advantage of parallel computing resources and data transferring channels to perform computation. In temporal computation, SDCs take advantage of dynamically reconfigurable resources to perform computation. Therefore, the mapping of an SDC is actually equivalent to identifying the spatial and temporal coordinates of every vertex and edge in the CDFG. Compilers are responsible for making this arrangement.

The combination of spatial and temporal computation provides a more flexible and powerful implementation framework for applications. Relative to architectures that enable only temporal computation (e.g., GPPs), SDCs can obviate costly deep pipelines and the centralized communication overhead. Relative to architectures that enable only spatial computation (e.g., conventional FPGAs, programmable array logic (PAL) architectures, and ASICs), SDCs can improve area efficiency. Therefore, combining spatial and temporal computation is one of the critical reasons for SDCs' high energy/area efficiency without reducing flexibility.

As opposed to sequential processors whose operations are driven by control flow (statically determined by compilers), SDCs have their operations driven mostly by configuration flow or data flow. The configuration of SDCs defines interconnections in addition to PE operations. All the PEs defined by one configuration execute in lockstep and under the same flow of control (thread). Although the configurations are also driven by control flow mostly, the operations in each configuration are in parallel or pipelined, which exploits compiler-directed parallelism. More importantly, configuration-driven SDCs can efficiently exploit explicit dataflow via interconnects, which is not supported in conventional instruction sets. A data-driven SDC is an implementation of an explicit dataflow machine, which abandons control flow execution completely. With all operations in one configuration as candidates, any

Table 2.1 Comparisons between SDCs and important computing fabrics

Architecture	Computation form		Execution mechanism				Flexibility
	Temporal computing	Spatial computing	Reconfiguration time[b]	Configuration-driven	Data driven	Instruction-driven	
Software-defined chip	√	√	ns–μs	√	√	×	Domain
FPGA	×[a]	√	ms–s	√	√	×	Standardization
ASIC	×	√	×	×	√	×	Fixed
Sequential CPU/VLIW	√	×	ns	×	×	√	Standardization
OoOE CPU	√	×	ns	×	√	√	Standardization
Multi-core	√	√	ns	×	×[c]	√	Standardization

[a]FPGAs can perform temporal computation, but it is not practical considering the overhead and effectiveness
[b]The reconfiguration time is not accurate, and the data come from recent works with technologies below 90 nm
[c]Dataflow mechanisms can be supported in software at the task/thread level, e.g., data-triggered multi-threading [43]

one that has its operands prepared will be executed. Here, data-driven SDCs follow an explicit producer–consumer data-dependent relationship.

Compared to control-flow-driven or instruction-driven execution, as occurs in, e.g., multi-core processors, configuration-/data-driven execution can avoid over-sequentialized execution of PEs, exploit fine-grained parallelism, and provide efficient synchronization among PEs. This execution style further supports explicit data communication, which can minimize the energy overhead of data movement. Therefore, configuration-/data-driven execution is one of the critical reasons for SDCs' high performance and energy efficiency.

2. Instantiations

The preceding comparison is made from the abstraction model layer, while the following makes a detailed comparison between SDCs and two mainstream computing architectures in terms of implementations and effects.

Figure 2.16 compares the specific implementations of SDCs and FPGAs at the circuit layer. (1) In terms of reconfiguration granularity, FPGAs use single-bit reconfiguration, while SDCs prefer coarse-grained reconfiguration, mostly mixed with fine- and medium-grained implementations. SDCs tend to use coarse-grained reconfiguration because mainstream applications mainly implement coarse-grained computation, and the use of coarse-grained reconfiguration dramatically improves the energy efficiency and performance of the circuit structure. In domains like cryptography and signal processing, the reconfiguration granularity of SDCs needs to become finer accordingly. The relatively coarse reconfiguration granularity gives SDCs a significant advantage in terms of the amount of configurations. The typical amount of configurations for FPGAs is 10–100 MB, while that for SDCs is 1 to 10 KB. The reduction in the amount of configurations will further shorten configuration time, making it possible to configure while computing and enabling the SDC to be dynamically reconfigurable. (2) In terms of fabrication process, FPGAs use a special process, while SDCs can use standard processes. This is essentially due to the fine reconfiguration granularity and low area efficiency of FPGAs, which require special processes to improve the design efficiency. (3) In terms of programming methods, FPGAs are mainly programmed in HDLs, or they can be programmed in high-level languages that require a lot of manual assistance. Moreover, the programmer needs to have sound knowledge of hardware circuits, and the electronic design automation (EDA) tools such as synthesis and P&R tools need to be used to covert the programs written into circuit descriptions, and generate configuration files. SDCs are mainly programmed with high-level languages, so the programmer does not need to have knowledge of circuits. In addition, the program written does not need to be converted into circuits, and configuration files can be generated directly. In fact, the problem of high programming difficulty of FPGAs can be considered to be caused by their fine programming granularity, because the configurations of FPGAs need to contain too much circuit-level information that cannot be contained in software, including logic functions of LUTs, register timing functions, crossbar interconnects, and interface definitions.

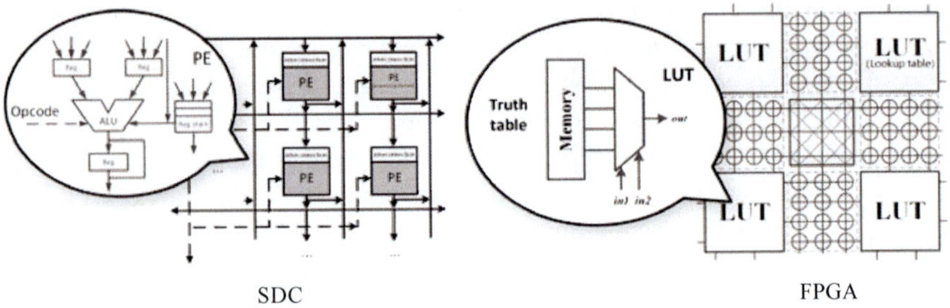

Fig. 2.16 Comparison of circuit structures between SDCs and FPGAs

Figure 2.17 shows the comparison of specific implementations between SDCs and multi-core processors at the system layer. (1) In terms of PE implementations, the multi-core architecture uses instruction-set processors, including OoO large cores with considerable overhead to ensure performance and latency, and energy-efficient sequential-execution small cores to ensure computational efficiency. However, the instruction-set processor has a low computational efficiency, because it integrates communication, control, and computation into a general-purpose module, with the computation occupying very less time and power in the overall operation. SDCs use data-driven ALUs that basically only need to perform computational functions. (2) In terms of communication between PEs, multi-core architectures have to use complex buses or NoCs for the unified memory of their programming models, while SDCs can use simple direct-connected wires and multiplexers to implement data communication. (3) In terms of parallel execution models, multi-core architectures rely heavily on multithreading to exploit thread-level parallelism, while SDCs can further exploit instruction-level parallelism and loop-level parallelism within threads.

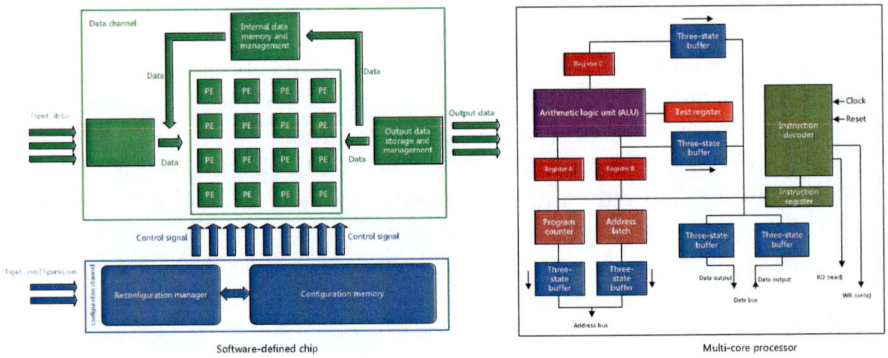

Fig. 2.17 Comparison between a typical SDC system and multi-core processor architecture (see the color illustration)

2.2 Characteristic Analysis

Section 2.1 introduces the basic principles of SDCs and demonstrates the comprehensive advantages of SDCs in terms of flexibility, energy efficiency, and ease of use by comparing them with mainstream computing chips such as ASICs, FPGAs, and CPUs. This section further discusses why SDCs can have all these advantages. We provides the key design methods that enable SDCs with high computational efficiency, low programming difficulty, unlimited capacity, and high hardware security. In addition to qualitative analysis, we also provides hardware modeling and quantitative analysis of SDCs.

2.2.1 High Computational Efficiency

For chip designs, broadly speaking, performance comes from parallelization and efficiency comes from specialization. The high performance and energy efficiency of SDCs comes from that the design model can combine specialization and parallelization better than before. As shown in Fig. 2.18, a typical SDC can be designed for specialization and parallelization optimization from multiple aspects.

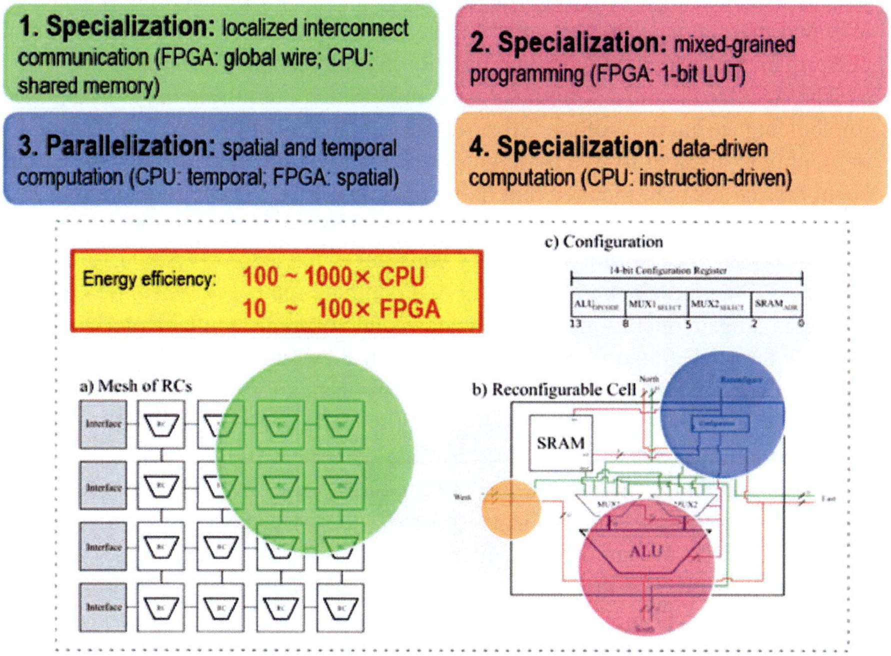

Fig. 2.18 Key architecture advantages of SDCs for high performance and energy efficiency (see the color illustration)

In terms of specialization, SDCs can adopt: (1) specialized granularity designs for application data, whether coarse, medium, fine, mixed or SIMD/SIMT, avoiding general-purpose single-bit granularity like FPGAs or 32-/64-bit fixed granularity like CPUs; (2) specialized interconnect design for application data movement patterns, avoiding the use of quite general-purpose and flexible crossbar switches with extremely high area overhead like traditional programmable devices, or the use of multi-level cache systems and shared memory mechanisms like CPUs; and (3) specialized control patterns for compute/data access in applications, such as dataflow-driven stream computing and sparse computation with indirect access, avoiding the general-purpose instruction-driven computation of CPUs.

In terms of parallelization, SDCs can concurrently adopt: (1) spatial computation, which fully expands tasks in two dimensions to provide maximum instruction-level parallelism while exploiting the locality of data (i.e., "producer–consumer" relationship), and to minimize the huge overhead of data movement; and (2) temporal computation, which folds tasks according to the hardware size, supports dynamic configuration switching, and forms a software pipeline to improve hardware resource utilization and avoid capacity limitations of traditional programmable devices.

To support the preceding qualitative analysis, the following part theoretically models the mainstream computing architectures to analyze the benefits from specialization and parallelization. This section refers to the modeling and analysis methods of the classic work [45]. Figure 2.19 shows the difference in implementations of the same function (full adder) with an ASIC, a GPP, an FPGA, and an SDC. It is assumed that the basic programmable PEs are 2.input LUTs, and the overhead and performance is evaluated when the function is extended to N full adders.

The ASIC is the most straightforward and efficient, requiring only seven gates to implement the functions of the full adder. Considering that the area of a 2-input LUT is roughly equivalent to five logic gates, the overhead of the ASIC implementation is slightly larger than the area and delay of a single LUT. For implementing N full adders, it is assumed that the area is $A_{\text{ASIC}} \approx N \times A_{\text{LUT}}$, the computation time is fixed to $D_{\text{ASIC}} \approx D_{\text{LUT}}$, and the energy consumption is $E_{\text{ASIC}} \approx N \times E_{\text{LUT}}$.

The GPP is implemented with temporal computation based on single LUTs, and the $7N$ logic gate operations of N full adder operations are implemented by sequentializing the LUT functions through instruction flow reconstruction. It has only one 2-input LUT for computation, but the instruction memory and data memory incur considerable overhead. In addition, instruction flow control is not yet reflected in the graph, but it is actually a high overhead. Therefore, $A_{\text{GPP}} = A_{\text{LUT}} + A_{\text{memory}} + A_{\text{controller}}$. In the area overhead of the memory, the area of instruction memory grows slowly due to the presence of instruction sharing mechanisms such as loops and SIMD. However, the area of data memory is closely related to the characteristics of the application. For those applications with a large memory footprint, the area of data memory will increase significantly with the size of the application, and vice versa. For a full adder example given here, the area of instruction memory and data memory is almost constant since it is implemented through iterative instruction loops and memory releases. The area overhead of the controller, although large, is largely unrelated to the application scale (N). Therefore, A_{CPU} is essentially a constant. The

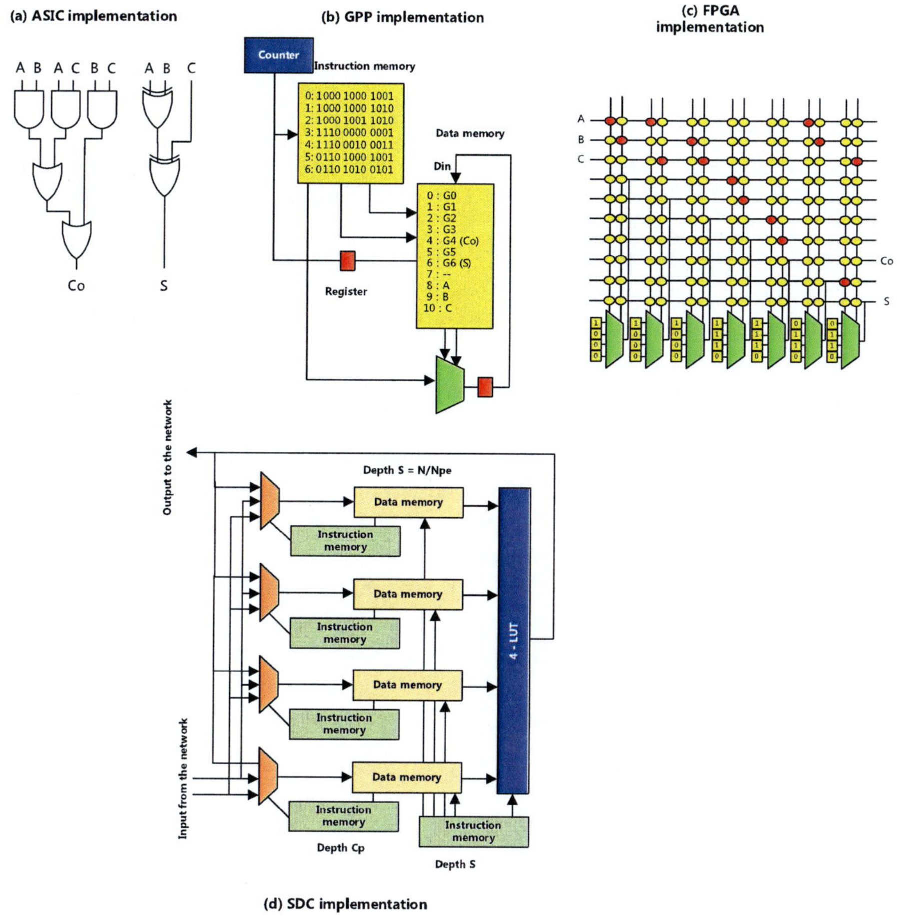

Fig. 2.19 Theoretical implementation of the full adder function

computation time of the GPP is $D_{\text{GPP}} = 7N \times (D_{\text{LUT}} + D_{\text{memory}})$, where **7** indicates that the full adder requires a total of **7** logic gate operations. Due to the sequential execution used, the total execution time is very long and becomes longer linearly as the problem size becomes larger. The energy consumption of the GPP is $\text{E}_{\text{GPP}} = 7N \times (E_{\text{LUT}} + E_{\text{memory}} + E_{\text{controller}})$.

The FPGA is implemented with statically reconfigurable and fully spatial computation. It uses $7N$ 2-input LUTs to implement $7N$ logic gate operations for N full adder operations, and uses crossbar switches to interconnect I/Os of all PEs. It can provide support for all 2-input logic gate network structures of size less than 7, but at the cost of considerable area and power overhead. Compared to ASICs, FPGAs have the main overhead coming from two parts: the memory for configurations and the crossbar interconnect logic (including a large number of crossbar switches and interconnection lines). The LUT requires $N \times 28$-bit configurations, and each input requires $N \times 2\log_2 N$ configurations. The size of the crossbar interconnect is $2N \times N$, so a total of $2N^2$ crosspoints are required, each with an area approximately twice the size of the memory. Therefore, the area of the FPGA is approximately $A_{\text{FPGA}} = N \times A_{\text{LUT}} + (28N + 2N \times \log_2 N) \times A_{\text{Mbit}} + 2N^2 \times A_{\text{crosspoint}} + N^2 \times A_{\text{wire}}$. The FPGA has the area proportional to the square of its size. The computation time of the FPGA is approximately $D_{\text{FPGA}} = 3 \times D_{\text{LUT}} + 2 \times D_{\text{crossbar}}$. The energy consumption of the FPGA is approximately $E_{\text{FPGA}} = 7N \times E_{\text{LUT}} + (28N + 2N \times \log_2 N) \times E_{\text{Mbit}} + 2N^2 \times E_{\text{crosspoint}} + N^2 \times E_{\text{wire}}$.

The SDC is implemented with a combination of spatial computation and temporal computation. It can use M ($7 < M < 7N$) 2-input LUTs to implement $7N$ logic gate operations with N full adder operations, while using a relatively simple mesh-based interconnect to connect adjacent PEs to each other. Although the SDC cannot efficiently support all possible logic gate networks, it can fully support the implementation of example circuits. The major differences between SDCs and FPGAs are the dramatic reduction in the amount of configurations and the complexity of the mesh interconnect logic. The LUT requires a total of $4 \times M$-bit configurations, and its I/O interconnects require $4 \times M$ configurations (only the interconnects of adjacent LUTs are considered). However, note that SDCs require dynamic switching of configurations. To complete $7N$ operations, up to $K = 7N/M$ reconfigurations are required. The area of the SDC is approximately $A_{\text{SDC}} = M \times A_{\text{LUT}} + 8M \times A_{\text{Mbit}} + M^2 \times A_{\text{wire}}$. The computation time of the SDC is $D_{\text{SDC}} = K \times (3 \times D_{\text{LUT}} + 2 \times D_{\text{mesh}})$. The energy consumption of the SDC is $E_{\text{SDC}} = KM \times E_{\text{LUT}} + 8MK \times E_{\text{Mbit}} + KM^2 \times E_{\text{wire}} = 7N \times E_{\text{LUT}} + 56N \times E_{\text{Mbit}} + 7NM \times E_{\text{wire}}$.

1. Performance Comparison

Based on the preceding modeling, performance is equal to the reciprocal of the task execution time, i.e., $1/D$. Note that the preceding analysis is oriented towards applications with fully parallel addition operations, where the exploitable parallelism is proportional to the task scale N. In fact, many applications cannot offer such a high degree of parallelism. For the above task of size N, assuming its parallelism is P, the performance of the ASIC is $1/((N/P) \times D_{\text{LUT}})$; that of the GPP is$1/(7N \times (D_{\text{LUT}} + D_{\text{memory}}))$; that of the FPGA is $1/((N/P)(3 \times D_{\text{LUT}} + 2 \times D_{\text{crossbar}}))$; that of SDCs becomes $1/(\max(K, N/P) \times (3 \times D_{\text{LUT}} + 2 \times D_{\text{mesh}}))$. Therefore, in the case of $K < N/P$, precisely the case where the task parallelism is smaller than the hardware resource size of the SDC, the SDC can enable near-ASIC performance.

The performance advantage of SDCs over GPPs comes from their higher degree of parallelism, while the performance advantage of SDCs over FPGAs comes from their shorter critical path.

2. Comparison of Energy Efficiency

Based on the modeling above, energy efficiency is performance divided by power consumption, and power consumption is equal to energy consumption divided by execution time, or E/D, so energy efficiency is $1/E$. Therefore, the energy efficiency of the ASIC is $1/(N \times E_{\text{LUT}})$; that of the GPP is $1/(7N \times (E_{\text{LUT}} + E_{\text{memory}} + E_{\text{controller}}))$; that of the FPGA is $1/(7N \times E_{\text{LUT}} + (28N + 2N \times \log_2 N) \times E_{\text{Mbit}} +$

$2N^2 \times E_{\text{crosspoint}} + N^2 \times E_{\text{wire}})$; that of the SDC is $1/(7N \times E_{\text{LUT}} + 56N \times E_{\text{Mbit}} + 7NM \times E_{\text{wire}})$.

Note that the computation granularity is not considered in the above analysis. All four computing architectures are modeled at a unified single-bit computing and interconnect granularity. But in fact, granularity specialization is a very important optimization factor, especially on SDCs. A parameter W is introduced below to characterize the degree of instruction sharing. For example, the parallel width of SIMD instructions in GPPs, and the bit width of the computation, communication and access data, can be expressed by this parameter. Increasing the instruction sharing width will significantly reduce the capacity to store instructions and configurations, as well as the complexity of the controller. Therefore, the energy efficiency of the GPP can be improved to $1/(7N \times (E_{\text{LUT}} + E_{\text{memory}}/W + E_{\text{controller}}/W))$, and that of the SDC can be improved to $1/(7N \times E_{\text{LUT}} + 56N \times E_{\text{Mbit}}/W + 7NM \times E_{\text{wire}})$.

The energy efficiency advantage of SDCs over FPGAs comes from their fewer configurations and less interconnect power consumption. The configuration power consumption of SDCs increases linearly as the problem size becomes larger, i.e., $O(N/W)$, while that of FPGAs increases at the level of $O(N\log N)$. The interconnect power consumption of SDCs increases linearly as the problem size becomes larger, i.e., $O(NM)$, while that of FPGAs increases exponentially. The energy efficiency advantage of SDCs over FPGAs comes from the fact that they avoid large amounts of data memory power and controller power.

3. Better than ASICs?

The modeling above quantifies the benefits and costs associated with the reconfigurable feature of hardware. The programmability requires additional area to store configurations, to implement functions other than those required, and to implement unused connections. The extra area makes the interconnections longer, the latency greater, and the performance lower. Longer interconnections mean more energy consumed on data communication. For implementations that are not fully spatially parallel, additional energy is consumed in reading configurations. Therefore, reconfigurable chips typically have lower logic density, poorer performance, and higher energy consumption than ASICs. Nevertheless, is it possible for reconfigurable chips to be better than ASICs in terms of energy efficiency and performance?

Reconfigurable SDCs can meet the runtime demands of tasks in a way that ASICs can not. In the above modeling analysis, if a task of fixed size N and fixed parallelism P is targeted, certainly ASICs are the most suitable. But if the load of the task changes at runtime, ASICs will face many problems. (1) ASICs may not be able to process variable-scale tasks at all. (2) Even though ASICs can handle task loads of different sizes, if the task load is reduced to a very low level at a certain time, there may be a serious waste of energy in ASICs, resulting in reduced energy efficiency, may even lower than that of SDCs. (3) ASIC chips cannot benefit from updates of algorithms, software and protocols, and this could also be an important advantage of SDCs over ASICs.

The second possible reason why SDCs outperform ASICs in terms of performance efficiency is the process. Specifically, reconfigurable chips are more robust and can use more advanced processes, thus achieving higher performance efficiency. In general, as integrated circuit process scales to smaller feature sizes and higher integration degree, production yields will become an important problem. The increasing impact of process variation is causing a larger percentage of transistors to be unavailable. Device characteristics change over time and a greater percentage of transistors are affected by aging failure. Post-fabrication programmability provides a way to mitigate process yields, fluctuations and aging problems. Therefore, SDCs and FPGAs are often available in smaller feature size and lower voltage process technologies.

Finally, for many emerging applications, the design can hardly be localized, and these applications are often limited by interconnect wires. In this case, the additional programmable logic becomes less important because the area of the chip is limited by the length of interconnect wires. Meanwhile, temporal reuse of long interconnects is better than specialized interconnection in terms of energy efficiency, which also makes reconfigurable designs better than ASICs.

However, ASICs can also introduce the above features to provide some limited reconfigurability to change task requirements, support scalable protocols and automatic restoration of faulty circuits, provide redundant resources to solve process problems, or even dynamically reuse its long interconnects. If so, ASICs will become reconfigurable SDCs.

2.2.2 *Low Programming Difficulty*

The programming difficulty depends on how many underlying details the programmer needs to understand and develop. For software programmers, the hardware is completely transparent. Billions of transistors in the CPU form countless logic gates, registers and memories, etc., which then form ALUs, pipelines, control state machines, cache queues, etc. However, the programmer only needs to focus on whether the high-level-language-based software implementation fulfills the specified function, and the compiler and hardware controller automatically map software functions to underlying hardware configurations. It is assumed that a certain amount of effort is required for optimization from application specification (natural language) to underlying hardware configuration (machine language). For CPUs, the whole process is natural language > (programmer) > high-level language > (compiler) > assembly > (hardware + dynamic optimization) > execution. In general, programmers need to do very little hardware optimization work for high-level languages, and many optimization like loop unrolling and affine optimization can be done by the compiler. Because high-level languages use a sequential execution model, the programming difficulty for CPUs is low. For FPGAs, the whole process is natural language > (programmer + static optimization) > HDL > (synthesis and P&R) > netlist > implementation. The complexity of FPGA programming comes from two aspects: (1) different from the model of software, which is sequentially executed, the DSL requires significantly more manual effort to implement the same algorithm,

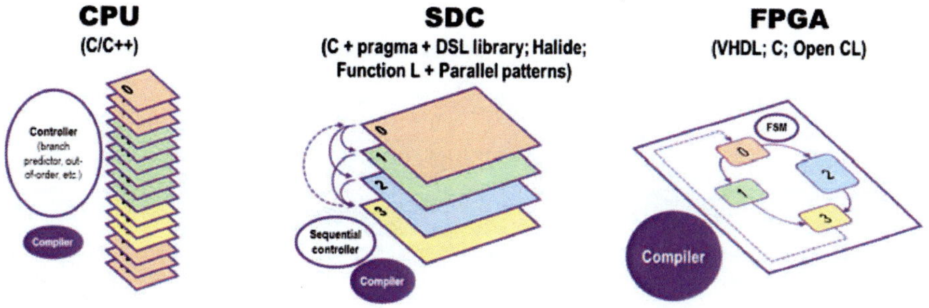

Fig. 2.20 Comparison of programming difficulties of CPUs, SDCs and FPGAs

and the programmer must have knowledge of hardware design; (2) the integrated P&R is basically a process of translating the HDL and optimizing a small amount of circuits, but not involving optimization of algorithms and systems, which results in a lot of optimization work laid on the programmer, and a hardware engineer only with extensive experience can complete the co-optimization of software and hardware well. Therefore, the programming difficulty for FPGAs is very high.

The process of implementing an SDC from application specification to configuration of underlying hardware is similar to that of software, i.e., natural language > (programmer + static optimization) > domain-specific high-level language > (compiler) > configuration > (hardware + dynamic optimization) > execution, as shown in Fig. 2.20. Domain-specific high-level languages are extensions of high-level languages, combining sequential and parallel execution, closer to the user's understanding of the application, which may slightly increase the learning difficulty for programmers, but will improve the efficiency of programmers and compilation effectiveness of programs. Therefore, the programming difficulty for SDCs is slightly higher than that of CPUs and much lower than that of FPGAs.

2.2.3 Unlimited Capacity

Mainstream programmable devices such as FPGAs have a capacity limit, while instruction processors such as CPUs and GPUs have almost no concept of program capacity. The reason for this is the huge difference in the configuration loading and functional reconfiguration time of the devices. The amount of FPGA configurations often reaches up to 100 MB, and interface programming is usually implemented through sequential configurations using protocols like the Joint Test Action Group (JTAG), resulting in a very long time for loading from the outside and function switching, reaching milliseconds or even seconds, which makes the device a long suspension upon each switch of configurations. The program segment of the instruction processor is generally very short, and can be loaded using a high-speed general-purpose interface, with the multi-level cache structure to further reduce

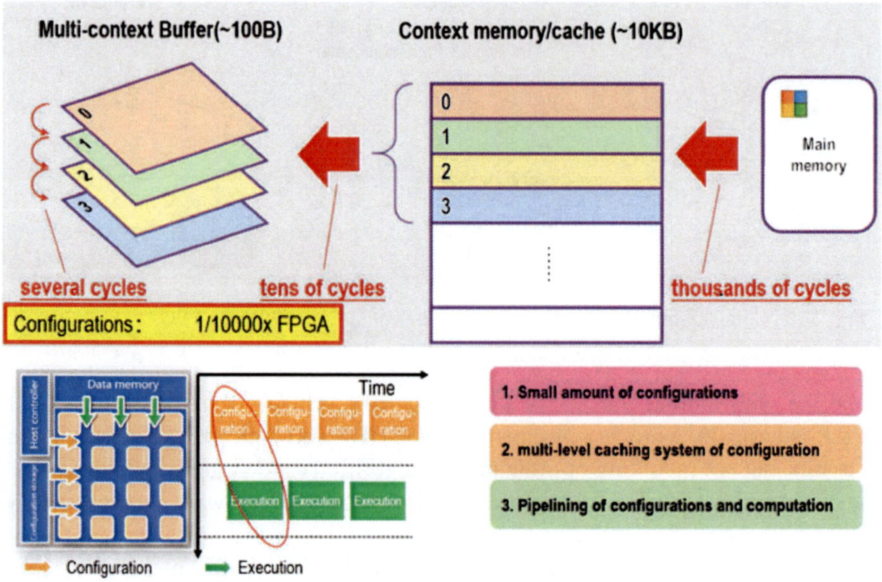

Fig. 2.21 Analysis of the dynamic reconfiguration capability of SDCs

access latency, making the loading of instruction flow hardly affect the normal operation of the processor. Therefore, the function reconfiguration time is a determinant of whether the capacity is limited, while the total amount of configurations is a determinant of the function reconfiguration time.

SDCs, as shown in Fig. 2.21, similar to CPUs, employ a variety of techniques to reduce configuration switching time, i.e., function reconfiguration time. (1) SDCs dramatically reduce the total amount of configurations by specialization, hierarchical design and even compression techniques, enabling less than 1/10,000 of FPGAs. (2) SDCs significantly reduce the loading latency of configurations through multi-level caching mechanisms such as buffers for fast multi-configuration switching and on-chip configuration SRAM, which can be as fast as switching configurations every cycle. (3) SDCs further reduce the impact of configuration switching on computing through interleaving and pipelining of configuration and computation, and the fastest configuration switching can have no impact on the computational process at all. Therefore, SDCs can enable programmable computing with unlimited capacity.

2.2.4 High Hardware Security

Hardware security is a very broad concept covering the entire lifecycle of a chip from design, verification, production, use, and even end-of-life. In particular, SDCs show unique resilience to physical attacks, a form of hardware security threat. The principle of SDCs against physical attacks is moving target defense (MTD) [46]. For a static system, it is relatively simple to run and use, but this also allows the attacker to gain an asymmetric advantage in the time domain. An attacker can spend an arbitrarily long time scouting the target system, studying its potential vulnerabilities, and choosing the best time to launch an attack. Once cracked, static systems cannot fix the problem in a short time either. As a result, static systems are relatively more vulnerable to attacks and cracking. The MTD is a design method for network security and is generally defined as the constant alteration of a system to reduce or move the attack surface (all resources accessible to an attacker that can be used to compromise system security, such as software, open ports, module vulnerabilities, or other resources that can be obtained through an attack). The SDC, as a computing system with dynamically reconfigurable functions, can be viewed as an implementation of the MTD.

With the continuous emergence of many new threatening attack methods like local electromagnetic attacks with even gate-level accuracy, multi-fault attacks that introduce several faults in each execution, and ultra low frequency (ULF) sound based on kilohertz (kHz) level signals, the dynamic reconfiguration feature of the SDC makes it more powerful to defense physical attack. Compared with traditional anti-attack techniques, dynamically reconfigurability can effectively reduce the performance, area and power overhead required for security improvement through resource reuse, and is also expected to resist existing new attacks that have not been effectively overcome through modifying designs. (1) SDCs can take advantage of local dynamic reconfiguration features to develop temporal and spatial randomization techniques that allow each iteration of software execution to be done at random temporal and spatial locations (ensuring the correctness of functions), making precise attacks extremely difficult. Similar to the MTD concept, when an attacker wants to attack a sensitive point implemented by the algorithm, the randomization method makes the temporal and spatial locations of the sensitive point change constantly, making it difficult for the attacker to carry out the attack even with a precise weapon. (2) SDCs can leverage redundant, dynamically reconfigurable resources to build defenses. Since SDCs are typically rich in compute and interconnect resources, developing anti-attack methods based on these resources will have little impact on the performance of normal applications. For example, computing resources can be used to implement physically unclonable functions for lightweight security authentication or secret key generation; randomization can be introduced to various topological properties of interconnection to resist physical attacks in addition to completing normal data transmission. In summary, SDCs have higher hardware security than GPPs, ASICs and FPGAs.

2.3 Key Research Problems

The ultimate research goal of the SDC is to design a computing chip architecture that can balance energy efficiency and flexibility. The ability to run software (applications) with different functions on a single fabricated IC chip while maintaining high performance and high energy efficiency is a worldwide challenge. Chip architectures are usually able to deliver good overall performance for certain types of applications, but not for others. In the more than 60 years since the invention of integrated circuits, general-purpose chips such as CPUs and FPGAs have been created to enable different application functions, but at the cost of low performance, high energy consumption, low efficiency and high cost. There is an urgent need to find new ways in which software (applications) can define chip functions in real time. With dynamically reconfigurable computing as the core, the SDC can meet the needs of changing software using intelligently changing hardware, so as to obtain absolute comprehensive advantages in energy efficiency, flexibility, design agility, hardware security and reliability and other key metrics. Therefore, it is an acknowledged development direction of computing chips, and is also an indispensable research direction of strategic importance for world powers.

The key scientific problem with the SDC is how to intelligently change hardware to meet the needs of changing software, and the main challenge is the huge gap between software applications and hardware chips. As shown in Fig. 2.22, the software and the chip use completely different paradigms when implementing the same function. The software is primarily based on imperative programming, which executes instruction flow by continuously changing the function of a general-purpose computing module, fulfilling the target function sequentially. The hardware is primarily based on declarative programming, which performs calculations by exhaustively declaring the functions and communications of each module, fulfilling the target function in parallel. The academia and industry have been exploring ways to bridge the gap between software and hardware, and have developed techniques such as HLS and VLIW processor compilation, which are still not yielding the desired results. HLS tools for FPGAs have gone commercial, but automation is not satisfactory. Manual effort is essential if a synthesis result of practical value is required. VLIW processors also rely on compilers to automatically schedule the execution of all instructions. However, in general-purpose computing, this technique has largely failed because for many irregular applications, compiler optimization is very difficult in principle.

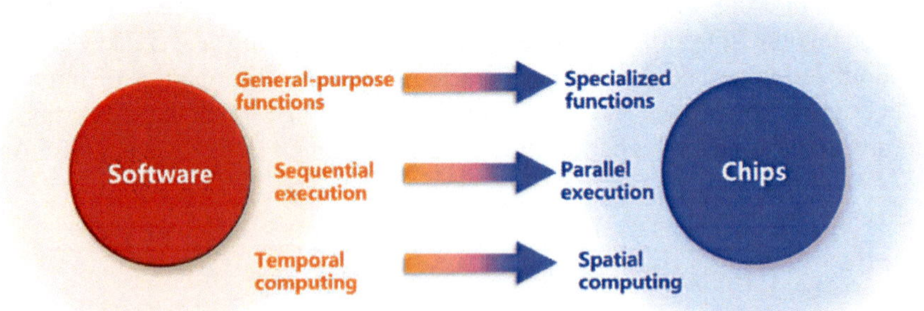

Fig. 2.22 Key challenge of SDCs: vast differences between software and chips

Further reflection on the above challenge reveals that the challenge can actually be summarized as the optimal conversion of software descriptions into hardware. Obviously, like the HLS problem, this is a nondeterministic polynomial-time complete (NPC) problem of extremely large size. It is not realistic to expect compiler software or EDA tools to solve this problem completely. Nowadays, the feasible solutions are heuristic, which significantly reduce the size of the optimization problem through manual or machine learning-assisted or dynamic hardware optimization; or are decoupling, which limits the size of the optimizing space by stripping functional implementation and modular optimization. Without extensive design experience as a guide, these solutions can easily fall into local optimal solutions.

The typical software and hardware development flow of SDCs has been described in detail previously. The review of the optimization problem of defining hardware with software indicates that the development flow of SDCs has actually done a lot of experience-based optimization in hardware design and software development. Figure 2.23 gives a typical design process for SDCs. (1) The SDC relies on empirical design to give a domain-specific programming model that identifies a common computing kernel for users, gives an application-specific programming paradigm, narrows the design space for user programming, and simplifies optimization in the programming phase. (2) The SDC adopts a combination of temporal computation and spatial computation, defines a hardware computing framework, sets up development methods for loop parallelization, pipelining and speculative execution, and specifies a standard synchronization method, thus narrowing the design space for compilation and optimization and simplifying the compilation phase. (3) The SDC supports specific dynamic optimization mechanisms and execution models and sets up dynamically balanced hardware scheduling policies, narrowing the design space for dynamic optimization and simplifying dynamic optimization. Therefore, with such a hierarchical design process, the SDC obtains effective guidance from experience in optimizing the problem solving process, significantly reducing the problem complexity compared to the HLS. The following discusses what optimization work is required to be done for the programming model, compilation techniques, and hardware model.

With the various needs of multi-domain applications and the overall data-centric development trend, for SDCs, the hardware model is the dominating factor impacting energy efficiency. A dynamically reconfigurable system can enhance the reconfiguration speed and thus improve energy efficiency by reducing resource redundancy; the programming paradigm is the basis of programmability, and a data-centric programming paradigm can enhance the programmability of hardware and simplify the development; software mapping is the bridge between software and hardware, and multi-task asynchronous co-mapping and dynamic scheduling techniques provide efficient abstract computation models for hardware architectures, while providing flexible hardware programming interfaces for programming paradigms. We believe that the following considerations must be taken into account in the optimization of hardware architectures, software mapping, and programming paradigms: (1) hardware architecture, an architecture design method that supports irregular control and data flow. Irregular parts of the application, such as access dependences and unbounded loops,

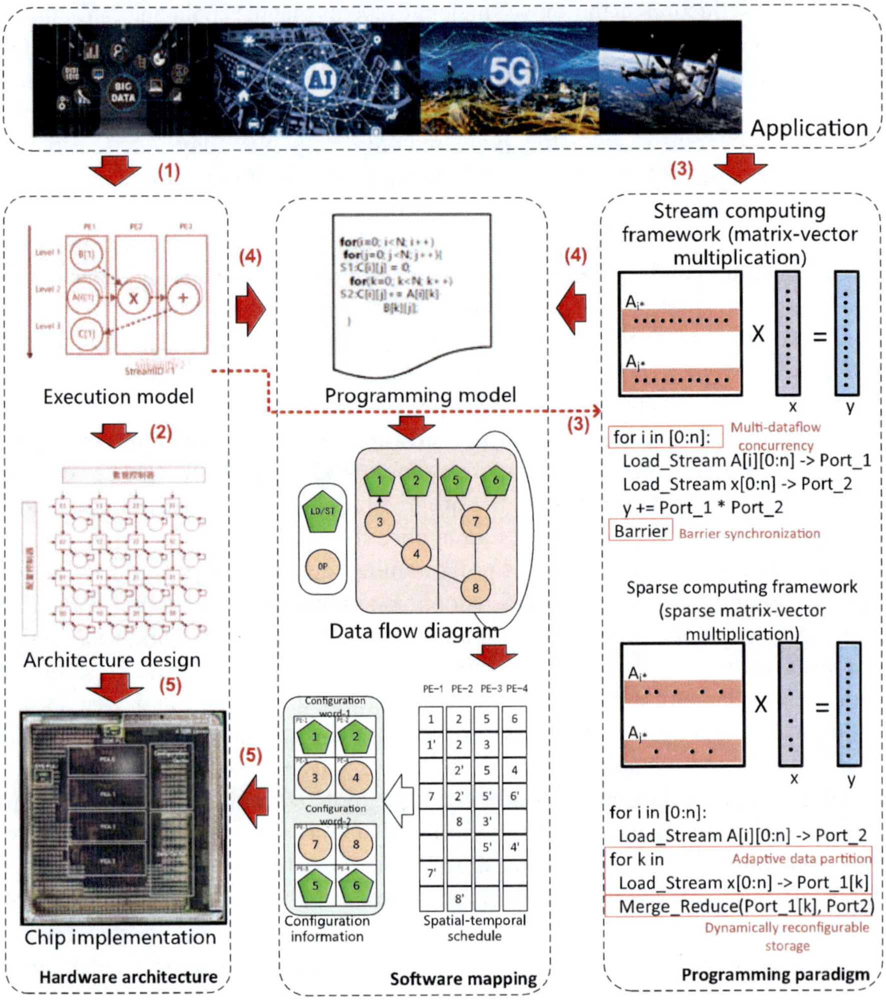

Fig. 2.23 Typical design process of an SDC

are difficult to be fully pipelined and parallelized, and the computational efficiency is low.

The fundamental reason is that the design of the configuration storage and management system in the SDC lacks sufficient temporal flexibility. (2) Software mapping, a dynamic mapping method oriented to reduce the cost of data communication. As applications and hardware resource enlarged, dynamic mapping becomes the key to reducing the complexity of mapping. As the performance and power cost of data communication has gradually become the bottleneck of most computing systems, simply employing methods like in-memory computing will incur higher data reuse costs, and thus how to effectively reduce the cost of data communication between the computation and memory hierarchy has become a key scientific problem to achieve dynamic mapping. (3) Programming paradigm, enables developers to efficiently exploit heterogeneous storage systems of the SDC. Hardware architectures of SDCs need to support heterogeneous storage to improve energy efficiency. Although simplifying the programming development framework, a transparent storage system, only partially improves access performance at runtime. The key to the design of programming paradigms is how to allow developers to characterize application's data access pattern through software so that the scheduler can automatically optimize the layout and movement of data within heterogeneous storage systems. The following sections describe the key research directions in detail from three aspects.

2.3.1 Programming Model and Flexibility

Current research on SDC programming paradigms focuses on how to express parallelism. Developers use the interfaces provided by the programming paradigm to describe data-level parallelism and task-level parallelism in the target application, so that hardware resources can be fully utilized when mapping. However, a programming paradigm that starts from a parallelism expression alone can make it difficult for developers to efficiently optimize the layout and movement of data. Currently, there are two major schemes for optimizing data access on SDCs: (1) using an on-chip multi-level cache that is transparent to the developer to buffer the data in main memory, and (2) requiring the developer to use the underlying hardware primitives to control data movement between off-chip main memory and on-chip scratchpad memory. The key problem with the two schemes is that the former has a significant power overhead in maintaining the cache state, while the latter is difficult to implement because it requires developers to understand the design details of the SDC storage architecture [47].

As shown in Fig. 2.24, developers need to study the programming model in terms of both parallelism and specialization, and study the development framework of data-centric applications, including programming paradigms for both regular and irregular applications. Regular applications can be abstracted into a series of operations on a data flow through stream processing. According to the research at the Stanford University, existing programming paradigms for stream processing on SDCs focus primarily on application parallelism and consider only continuous and fixed-step accesses when dealing with dataflow accesses [48]. The core difficulty of the programming paradigm of regular applications is how to extend the existing stream processing programming paradigms for the complex dataflow access behaviors in regular applications, combined with the characteristics of dynamically reconfigured storage systems. In contrast, when dealing with irregular applications typified by graph computing, research at the University of California, Los Angeles (UCLA), has shown that stream processing introduces a wide scope of random accesses that severely affect system performance and require the consideration of data partitioning [49]. The core difficulty of the programming paradigm of irregular applications is

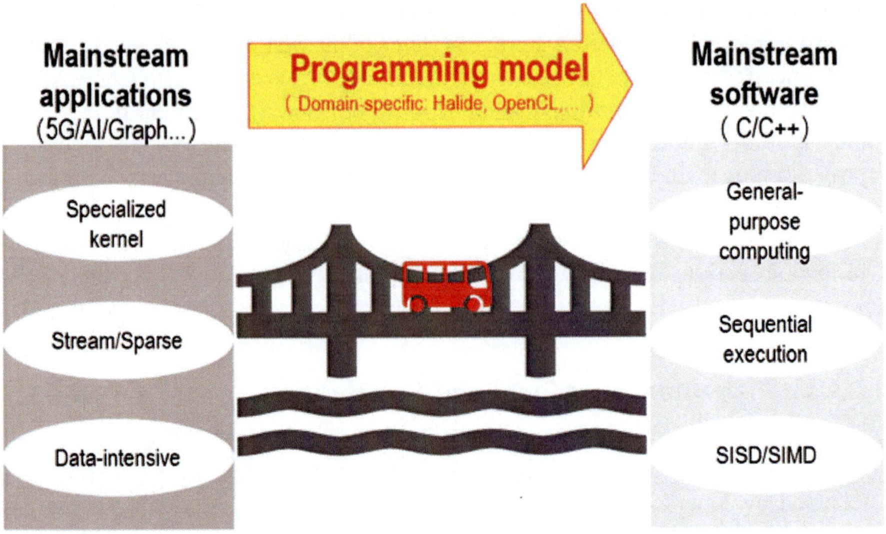

Fig. 2.24 Programming model design—Applications

how to use data partitioning to reduce the scope of random accesses and fully reuse partitions through dynamic reconfiguration of the storage system.

Specifically, research can be conducted on programming paradigms of stream processing for regular applications and developing frameworks of sparse computing for irregular applications. The design of the heterogeneous storage system of the SDC may include multi-level cache, on-chip scratchpad memory, off-chip compute and memory units. To enable more efficient layout and movement of data in the system, while avoiding having developers explicitly manage data in the underlying hardware primitives, application development frameworks need to provide a data-centric programming paradigm at a high level of abstraction. Data access patterns for stream processing and sparse computing can be analyzed to design programming paradigms applicable to SDCs. For the programming paradigm of stream processing for regular applications, the concurrent extension of continuous and fixed-step access data flow on heterogeneous storage systems will be explored to improve the performance of stream computing on SDCs; meanwhile, the dynamic reconfiguration characteristics of storage systems will be exploited to incorporate new data access primitives such as indirect access and barrier synchronization to improve the functional flexibility of SDCs. Based on the distributed graph computing framework, sparse computation for irregular applications narrows the scope of random accesses through data partitioning, and uses the flexible storage system of SDCs to dynamically switch between the cache and scratchpad memory to maximize the reuse of each data partition.

2.3.2 *Hardware Architecture and Efficiency*

The hardware architecture of the SDC mainly consists of computing units, interconnects, configurations and memories. To better design hardware architectures, architects need to study key technologies such as spatial parallelization, pipelining, distributed communication, and dynamic reconfiguration respectively, as shown in Fig. 2.25. Current research is focused on the exploration of computation and interconnection. However, as research efforts have progressed and semiconductor process technologies have advanced, configuration and storage systems have become bottlenecks in hardware performance and efficiency, and are a major area of concern for architecture research. The configuration of SDCs is different from both the configuration bitstream of FPGAs and the instruction flow of CPUs. It usually takes the fixed form of multi-context memory or cache memory, and contains the configuration of operations, control, and explicit data flow, but the loading and switching of configurations cannot be optimized according to the computational and data access characteristics of the application. Therefore, the problem of configuration is actually a matter of configuration storage subsystem. The key problems of traditional storage systems with fixed patterns are: difficulty in adapting to irregular control flow and access patterns in applications, resulting in low system operation efficiency. We need to conduct research on structurally flexible storage system architectures and co-optimization techniques of reconfiguration and computation, mainly including fast reconfiguration methods for storage subsystem and hardware data-prefetch and data-replacement mechanisms optimized for runtime characteristics. The structural flexibility of the storage subsystem is the basis for efficient acceleration of irregular applications, while the co-optimization of reconfiguration and computation is the key to dynamic reconfiguration of the system according to application patterns. According to the research at the Stanford University, the core of storage system flexibility design is how to support control and data patterns for diverse application execution [22]. According to a collaborative research team from Intel and Texas A&M University, the key to co-optimization of reconfiguration and computation is how to efficiently characterize and extract diverse behavioral patterns of applications at runtime [50]. The current configuration system for SDCs mainly uses a fixed-pattern structure and has not yet solved these two core difficulties.

Specifically, the hardware architecture should be different from the traditional computation-unaware storage subsystem design, and allow the storage subsystem to have stronger dynamic reconfiguration ability to combine with and adapt to the computation datapath. The storage subsystem changes the structural form of computing and memory at runtime through rapid reconfiguration, possibly forming a new architecture that integrates memory and computing together, accomodates the irregularity of control flow, and avoids the memory accesses from becoming irregular and fragmented. Meanwhile, the memory subsystem uses a distributed, locally configurable and redistributable design that provides a flexible hardware interface for software programming and dynamic optimization. Fast local reconfiguration is implemented by reallocation of configuration path resources based on computational

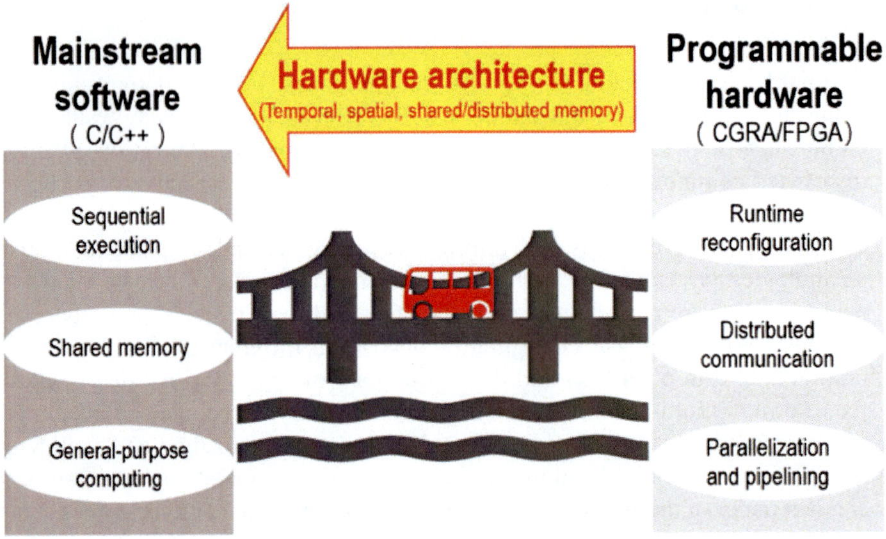

Fig. 2.25 Design of a computation model—Hardware abstraction

and dataflow patterns through runtime analysis of key characteristics of the application, such as data reuse distance and frequency, to better accommodate the diversity of software algorithms.

2.3.3 Compilation Methods and Ease of Use

Software mapping for SDCs includes static mapping and dynamic mapping. The current research focuses on static mapping, which determines the operation scheduling, access latency, and interconnection before runtime to reduce the hardware computation overhead. The key problems with static mapping are: always generating the most conservative design without adapting to changes in runtime requirements; mapping complexity grows exponentially with the application and hardware size, and thus can only be oriented to a small amount of code, such as loop bodies. As emerging applications become more and more diverse, simple static mapping is no longer sufficient, and dynamic mapping is gradually becoming the focus of future research, as shown in Fig. 2.26.

Dynamic mapping techniques oriented for data communication need to be studied, mainly including IRs to support asynchronous task communication, and dynamic task scheduling techniques that optimize data communication. IRs that support asynchronous task communication are extensions to static mapping and are the basis for reducing mapping complexity and enabling dynamic mapping, while dynamic task scheduling to optimize data communication is a key strategy for optimizing system performance and power consumption. According to a collaborative study by Simon

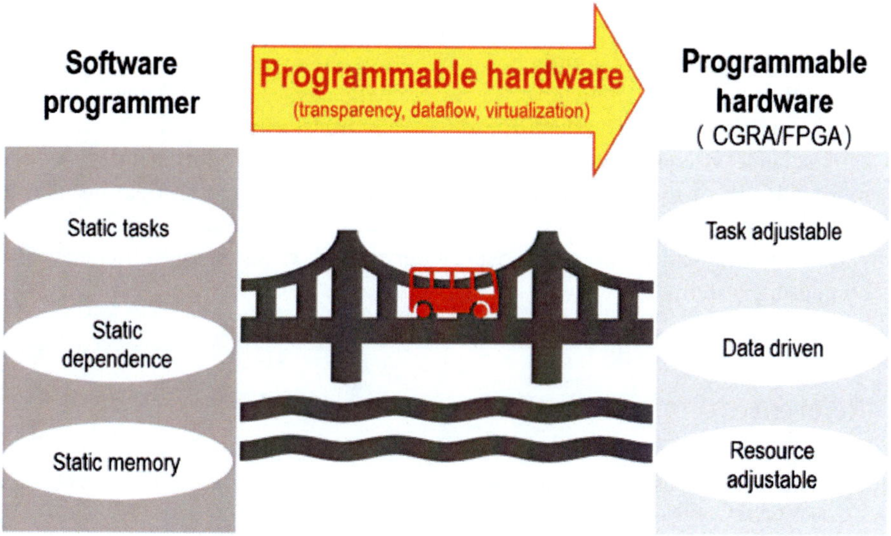

Fig. 2.26 Development of compilation techniques—Transparency and automation

Fraser University and the UCLA, the key to the design of IRs for large-scale application acceleration is how to implement a divide-and-conquer hierarchical mapping method that partitions static mappings constrains into mutually independent tasks while implementing task-level mappings based on a dataflow model [12]; according to a study by Carnegie Mellon University, the key to dynamic scheduling of tasks in modern computing systems is how to achieve an effective trade-off between data reuse and task scheduling [51]. The current dynamic mapping technique for SDCs still aims to improve the utilization of computing resources, and cannot solve these two core difficulties.

Specifically, research can be conducted on multitask co-mapping methods for asynchronous data communication and co-scheduling of tasks and data to reduce the cost of data communication. In terms of application mapping methods, applications can be represented as task graphs with data and control dependences, providing a hierarchical task IR with asynchronous data communication and significantly reducing the scale and complexity of application mapping. Tasks can use stream computing model or fork-join model, depending on the characteristics of the application. The mapping method needs to provide a uniform timing-insensitive asynchronous communication interface for each task to avoid overly strict constraints on the timing scheduling and resource mapping of tasks, thus relaxing the co-scheduling flexibility and optimization space for tasks and data. In terms of dynamic scheduling techniques, as computational power consumption and execution time overhead have become several orders of magnitude smaller than the overhead of data movement in modern computing systems, system design is gradually becoming data-centric and reducing the cost of data access is becoming a key optimization goal. This requires not only the support of hardware architecture, but also the support of task scheduling techniques. We can conduct a comprehensive analysis and dynamic scheduling scheme on computing resources, multi-level cache, scratchpad memory, and new storage and computing devices that varies in terms of functions, efficiency, and performance. This scheme strives to avoid unnecessary data movement during task execution, make full use of the characteristics of different memory and computing modules, enable computing tasks and their data to be executed on the most appropriate hardware structure, and improve the overall performance and energy efficiency of the computing system.

References

1. Krizhevsky A, Sutskever I, Hinton GE. ImageNet classification with deep convolutional neural networks. Commun ACM. 2017;60(6):84–90.
2. Dally WJ, Turakhia Y, Han S. Domain-specific hardware accelerators. Commun ACM. 2020;63(7):48–57.
3. Jiangyuan G. Research on compilation mapping techniques for reconfigurable processing arrays. Beijing: Tsinghua University; 2020.
4. Farooq U, Marrakchi Z, Mehrez H. Tree-based heterogeneous FPGA architectures: FPGA architectures: an overview. New York: Springer; 2012.
5. Norrie T, Patil N, Yoon DH, et al. Google's training chips revealed: TPUv2 and TPUv3. In: IEEE Hot chips 32 Symposium. 2020. p. 1–70.
6. Turakhia Y, Bejerano G, Dally WJ. Darwin: a genomics coprocessor. IEEE Micro. 2019;39(3):29–37.
7. Olofsson A. Intelligent design of electronic assets (IDEA) posh open source hardware (POSH). Mountain View: DARPA; 2017.
8. Hennessy JL, Patterson DA. A new golden age for computer architecture. Commun ACM. 2019;62(2):48–60.
9. Ragan-Kelley J, Adams A, Sharlet D, et al. Halide: Decoupling algorithms from schedules for high-performance image processing. Commun ACM. 2017;61(1):106–15.
10. Lattner C, Adve V. LLVM: a compilation framework for lifelong program analysis & transformation. In: International symposium on code generation and optimization. 2004. p. 75–86.
11. Bachrach J, Vo H, Richards B, et al. Chisel: constructing hardware in a scala embedded language. In: DAC design automation conference. 2012. p. 1212–21.
12. Sharifian A, Hojabr R, Rahimi N, et al. uIR-an intermediate representation for transforming and optimizing the microarchitecture of application accelerators. In: IEEE/ACM international symposium on microarchitecture. 2019. p. 940–53.
13. Weng J, Liu S, Dadu V, et al. Dsagen: synthesizing programmable spatial accelerators. In: ACM/IEEE 47th annual international symposium on computer architecture. 2020. p. 268–81.
14. Flynn MJ. Some computer organizations and their effectiveness. IEEE Trans Comput. 1972;100(9):948–60.
15. Liu L, Zhu J, Li Z, et al. A survey of coarse-grained reconfigurable architecture and design: taxonomy, challenges, and applications. ACM Comput Surv. 2019;52(6):1–39.
16. Budiu M, Venkataramani G, Chelcea T, et al. Spatial computation. In: Proceedings of the 11th international conference on architectural support for programming languages and operating systems. 2004. p. 14–26.
17. Mei B, Vernalde S, Verkest D, et al. ADRES: an architecture with tightly coupled VLIW processor and coarse-grained reconfigurable matrix. In: International conference on field programmable logic and applications. 2003. p. 61–70.

18. Mishra M, Callahan TJ, Chelcea T, et al. Tartan: Evaluating spatial computation for whole program execution. ACM SIGARCH Comput Architect News. 2006;34(5):163–74.
19. Parashar A, Pellauer M, Adler M, et al. Triggered instructions: a control paradigm for spatially-programmed architectures. ACM SIGARCH Comput Architect News. 2013;41(3):142–53.
20. Swanson S, Schwerin A, Mercaldi M, et al. The wavescalar architecture. ACM Trans Comput Syst. 2007;25(2):1–54.
21. Nikhil RS. Executing a program on the MIT tagged-token dataflow architecture. IEEE Trans Comput. 1990;39(3):300–18.
22. Prabhakar R, Zhang Y, Koeplinger D, et al. Plasticine: a reconfigurable architecture for parallel patterns. In: ACM/IEEE 44th annual international symposium on computer architecture. 2017. p. 389–02.
23. Govindaraju V, Ho C, Nowatzki T, et al. Dyser: Unifying functionality and parallelism specialization for energy-efficient computing. IEEE Micro. 2012;32(5):38–51.
24. Clark N, Kudlur M, Park H, et al. Application-specific processing on a general-purpose core via transparent instruction set customization. In: The 37th international symposium on microarchitecture. 2004. p. 30–40.
25. Sankaralingam K, Nagarajan R, Liu H, et al. TRIPS: A polymorphous architecture for exploiting ILP, TLP, and DLP. ACM Trans Architect Code Optim. 2004;1(1):62–93.
26. Bouwens F, Berekovic M, Kanstein A, et al. Architectural exploration of the ADRES coarse-grained reconfigurable array. In: International workshop on applied reconfigurable computing. 2007. p. 1–13.
27. Voitsechov D, Etsion Y. Single-graph multiple flows: energy efficient design alternative for GPGPUs. ACM SIGARCH Comput Architect News. 2014;42(3):205–16.
28. Chin S A, Anderson JH. An architecture-agnostic integer linear programming approach to CGRA mapping. In: Proceedings of the 55th annual design automation conference. 2018. p. 1–6.
29. Chen T, Srinath S, Batten C, et al. An architectural framework for accelerating dynamic parallel algorithms on reconfigurable hardware. In: Annual IEEE/ACM international symposium on microarchitecture. 2018. p. 55–67.
30. Nowatzki T, Gangadhar V, Ardalani N, et al. Stream-dataflow acceleration. In: ACM/IEEE international symposium on computer architecture. 2017. p. 416–29.
31. Watkins MA, Nowatzki T, Carno A. Software transparent dynamic binary translation for coarse-grain reconfigurable architectures. In: IEEE international symposium on high performance computer architecture. 2016. p. 138–50.
32. Liu F, Ahn H, Beard SR, et al. Dynaspam: dynamic spatial architecture mapping using out of order instruction schedules. In: ACM/IEEE international symposium on computer architecture. 2015. p. 541–53.
33. Park H, Park Y, Mahlke S. Polymorphic pipeline array: a flexible multicore accelerator with virtualized execution for mobile multimedia applications. In: IEEE/ACM international symposium on microarchitecture. 2009. p. 370–80.
34. Pager J, Jeyapaul R, Shrivastava A. A software scheme for multithreading on CGRAs. ACM Trans Embedded Comput Syst. 2015;14(1):1–26.
35. Man X, Liu L, Zhu J, et al. A general pattern-based dynamic compilation framework for coarse-grained reconfigurable architectures. In: Design automation conference. 2019. p. 1–6.
36. Josipovi C L, Ghosal R, Ienne P. Dynamically scheduled high-level synthesis. In: ACM/SIGDA international symposium on field-programmable gate arrays. 2018. p. 127–36.
37. Nicol C. A coarse grain reconfigurable array (CGRA) for statically scheduled data flow computing [EB/OL]. https://wavecomp.ai/wp-content/uploads/2018/12/WP_CGRA.pdf. Accessed 25 Dec 2020
38. Li F, Pop A, Cohen A. Automatic extraction of coarse-grained data-flow threads from imperative programs. IEEE Micro. 2012;32(4):19–31.
39. Maggs BM, Matheson LR, Tarjan RE. Models of parallel computation: a survey and synthesis. In: Proceedings of the twenty-eighth annual hawaii international conference on system sciences. IEEE; 1995. p. 61–70.

40. Asanovic K, Bodik R, Catanzaro BC, et al. The Landscape of parallel computing research: a view from Berkeley. Berkeley: University of California; 2006.
41. Svensson B. Evolution in architectures and programming methodologies of coarse-grained reconfigurable computing. Microprocess Microsyst. 2009;33(3):161–78.
42. Stanier J, Watson D. Intermediate representations in imperative compilers: a survey. ACM Comput Surv. 2013;45(3):1–27.
43. Tseng H, Tullsen DM. Data-triggered threads: Eliminating redundant computation. In: IEEE international symposium on high performance computer architecture. 2011. p. 181–92.
44. Liu L, Deng C, Wang D, et al. An energy-efficient coarse-grained dynamically reconfigurable fabric for multiple-standard video decoding applications. In: Proceedings of the IEEE custom integrated circuits conference. IEEE; 2013. p. 1–4.
45. Dehon AE. Fundamental underpinnings of reconfigurable computing architectures. Proc IEEE. 2015;103(3):355–78.
46. Zhuang R, DeLoach SA, Ou X. Towards a theory of moving target defense. In: Proceedings of the first ACM workshop on moving target defense. 2014. p. 31–40.
47. Nowatzki T, Gangadhan V, Sankaralingam K, et al. Pushing the limits of accelerator efficiency while retaining programmability. In: IEEE international symposium on high performance computer architecture. 2016. p. 27–39.
48. Thomas J, Hanrahan P, Zaharia M. Fleet: a framework for massively parallel streaming on FPGAs. In: International conference on architectural support for programming languages and operating systems. 2020. p. 639–51.
49. Zhou S, Kannan R, Prasanna VK, et al. HitGraph: high-throughput graph processing framework on FPGA. IEEE Trans Parallel Distrib Syst. 2019;30(10):2249–64.
50. Bhatia E, Chacon G, Pugsley S, et al. Perceptron-based prefetch filtering. In: ACM/IEEE 46th annual international symposium on computer architecture. 2019. p. 1–13.
51. Lockerman E, Feldmann A, Bakhshalipour M, et al. Livia: data-centric computing throughout the memory hierarchy. In: International conference on architectural support for programming languages and operating systems. 2020. p. 417–33.

Chapter 3
Hardware Architectures and Circuits

Accelerator design is guided by cost—Arithmetic is free (particularly low-precision), memory is expensive, communication is prohibitively expensive.
—Bill Dally, MICRO 2019

The hardware design fundamentally determines the essential properties of the chip, such as performance, energy efficiency, parallelism, and flexibility. Both compilation and programming methods are essentially designed to make it more efficient and convenient for the user to exploit the potential of the hardware. In recent years, new application scenarios have emerged one after another, and the rapid development of big data computing, neural network acceleration, edge computing and other fields have imposed higher requirements for hardware architectures in terms of computational performance and power consumption and other metrics. The software-defined architecture provides a multi-dimensional complex design space for hardware architectures, and each dimension allows a variety of design choices. Moreover, different solutions can often provide very different metrics. Therefore, if designers can come up with the optimal architectural design solutions, the software-defined architecture can meet the needs of various domains.

After more than two decades of research, the architecture-level design space of the software-defined architecture has been relatively complete, with the design direction of each dimension extensively explored. Moreover, the focus of research on the hardware of the software-defined architecture has gradually shifted from the design of architecture-level arrays. Currently, the research hotspot has gradually shifted to: (1) Establishing agile hardware development framework. Domain-specific software-defined architectures are generated automatically using high-level programming languages, and the design space is explored. (2) Integration of novel computing circuits with traditional array structures to fully exploit the potential of high-performance and energy-efficient computation models of software-defined architectures and new computing circuits. With the gradual improvement of agile development tools, the software-defined architecture, a fast-growing efficient computing architecture, allows a significantly shortened design cycle and can be integrated with new computing circuits to secure an increasingly important position in various domains.

S. Wei et al., *Software Defined Chips*,
https://doi.org/10.1007/978-981-19-6994-2_3

In this chapter, we will discuss the architecture design methods of the SDC systematically and explore how to design an excellent SDC, from the perspective of architecture design primitives, hardware design space, and agile development methods. Section 3.1 presents the architecture-layer design primitives of SDCs, consisting of computation, memory, interconnect, interface, and configuration primitives. These hardware primitives are the "building blocks" constituting the SDC, and they can be abstracted in software. The tradeoffs between specialization and flexibility can be made to suit the needs of the user. Section 3.2 presents the current typical agile development frameworks in academia, discusses how to quickly build software-defined architectures using architecture design primitives, and introduces the instructional methods and exploration findings of architecture designs that are currently available. Section 3.3 describes the research on cutting-edge circuit-level computation models, including adjustable circuits, analog computing, approximate computing, and probabilistic computing, and discusses the potential advantages and application prospects of new computing circuits in software-defined architectures.

3.1 Design Primitives for Software-Defined Architectures

The SDC architecture consists of hardware primitives such as computation, memory, interconnect, interface, and configuration primitives. From the software perspective, the compiler abstracts the hardware implementation model using the design primitives and provides the corresponding programming model based on the design primitives, thus eliminating the need to consider details on hardware architectures during programming. From the hardware perspective, design primitives are modular expressions of the hardware structure, and different design primitives correspond to different hardware modules. In an SDC, as shown in Fig. 3.1, the computation primitive corresponds to the structure of the PE in the PEA; the memory primitive corresponds to the on-chip memory structure; the interconnect primitive corresponds to the interconnect structure between the PEs; the interface primitive corresponds to the module-level and chip-level interfaces; the configuration primitive corresponds to the on-chip configuration system. Usually, each design primitive corresponds to multiple implementations in hardware. Different implementations can provide different metrics. For example, some implementations allow higher performance, while others focus more on lower power consumption. When designing a software-defined architecture, the designer is required to select the corresponding hardware model for each design primitive depending on scenario requirements and application characteristics, and combine different hardware models to constitute a software-defined architecture that adapts to different scenarios. This section provides a broad introduction to the design primitives of software-defined architectures, classifies and summarizes the implementations of each design primitive, and comprehensively compares the characteristics of different implementations using a typical architecture as an example.

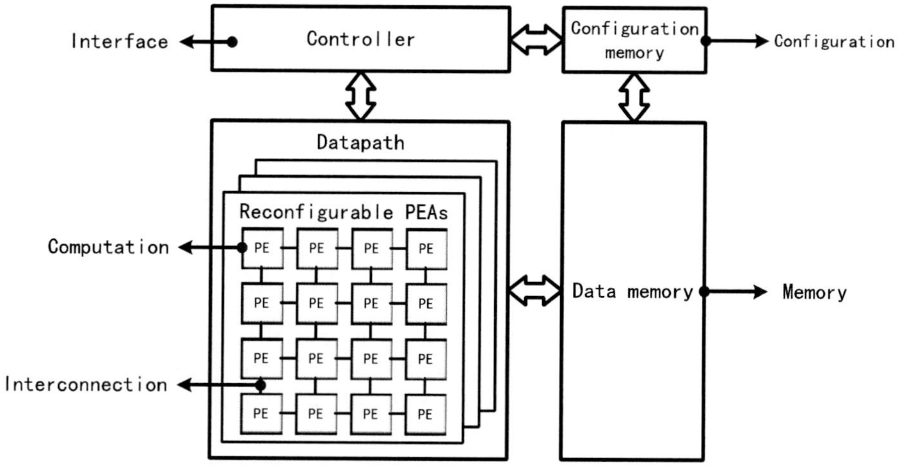

Fig. 3.1 Module composition of a typical software-defined architecture and the corresponding hardware design primitives of each module

3.1.1 Computation and Control

3.1.1.1 Classification of Software-Defined Computation

The computation primitive in a software-defined architecture is implemented by a reconfigurable processing element array (PEA). Usually, a software-defined architecture contains one or more PEAs as the computing engine, with each PEA consisting of multiple PEs and interconnects between PEs. PEs are for executing computing instructions, and interconnects are used for communication between PEs. The computation models of the PEA include spatial computation and temporal computation, which are the keys to differentiate the software-defined architecture from other architectures and to achieve energy-efficient and high-performance computation. Currently, there are various ways to design the internal structure of PEs. A simple PE may contain only one function unit (FU), while a complex PE may contain mechanisms such as instruction scheduling and flow control. The difference in the instruction scheduling mechanism is the main factor contributing to the difference in the internal structure of PEs. Instruction scheduling in PEs can be divided into static scheduling and dynamic scheduling. Dynamic scheduling can be divided into static dataflow and dynamic dataflow, depending on whether it is multi-dataflow-driven. In this section, the concepts related to PE computation models are introduced, and the design of PE structures corresponding to each computation model is discussed in the subsequent sections.

1. Spatial and Temporal Computation

Spatial computation and temporal computation are two implementations of parallel computing in PEAs. Spatial computation means that different computing instructions are mapped to PEs at different spatial locations, and the data transfer between different instructions is completed by interconnects between PEs. In the software-defined architecture, temporal computation has two meanings. One is that instructions are executed in the pipeline constituted by multiple PEs. The other one is that multiple instructions are executed in a single PE in a TDM manner. The software-defined architecture supports both spatial computation and temporal computation efficiently, and is therefore able to exploit multiple parallel mechanisms in applications, including ILP, DLP, and TLP.

(1) Spatial Computation

As mentioned above, spatial computation is the computation model that unfolds multiple operators spatially, maps them to different PEs, and completes data transfer between instructions using interconnects. Spatial computation is one of the key features of software-defined architectures due to the large number of spatially distributed PEs and interconnect resources in the PEA.

Figure 3.2a shows a typical implementation of spatial computation in a software-defined architecture, using the vector dot product as an example. Loop unrolling is a common method in spatial computation to exploit ILP inside loops. In the example above, the loop body, in which the vector dot-product calculation is performed with a factor of 4, is unrolled. In the unrolled loop, four sets of data are calculated in each iteration, and the corresponding dataflow graph is shown in Fig. 3.2b. The nodes in the dataflow graph indicate instruction computation and the arcs indicate data dependence between instructions. In the PEA configuration, the nodes in the graph are spatially mapped to different PEs, and the operations on the nodes correspond to instructions in the PEs; the arcs in the graph are mapped to the interconnects in the PEA. The interconnect structure is very simple in this example, and some data needs to be forwarded by PE (such as operation R in Fig. 3.2c). When the execution starts, the data is loaded from memory to the corresponding PE and transferred along the interconnects in the PEA, and the result is finally written back to memory by the output unit after the execution.

```
input a[N];
input b[N];
output c = 0;

#pragma unroll 4
for i = 1:N {
  c += a[i] * b[i]
}
```

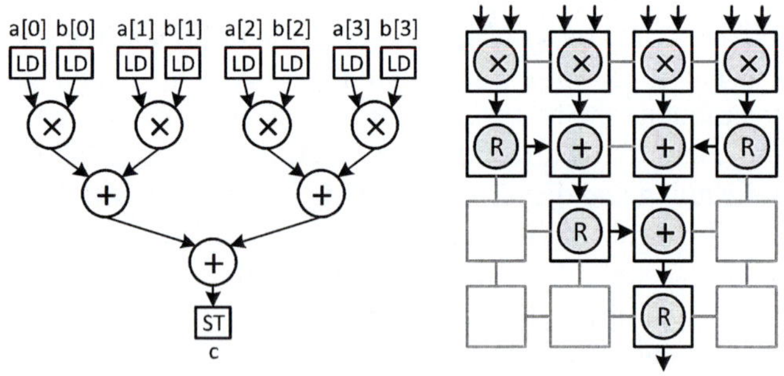

(a) Vector dot-product pseudocode (b) Data flow graph inside the loop (c) Data flow graph of spatial computation

Fig. 3.2 Example of spatial computation for software-defined architectures (LD in Figure (**b**) represents loading data from memory; ST stands for memory; operation R in Figure (**c**) represents the corresponding PE as a data route)

Note that Fig. 3.2 only shows a typical implementation of spatial computation, with some details of the discussion ignored for simplicity. The example contains inter-loop dependence. Specifically, the variable c needs to be synchronized in different iterations, but the corresponding dependence is not drawn in the dataflow graph. To support such dependence is related to the control flow of the PEA, which will be described in detail in subsequent sections. Meanwhile, this mapping, although straightforward, is not optimal. For example, there are problems such as unbalanced dataflow paths and low utilization of PEs in Fig. 3.2c.

In traditional out-of-order (OoO) processors, inter-instruction dependence is usually recorded dynamically by mechanisms such as the scoreboard; the size of the instruction window limits the degree of ILP, and inter-instruction data transfer is done through registers. In comparison, the spatial computation model has the following advantages:

① The PEA configuration directly contains the degree of ILP and dependence between instructions. The degree of the ILP is determined by the size of the PEA, and there is no need to use a scoreboard with high power consumption to record the dependence.

② Multiple PEs execute instructions in parallel, allowing a higher computing bandwidth than the processor.

③ The inter-instruction data dependence is satisfied directly by explicit data transfer on the distributed interconnect, which provides a higher data bandwidth compared to registers used as intermediate memory.

Therefore, the spatial computation model is a key feature that distinguishes software-defined architectures from general-purpose computing architectures and is a major factor to achieve high-performance and energy-efficient computation.

(2) Temporal Computation

The software-defined architecture can support multi-level temporal computation models. First, the I/O interfaces of PEs usually contain registers or data buffer units that can be used as pipeline registers to implement the instruction-level pipeline computation model. In Fig. 3.2c, each PE can be viewed as a stage of the pipeline; computations are performed in the entire PEA in a fully pipelined manner; four elements from arrays a and b can be received respectively in each cycle. The pipeline is the most prevalent temporal computation model in software-defined architectures.

In the meanwhile, the PEA can be reconfigured at runtime to enable temporal computation. For example, to calculate the dot product of vectors with a length of 32, the PEA can calculate the dot product of vectors with a length of 4 eight times separately using the previous configuration to obtain eight intermediate results from c0 to c7, as shown in Fig. 3.2. To get the final dot-product result, the PEA needs to accumulate the intermediate results from c0 to c7. Since these two steps require different functions of the PEA, the PEA needs to be reconfigured to enable accumulation of eight inputs after calculating c0 to c7. Figure 3.3 shows the process of PEA reconfiguration, after which the function of the four PEs at the input changes from the multiplication operator to addition operator.

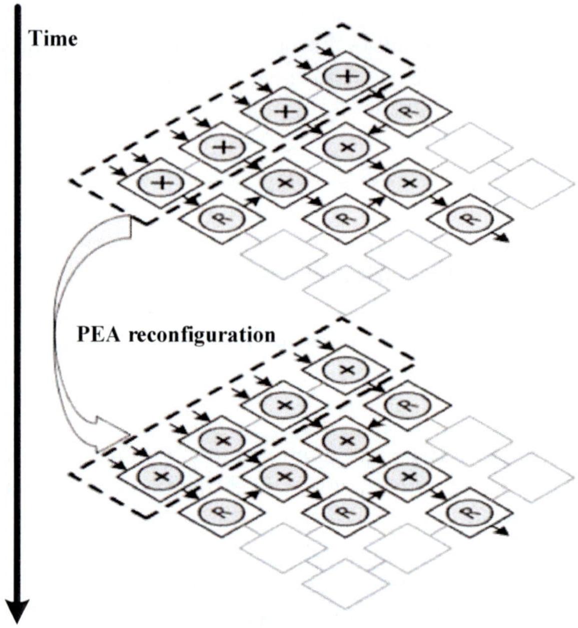

Fig. 3.3 PEA reconfiguration process

In addition, a single PE can contain multiple instructions, and the dynamic switching of instructions enables temporal computation. In this case, multiple instructions can share the internal resources of the PE in a TDM manner at runtime. For example, as shown in Fig. 3.3, if both addition and multiplication instructions are available in the four PEs at the input (inside the dashed box) and the scheduling logic enables them to execute multiplication instructions when calculating dot products and automatically switch to addition instructions when performing accumulation, the functions of the entire PEA change with computation time without the need for the overall PEA reconfiguration.

As a more typical example, Fig. 3.4 shows how to perform a complete vector dot-product calculation using only one multi-instruction PE. The input data of the PE is switched between external inputs (a[i], b[i]) and internal registers (r, c), and the computing instructions of the FU are switched directly between addition and multiplication. At runtime, the PE calculates a[0] $\times$ b[0] in one cycle, stores the result in register r, and calculates $c = c + r$ in the subsequent cycle. Therefore, the throughput is a set of input data every two cycles. If multiple PEs are used simultaneously to calculate the dot products of different vectors, the computational throughput will increase with the number of PEs.

Based on the discussion above, there are three methods of implementing temporal computation for the software-defined architecture: pipelined execution between PEs, dynamic reconfiguration of PEAs, and switching between multiple instructions within PEs. Pipelined execution is the principal method of implementing efficient temporal computation for all software-defined architectures, while the other

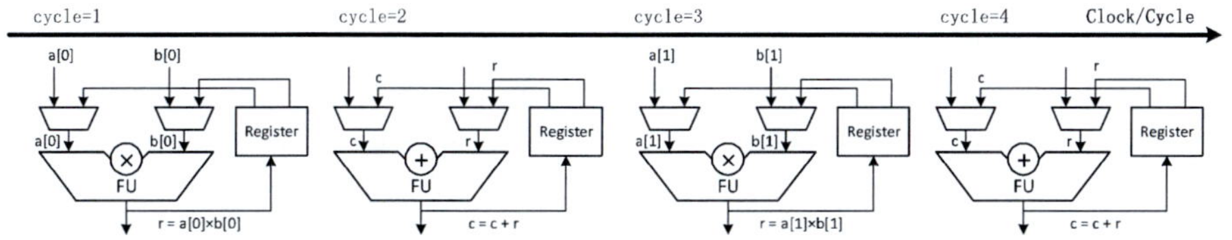

Fig. 3.4 Example of vector dot-product calculation in temporal computation by a multi-instruction PE

two require additional control and scheduling logic. The design of the configuration system and instruction scheduling will be discussed in detail in subsequent sections. The combination of temporal computation and spatial computation can form a variety of flexible computation patterns, which should be selected depending on the application characteristics and scenario requirements.

2. Static Scheduling and Dynamic Scheduling of PEs

In software-defined architectures, scheduling means to determine the sequence of computing instruction execution and data transfer. For example, as shown in Fig. 3.4, instruction scheduling needs to determine the cycle in which the PE performs multiplication or addition, and when the data arrives at the input port of the PE. Instruction scheduling can be done completely statically by the compiler at compile time, like the static analysis of the compiler in VLIW processors determining the order of instruction execution [1], which is known as static scheduling. Instruction scheduling can also be determined dynamically by the PE at runtime, like the OoO processor [2] that records the readiness of each instruction operand at runtime and dynamically determines the instructions to be issued through a hardware mechanism, which is called dynamic scheduling.

(1) Static Scheduling

Static scheduling is applicable to regular loop bodies, including regular computations, memory accesses, and communications. The compiler needs to know in advance the time required for each operation, such as the number of cycles required by computing instructions, the latency of memory access operations, and the path and latency of data communications, in order to develop an efficient scheduling strategy. In the vector dot-product calculation in Figs. 3.2 and 3.3, all operations within the loop body are regular. Specifically, during the pipelined operation of the PEA, the compiler can determine that a total of eight sets of data from a[i] to a[i + 3] and b[i] to b[i + 3] can be read from memory in each cycle, and that each PE can receive a new set of data for computation in each cycle. In the case of vectors with a length of 32, a reconfiguration is required after eight intermediate results are obtained, and the results are written back to memory after being accumulated after the reconfiguration. In this case, all hardware behaviors at runtime are predictable by the compiler, and the compiler can adopt technologies such as loop unrolling and software pipelining [3–5] to optimize the scheduling strategy [6, 7] and improve the computational throughput and efficiency.

However, when there are irregularities in the loop body to be accelerated, such as the presence of branches or variable memory access latency, the static scheduling policy carried out by the compiler is usually conservative, leading to performance degradation, low hardware utilization, and etc. This is primarily because the compiler usually needs to estimate the latency of various operations in a worst-case scenario to ensure the correctness of the loop at any given time. For example, when a loop body contains branches, a typical practice is to map all branch paths to the hardware, execute all branches in each loop iteration, and finally select the data of the corresponding branch based on the branch conditions. In this case, the execution time of branches will always be determined by the branch with the longest execution time. As shown in Fig. 3.5, a loop body contains two branches, where the execution time of branch path 1 is longer than that of branch path 2, but the branch 2 is biased. Therefore, there is a much higher probability that path 2 is executed. In this case, the compiler will still map both path 1 and path 2 to the PEA and execute both branches in each iteration, with the multiplexer eventually selecting the correct data for the current iteration. Obviously, the execution time of branches in each iteration is determined by path 1, although most of the time the actual data is provided by path 2. This will cause significant performance degradation and additional power consumption on the execution of path 1, thus reducing the hardware utilization and energy efficiency. Similarly, in the case of irregular memory access characteristics, such as latency uncertainty due to use of structures like a cache, the compiler usually needs to estimate the time required for data loading and saving in a worst-case scenario, which can also lead to a significant reduction in hardware execution efficiency.

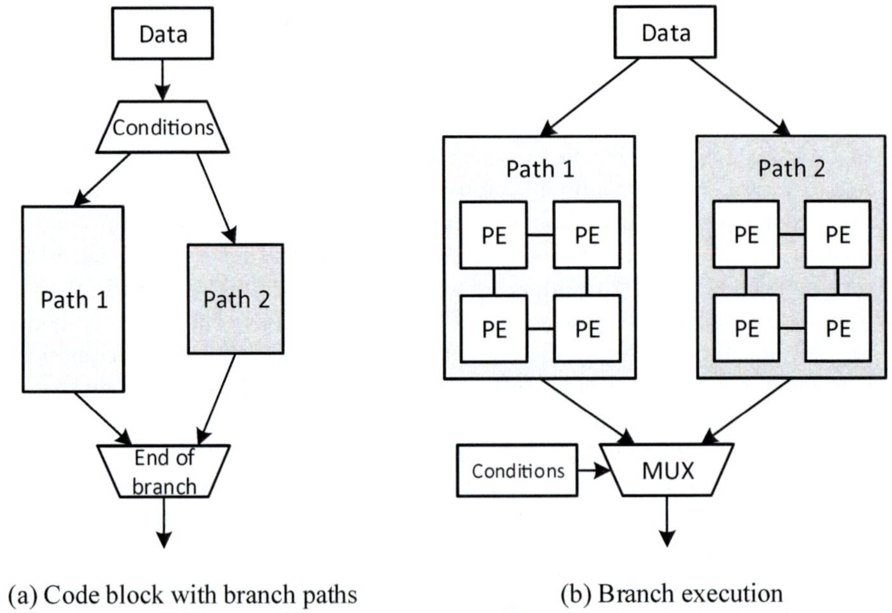

Fig. 3.5 Execution of applications with branches

(2) Dynamic Scheduling

Dynamic scheduling in software-defined architectures is a mechanism where the hardware determines the sequence of instruction execution at runtime. Dynamic scheduling usually uses the dataflow computation model. In the traditional control-flow computation, the computation sequence is determined based on the relative address sequence determined by the compiler. In the dataflow computation, the PE contains a data detection mechanism, which checks the readiness of operands. Specifically, an instruction will be executed when all the required operands are ready. Multiple instructions with ready operands can be executed in parallel or asynchronously. The relative sequence of instruction execution is determined entirely by the flow of data, allowing full exploitation of ILP without additional control mechanisms.

It is not complicated to enable software-defined architectures to support dataflow mechanisms, which only requires the addition of data detection mechanisms inside each PE to check for data arrivals in all input channels corresponding to instruction operands. At runtime, each PE dynamically decides when a local instruction is triggered based only on the readiness of its own local data and sends the result to other PEs through the interconnect to activate subsequent instructions. This process is cyclic, and all PEs execute instructions asynchronously and collaborate with each other through explicit data transfer until the entire program is executed. The dataflow mechanism is also used to enhance performance in an OoO processor, which records data dependence of instructions through a global scoreboard and uses a centralized register file for inter-instruction data communications. In contrast, the dataflow implementation in the software-defined architecture is distributed and asynchronous, without the design of centralized control. Therefore, it is more in line with the semantics of the dataflow computation, and is an efficient implementation.

The dataflow computation model can be divided into the static dataflow and dynamic dataflow computation model. The main difference between them is whether they are running in a blocking manner and whether they allow OoO execution between multiple threads[1] of a loop. In static dataflow computation, threads are executed sequentially, meaning that one iteration must be completed before moving onto the next iteration. Moreover, the static dataflow computation model is running in a blocking manner. All communication paths in static dataflow computation do not contain buffers, which means that the producer[2] can send data for the next iteration only after the data of the previous iteration has been consumed by the consumer; otherwise the operation would be blocked due to the output channel being occupied even if the input data is ready. In contrast, the dynamic dataflow computation model adds additional thread tags to each data token to identify data from different threads. It has a tag matching mechanism to ensure that multiple data sets required for an operation come from the same thread, so it can support OoO execution of multiple threads. It means that any iteration can be executed as soon as the data is ready and the

[1] Each iteration of a loop is called a thread.

[2] In the dataflow computation model, the operation generating the data token is generally referred to as the producer, and the operation receiving the data is referred to as the consumer.

tags of the input data match, and that subsequent threads can be executed before the previous one is completed. Meanwhile, the communication channel in the dynamic dataflow computation contains a buffer, and all data is first stored in the buffer before being read by the PE. This allows any operation to be executed if the buffer in its output channel is not full, so dynamic dataflow computation is often non-blocking. Of course, if the buffer in the output channel of an operation is fully occupied, the operation can be executed only after some data in the buffer is consumed.

Figure 3.6 shows the differences between static and dynamic dataflow computation. The example contains two sequential operations (OP1 and OP2). OP1 receives data a#i and b#i loaded from memory, where i is the iteration number of a loop. In this example, access latency occurs when OP1 is loading b#1 due to cache misses, port conflicts or other errors. In the execution of the static dataflow computation in Fig. 3.6a, OP1 must wait for b#1 to arrive before it can start because the operand is not ready. In contrast, in the execution of the dynamic dataflow computation in Fig. 3.6b, the input channel of each operation has a buffer, and the data of different iterations will be stored in different locations of the buffer. When b#1 is not hit, OP1 can receive the data for the next iteration (i.e., a#2 and b#2) and start, thus shortening the waiting time of OP1. In this way, v#2 will be before v#1 in the output result. In addition, if OP2 takes longer to complete an operation than OP1, that is, the data throughput of OP1 is greater than the data consumption rate of OP2, OP1 will be blocked by OP2 in the static dataflow computation because the data in the output channel is not consumed. In contrast, for dynamic dataflow, OP1 does not need to wait for OP2 to consume the data due to the presence of the buffer. Instead, it can send data and start subsequent operations if the buffer is not full. By supporting multithreaded OoO execution and non-blocking operations, the dynamic dataflow computation can exploit a higher degree of parallelism and achieve higher performance than the static dataflow computation.

(3) Comparison of Static Scheduling and Dynamic Scheduling

As mentioned above, the main advantage of static scheduling lies in that its hardware architecture is designed to be simple and straightforward, with high computational

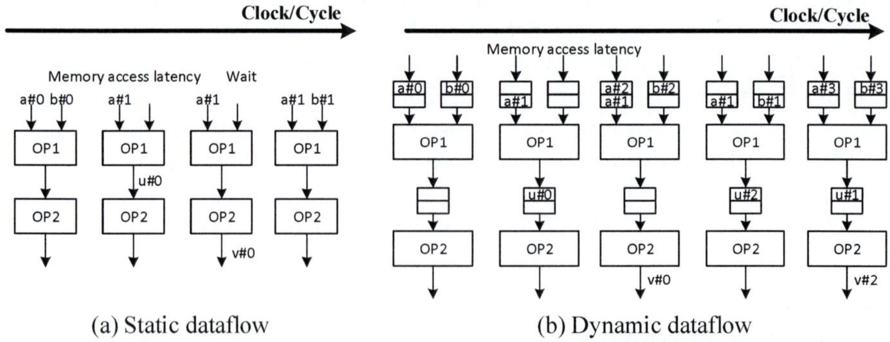

Fig. 3.6 Execution process of static dataflow computation and dynamic dataflow computation

performance and energy efficiency for regular applications. However, it is less efficient for control-flow processing due to the simplicity of the control logic within the PE. Relatively, dynamic scheduling is more flexible for control-flow processing and can tolerate dynamic characteristics in applications, including uncertain access and communication latency and branches in computing instructions. Therefore, dynamic scheduling can achieve higher performance than static scheduling for irregular applications. The disadvantage is that additional complex logic needs to be supported in the PE. Specifically, the dynamic dataflow computation model needs to have a data detection mechanism to dynamically check whether the data in the input channel is ready and whether the output channel is occupied; in addition, it needs to add an extra buffer in each communication path and increase the bit width of the data token to store additional tags, the corresponding tag matching mechanism, etc. These additional costs make dynamic scheduling typically less computationally efficient than static scheduling.

Table 3.1 compares the characteristics of static scheduling and dynamic scheduling, where flexibility refers to the range of applications to which it can be adapted. Since different scheduling mechanisms have different design metrics and costs, when designing a software-defined architecture, the appropriate PE scheduling mechanism must be selected based on the characteristics and requirements of the target application. For example, for applications that are computationally regular and statically predictable, such as matrix multiplication, static scheduling can be used to achieve the highest performance and energy efficiency, while for applications that are computationally irregular and have uncertain behaviors, such as graph computing, dynamic scheduling can provide higher performance.

3. Other Considerations in Designing PEs

Table 3.1 Comparison of static scheduling and dynamic scheduling

		Static scheduling	Dynamic scheduling	
			Static dataflow	Dynamic dataflow
Instruction scheduling method		Compiler	PEs	PEs
Hardware design complexity		Low	Medium	High
Whether to support multithreaded OoO execution		No	No	Yes
Flexibility of target applications		Low	High	High
Performance	Regular applications	High	High	High
	Irregular applications	Low	Medium	High
Energy efficiency	Regular applications	High	Medium	Low
	Irregular applications	Low	Medium	High

In the design of PEs in a software-defined architecture, other considerations need to be considered, and some noteworthy design points are briefly discussed below.

(1) Types of Computations Supported by FUs

The FU refers to the computation module in a PE. Since software-defined architectures are usually intended for algorithm-specific and domain-specific acceleration, PEs do not need to support general-purpose computing functions like ALUs in GPPs. Only software-defined architectures intended for general-purpose acceleration need to employ ALUs as FUs. Usually, the simplest FU, with only a very small number of functions supported, can efficiently complete computational tasks. For example, to accelerate matrix algorithms (e.g. matrix multiplication), the FU supporting multiplication and addition operations, can meet the requirements. In addition, PEs can be customized for specific applications. For example, when accelerating the inference algorithm of a neural network, the softmax computing function can be added to the PEA to efficiently complete the computational tasks of the neurons. In some floating-point computation applications, the FU needs to support the computation of the floating-point number.

When designing a PE, the operations supported by the FU need to be determined by the requirements of target algorithms and domains. The fewer the types of operations required by the application, the simpler the FU structure, and the higher the computational performance and energy efficiency achieved. Conversely, the more operations supported by the FU, the more complex the FU structure, and the higher the power consumption at runtime, making the computation less energy efficient.

(2) Data Granularity

The data width of software-defined architectures intended for multi-algorithm acceleration is typically 32-bit, but the data granularity should be chosen based on application requirements in many cases. The data of PEAs can be classified into four types by granularity, namely bit, word, vector, and tensor. The bit type is generally used to pass predicates for control flow; and the word type can be chosen flexibly. For example, many encryption algorithms (such as AES) performs computation on 8-bit data, and here the PEA can support only 8-bit data. The vector type is mainly used to implement parallel computing in SIMD pattern, and many architectures will use vectors with variable word length. For example, a 128-bit vector may contain 8 $\times$ 16-bit data or 16 $\times$ 8-bit data, and for typical applications such as neural network accelerators, the algorithm itself is robust. Discarding certain accuracy by using smaller data granularity can significantly improve the computation and communication throughput. The tensor type, that is, matrix or multidimensional array, is usually transformed into vector operations during computation.

In general, the selection of data granularity requires a trade-off between computational accuracy and efficiency. The smaller the data granularity, the lower the computational accuracy. However, the computation, communication and memory units can process massive data simultaneously with high throughput and reduced power consumption. Therefore, the data granularity needs to be determined according to the accuracy requirements of the algorithm. For applications with low requirements for data accuracy (e.g., neural networks), a lower bit width can be used. For algorithms with high accuracy requirements (e.g., scientific computing), the loss of accuracy due to bit width reductions can lead to significant computational errors, and a higher bit width is required.

(3) Design of Registers

The PE can contain a certain number of registers for storing local data, or it can contain no registers to keep the structure simple. In general, the PEs allowing multi-instruction execution, are required to contain registers for data sharing between internal instructions. Single-instruction PEs are not required to contain registers, and data transfer is mainly done through external interconnects. There are also structures that use function-specific registers in the PE. A typical example is the output register that stores only the results of the last computation. In the example in Fig. 3.4, a single output register is used to save the intermediate results during the vector dot-product operation.

Local registers allow PEs to store data locally, thus reducing the frequency of communication with other PEs and the global memory access for storing intermediate results. In addition, registers can increase the flexibility of the compiler, providing more options for multi-instruction mapping and data sharing. However, registers will complicate the structure of the PE because the operands of the FU may come from external interconnects or internal registers. Therefore, the specific selection logic is required to ensure the correct source of the operands. Meanwhile, registers allow more flexible instruction mapping, but make the exploration space significantly larger and compiling more difficult.

4. Summary

The preceding sections introduced the common concepts in the design of software-defined architectures, including spatial computation, temporal computation, static scheduling and dynamic scheduling. Accordingly, there are two most important dimensions in the PE design space: ① single-instruction & multi-instruction, and ② static scheduling & dynamic scheduling. For the first dimension, the support of multi-instruction execution directly determines that the PE can support temporal computation internally. Meanwhile, there are significant differences between single-instruction and multi-instruction PEs. Specifically, the former only needs to contain simple FUs to improve performance and energy efficiency for regular applications; the latter needs to contain additional elements such as instruction cache, internal instruction scheduling, and registers to allow a more flexible computation model and efficiently support irregular applications. Therefore, the choice between single-instruction and multi-instruction PEs is a trade-off between the design complexity, computational efficiency, flexibility, and other metrics of PEs. For the second dimension, the comparison of static scheduling and dynamic scheduling in multiple perspectives has been presented in Table 3.1. Static scheduling enables the

Table 3.2 Design space and typical architecture of PEs

Scheduling method	Single-instruction PE	Multi-instruction PE
Static scheduling	Systolic arrays such as Warp [8], FPCA [9], Softbrain [10], Tartan [11], and PipeRench [12, 13]	Coarse-grained reconfigurable architectures such as MorphoSys [14], Remarc [15], ADRES [16], MATRIX [17], and Waveflow [18]
Dynamic scheduling	Static dataflow architectures such as DySER [19], Plasticine [20], and Q100 [21]	Dynamic dataflow architectures such as TRIPS [22], SGMF [23], TIA [24], WaveScalar [25], and dMT-CGRA [26]

computation of regular applications in an efficient and simple way, while dataflow-based dynamic scheduling is more flexible and has better adaptability to irregular applications, but with higher design complexity.

These two design dimensions largely determine the design method, computation model, and computational performance and efficiency of PEs, as well as types of applications supported. Therefore, these two dimensions are the most important factors in the design space of software-defined architectures, which set the boundaries of the architecture design space to some extent. Table 3.2 lists the design space consisting of these two dimensions and typical architectures for each structure. The subsequent sections in this chapter will discuss in detail the differences in hardware design and application scenarios resulting from the different number of instructions and scheduling methods of these architectures.

In addition to these two main dimensions, other dimensions need to be considered during design, such as data granularity and the operations supported by FUs. In short, these dimensions make up the vast design space of software-defined architectures. In this design space, there is no universally optimal architecture, and there are different optimal architectures for different scenarios. The most important principle in designing an architecture is to select dimensions according to the needs of the specific application scenario and weigh the advantages and cost of each mechanism in the target application. The most suitable software-defined architecture can be found according to the scenario requirements and application characteristics.

3.1.1.2 Functional Reconfiguration of PEs

This section will introduce the differences in hardware architectures between single-instruction and multi-instruction PEs using typical architectures as examples.

1. Single-Instruction PEs

Single-instruction PEs mainly take advantage of spatial computation and pipeline computation to fully exploit DLP in regular applications. They contain only single instructions, without additional instruction cache or scheduling logic, and therefore usually provide only the key computing functions that can meet the application requirements. This ensures the simple structure of PEs and the highest performance and energy efficiency that can be achieved in regular applications.

There are many software-defined architectures that use single-instruction PEs, such as Softbrain [10], DySER [19], HReA [27], and Tartan [11], which use simple and similar PEAs. Here, the dynamically specializing execution resources (DySER) architecture is taken as an example to introduce the hardware architecture, and its structural diagram is shown in Fig. 3.7. Overall, the design idea of the DySER architecture is to integrate the PEA into the processor pipeline as a more flexible and efficient coarse-grained execution stage, while other pipeline stages of the processor are to load instructions and data at runtime. Data communication is carried out between the processor pipeline and the PEA via FIFO buffers, and the processor configures DySER at runtime via an extended instruction set. The design of each module is described in detail below.

(1) PEs and PEAs

The PEA is the computing engine of the entire DySER architecture, and the PE in the PEA contains only one FU that can be configured to support multiple computing instructions, but not any control logic. Each FU is connected to four adjacent switches, which form a static interconnection network. FUs can read input data from adjacent

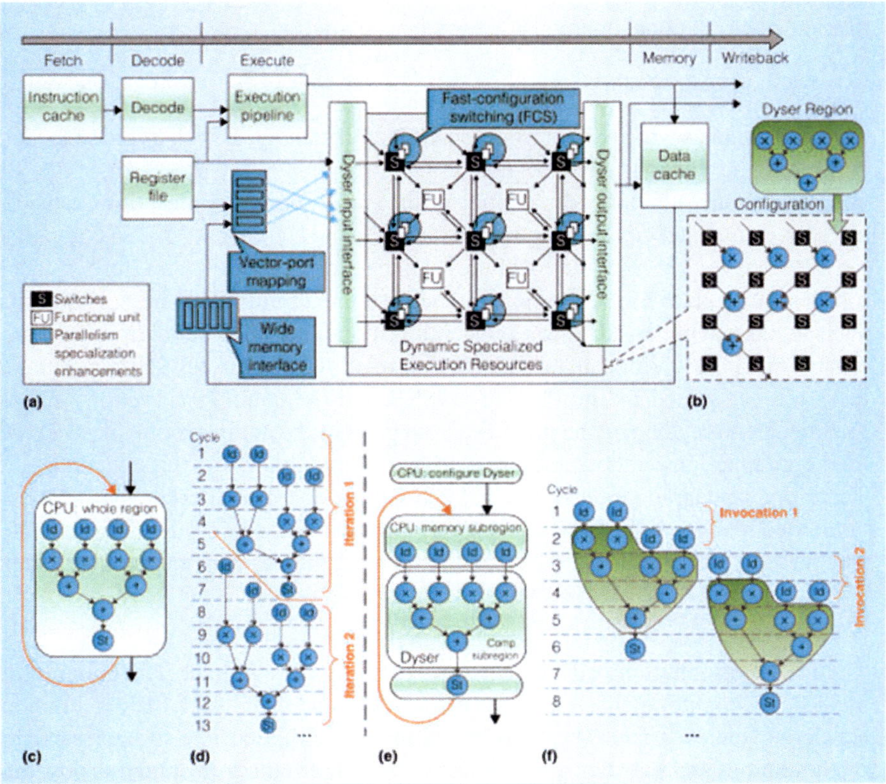

Fig. 3.7 Structural diagram and operational flowchart of DySER

switches and send the output to the corresponding switches for transmission to FUs. In the computation process, DySER uses a typical static dataflow mechanism. Specifically, it checks whether all operands are ready (valid signal) before each operation and whether the output can send data after the computation is completed (credit signal), and executes instructions to complete the computation and send the result only when the two conditions are satisfied, otherwise the computation of the FU will be blocked.

Figure 3.7b shows the configuration of the dot-product operation, in which the configuration of FUs and switches is similar to that in Fig. 3.2. The computation can be performed in a fully pipelined manner between different iterations of the dot product. In Fig. 3.7d, the dual-issue OoO CPU can contain only 2 computations per cycle with an average interval of 6 clock cycles between iterations, while in Fig. 3.7e, the initialization interval of the dot-product operation is reduced to 2 clock cycles due to the higher degree of parallelism of the DySER computation. When multiple iterations are performed alternately, the throughput of the pipeline is increased by nearly 3 times. Meanwhile, since DySER can contain many computing instructions in one configuration, these instructions do not require repeated operations like instruction fetch, decoding, and submission by the CPU, and the intermediate data does not need to read/write registers each time. Therefore, DySER has a higher average computational energy efficiency per instruction.

(2) Memory Access and Control

The loading and saving of data from registers and memory, as well as the sequence of computing instructions, are completed by the pipeline of the processor, so the PEA can support only efficient computation without any form of control flow. Conceptually, this design decouples computation, control, and memory access. In comparison, the CPU has higher computational performance for control flow because of mechanisms such as predictive execution and better compatibility with memory structures such as the cache. DySER has outstanding computational throughput and parallelism, but is less efficient to support control flow. Therefore, the computational tasks are completed by the PEA of DySER and the control instructions and data loading tasks are done by the pipeline. Both parts of the hardware only need to retain their core functions without additional logic to compensate for their own drawbacks. Therefore, remapping after functional decoupling enables different architectures to work together, giving full play to their strengths while compensating for each other's weaknesses, thus forming a new architecture with greater computational efficiency.

(3) PEA Configuration

The PEA configuration of DySER is generated by the compiler. The compiler first identifies compute-intensive regions of the code and enables DySER to efficiently accelerate such regions. These code regions are compiled into the corresponding configurations and an extended instruction is inserted in the corresponding position of the instruction flow. When the CPU executes this extended instruction, it will trigger the PEA configuration and load the corresponding data to start the computation on

DySER PEA. The performance overhead of DySER at runtime mainly comes from the time consumed by PEA configuration. As shown in Fig. 3.8a, if all computations can be mapped to the PEA, the PEA can be executed in a pipelined manner between multiple iterations after the initial configuration (such as the vector dot product), thus achieving the highest computational efficiency. If the size of the computation region is larger than that of the PEA, it will be compiled into multiple configurations, which need to be executed sequentially, as shown in Fig. 3.8b. In this case, the high serialization of configurations and computations makes it difficult to take full advantage of pipelined execution, resulting in a significant reduction in the utilization and computational efficiency of PEAs.

DySER uses several mechanisms to avoid serialization problems with configurations and computations. First, the compiler treats the computation region differently depending on its size at compile time. Specifically, for small computation regions, it adopts loop unrolling to make the number of instructions match the size of the PEA. For large computation regions, the DySER compiler makes the dataflow graph match

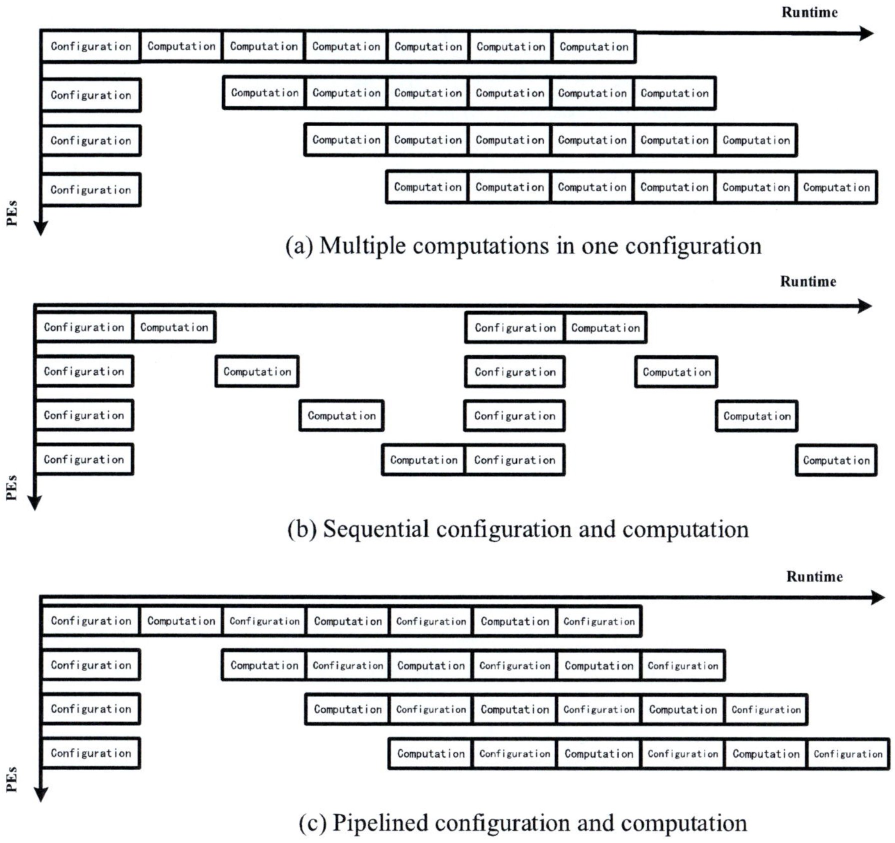

Fig. 3.8 Configuration and computational flowchart of a typical single-instruction PE (assume that data loading and storage are done by the CPU)

the subgraph, combines common subgraphs, reduces the number of instructions, and reuses the PEs corresponding to these regions at runtime, thus reducing the number of reconfigurations. Second, DySER provides a fast configuration switching (FCS) mechanism to reduce the cost required for dynamic configuration switching. That is, multiple configurations and an FSM are stored inside each switch. At runtime, the FSM checks whether the computations of this configuration are complete, and automatically switches its local configuration for subsequent computations upon the completion. The FCS mechanism allows each switch and PE to be reconfigured independently at runtime without affecting the operations of other PEs. It means that configurations and computations can be performed in parallel, thus hiding the configuration cost. As shown in Fig. 3.8c, the pipelined execution of configurations and computations allows much higher hardware utilization than the sequential execution. This mechanism enables high-throughput computations even when the application size exceeds the PEA size.

(4) Design Considerations of Single-Instruction PEs

The above describes the PEA structure of DySER and its solutions for memory access, control flow and configuration. In fact, since software-defined architectures that employ single-instruction PEs typically have similar PEAs, other structures require corresponding mechanisms to handle control flow and configuration problems as well.

To process control flow efficiently, the software-defined architecture decouples control from computation and uses the PEA as the computing engine and other architectures for control flow. For example, SGMF [23] uses PEAs to replace PEs in GPUs; MANIC [28] uses PEAs for efficient and flexible vector processing; architectures such as ADRES [16] and RAW [29] use PEAs as peripheral coprocessors of CPUs. They all adopt traditional sequential execution based on the program counter (PC) to process control flow. Softbrain has stream abstractions-based scheduling logic at the periphery of the PEA. All control flows between tasks are converted into data dependence to achieve task control by data transfer between streams. In these architectures, the software-defined architecture performs only data-parallel and regular operations, taking full advantage of its energy-efficient and high-throughput computation.

For overhead of frequent configurations of single-instruction PEs, the parallel execution of partial reconfiguration of the PEA and computations is a major solution, as shown in Fig. 3.8c. The large-scale applications are divided into multiple smaller subgraphs at compile time, each of which should be able to match the size of the PEA. In terms of hardware, the PEA is divided into multiple subarrays, that can be reconfigured independently and start executing the next subgraph after completing the current one. For example, HReA [27] allows the PEA to be configured independently per row to achieve pipelined configurations and computations, and provides a configuration information compression mechanism to reduce the time required for a single configuration. Tartan [30] provides a hardware virtualization mechanism that dynamically maps subgraphs to optimal hardware resources based on the predefined target function, and subgraphs are dynamically switched at runtime to complete large-scale applications. In short, when applications cannot be fully mapped, how to reduce the configuration overhead should be considered in single-instruction PEs; otherwise the hardware computing power will not be fully utilized.

2. Multi-instruction PEs

The internal structure of a multi-instruction PE is much more complex than that of a single-instruction PE. In addition to the basic FUs, a typical multi-instruction PE requires an instruction cache to store local instructions, an instruction scheduling mechanism to dynamically determine instruction order, and a register file to store temporary data between instructions. Despite the reduced computational efficiency and increased hardware overhead of the PE, the PE can support temporal computation internally, flexibly adapt to the control flow in the program, and execute applications much larger than the hardware scale without reconfiguration. In addition, the multi-instruction PE can support multithreading if different input data can choose instructions.

The most critical module is the instruction scheduler, which directly determines the instruction execution order. Considering energy efficiency and complexity, instruction scheduling within a PE is usually lightweight. In general, the scheduler can be considered as an FSM, and each state corresponds to the execution of one instruction. The conditions for state transfer (i.e., instruction switching) usually depend on the input data, output channels, and other structural state identifiers. The typical software-defined architectures using multi-instruction PEs are TRIPS, WaveScalar, dMT-CGRA, and TIA. This section introduces the design consideration of multi-instruction PEs with the triggered instruction architecture (TIA) as an example.

(1) Definitions of Instruction Sets

Figure 3.9a shows the internal structural diagram of PEs of TIA. TIA uses a trigger-based instruction scheduling mechanism. Every instruction specifies the structural state required for instruction triggering in addition to the operations. In TIA, the structural state of each PE consists of the following parts: ① data registers (reg0 to reg3), ② predicate registers (P0 to P3), ③ input channel data (in0 to in3), and ④ output channel status (out0 to out3). Each predicate register stores 1-bit data, and conditional instructions (such as comparison) use the predicate registers as outputs. The trigger condition in the instruction can specify all the structural states except the data register, and the operation in the instruction can change these states. As shown in Fig. 3.9b, the keyword when is followed by the trigger condition of the instruction, which can specify the value of a certain predicate register or specify an input port that contains data with a specific tag. For example, instruction 1 specifies that P0 in the predicate register is 0 and requires that the input channels in0 and in1 contain data. The keyword do is followed by the specific operations of the instruction, which can perform operations on data and change the structural state. For example, instruction 2 moves data from input channel in0 and sets the predicate register P0 to 1. In fact, the operation of the instruction also implies the requirements for the output channel. For example, instruction 2 needs to write data to channel out0, and the instruction implies the trigger condition that channel out0 is not full and can receive data.

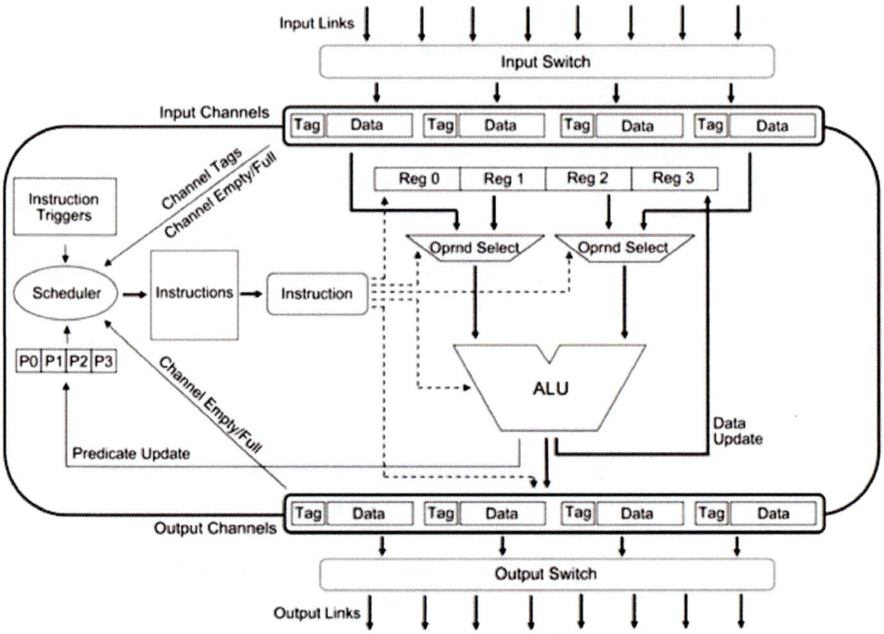

(a) PE structural diagram

```
//when condition do operation

//instruction 1
when (!p0 && %in0.tag != EOL
          && %in1.tag != EOL) do
    cmp.ge p1, %in0.data, %in1.data (p0:=1)
//instruction 2
when (p0 && p1) do
    enq %out0, %in0.data (deq %in0, p0:=0)
//instruction 3
when (p0 && !p1) do
    enq %out0, %in1.data (deq %in1, p0:=0)
```

(b) Instructions example

Fig. 3.9 PE structural diagram and instruction examples of TIA

(2) Instruction Scheduling

The instruction definition of TIA explicitly specifies the instruction trigger conditions, and the scheduler matches the current structural state of the PE with the trigger conditions of all instructions in each cycle. All instructions whose trigger conditions are satisfied can be executed. For example, as shown in Fig. 3.9b, instruction 2 can be triggered when predicate registers P0 and P1 are 1 and out0 is not full. Essentially, the values of the predicate registers can be considered as states of the FSM, while the requirements for the input and output channels are prerequisites for the execution of dataflow. A change in the value of the predicate register means that a state transition occurs on the FSM, indicating that the next instruction to be executed has changed. For example, after instruction 1 is executed, P0 is set to 1. P1 is determined by comparison results of channels in0 and in1, and the value of P1 will determine whether to trigger instruction 2 or instruction 3. After instruction 2 or instruction 3 is executed, P0 is reset to 0. After the input data is ready again, instruction 1 will be triggered to execute. The structural state is constantly changing, and instructions can be triggered to execute in cycles according to the specified logic.

Since the instruction itself can explicitly specify the value of the predicate register or write the result of conditional computation into the predicate register, it means that the current instruction can specify the next instruction to be executed. Therefore, instruction scheduling in TIA can be thought of as an FSM model in which state transfer is determined by the instructions. In this way, multiple instructions inside the PE can be executed in a predetermined order, while the branch computation is converted to predicate execution, thus dynamically determining the instruction order and enabling internal temporal computation.

(3) Multithreading

In Fig. 3.9a, the data in the input and output channels of TIA has a corresponding tag value, and the trigger condition of the instruction can also specify the tag value of the data. The instruction can be triggered only when the tag value specified in the instruction matches that of the input data. The tag values are statically specified in programming and can have a variety of meanings, which enables TIA to support various multithreading mechanisms. For example, if the tag is specified as different iterations of the loop, TIA can support OoO execution of multiple iterations. In addition, the tag can be used to identify data of different tasks, allowing the PE to contain instructions of multiple tasks at the same time and select the corresponding task for execution at runtime according to the tag of the input data. This allows multiple independent tasks to multiplex the same PEA using TDM. From the FSM perspective, this indicates that the PE contains multiple independent FSMs, and different data will select corresponding states in different FSMs. Note that since multiple tasks share the same hardware state (e.g., predicate registers), the instructions of multiple tasks should avoid conflicts when reading and writing the hardware state; otherwise the instructions will be executed in a wrong order.

(4) Design Considerations of Multi-instruction PEs

Compared to single-instruction PEs, multi-instruction PEs contain more complex modules. The design objective is no longer keeping the structure simple, but adding internal scheduling logic to make the PE more flexible. The PE of TIA is a typical multi-instruction structure, like WaveScalar, which also has a tag-like mechanism to select the instruction to be executed based on the tag of the input data. Moreover, TRIPS uses a new instruction set that explicitly specifies data communication between instructions to trigger the instruction. In summary, there are many implementations of instruction scheduling within a PE, all of which focus on constructing instruction FSMs and dynamically transferring states with input data or instructions to drive instruction switching.

A great advantage of multi-instruction PEs is that the temporal computation model can support larger-scale applications using a small number of computing resources without frequent configuration switchings. In general, the total number of operations that can be supported by a multi-instruction PEA is the product of the number of PEs and the maximum number of instructions within a single PE. The above analysis shows that the multi-instruction PE can support more flexible computations, use predicate mechanisms within the PE to support control flow, and support various multithreading mechanisms through a tag matching mechanism. However, the instruction scheduling logic makes the PE functions much more complex, thus increasing the overall power consumption of the PEA, and the energy efficiency is often lower than that of a single-instruction PE. The key to reducing scheduling overhead is to ensure that the scheduling logic is implemented in a lightweight manner. Since a software-defined architecture is usually applied to one domain or a limited number of applications, it is often possible to perform effective instruction scheduling using simple logic without using general-purpose scheduling logic which can make PEs very complicated.

3.1.1.3 Instruction Scheduling of PEs

The preceding sections introduced the design considerations of single-instruction and multi-instruction PEs with typical architectures as examples. The concept of instruction scheduling within a PE is mentioned in the design of a multi-instruction PE, which refers to determining the execution sequence of multiple instructions within a PE at runtime. Instruction scheduling as described in this section refers to determining the relative execution sequence of instructions between different PEs. Moreover, the preceding sections introduced the concepts of static and dynamic scheduling and compared their advantages and disadvantages, and the design of their corresponding PE structures is presented below.

1. Static Scheduling

Like the single-instruction PE, the main design principle of static scheduling is to make the internal structure of the PE as simple as possible. In such software-defined architectures, instruction scheduling is mainly done by the compiler. For regular loop bodies, if their internal computation does not contain any control flow, the compiler can use optimization algorithms such as modulo scheduling [4, 31] and integer linear programming (ILP) to compute a software pipeline with the minimum initiation interval [32, 33] while satisfying the hardware resource constraints and the inter-loop dependence in the application. Then, it maps the pipeline to the PEA, thus obtaining the instruction scheduling scheme with the highest throughput. For example, as shown in Fig. 3.2, there is an inter-loop dependence on the accumulation of variable c with a length of 1. It means that the next loop must start computation 1 clock cycle after the previous one, to make sure that the result is correct. Therefore, the minimum initialization interval for the vector dot-product in this example is 1. After the initialization interval is statically computed by the compiler, the hardware cycle in which all instructions in the loop start to be executed can all be determined. When the pipeline is established, the operations need to wait for a certain time before starting the first operation. Once the pipelined execution starts, all operations are repeated every II cycles. Therefore, the execution cycle of any instruction in the PE can be expressed by cycle $= \mathrm{II} \times n + d, n = 0, 1, \ldots$, where d represents the initialization latency when establishing the pipeline and n means the number of running iterations. Similarly, it is possible to statically configure the switch by calculating the moment at which the switches in the interconnection network transfer data each time.

(1) Static-Scheduling PEs

By pre-computing the initialization interval, the execution timing of all operations can be fully determined before running, and all behaviors of the hardware can be statically predicted. Therefore, static scheduling does not require complex control logic in hardware but requires the PE to configure d and n or similar parameters. In this way, computing instructions can be executed once every fixed cycle after the configuration is complete. Figure 3.10 shows the structure of a typical static-scheduling PE, in which the context register stores the parameters related to the execution timing. Of course, some other special information related to the computation can be stored according to the requirements of different software-defined architectures. For multi-instruction PEs, the context register generally needs to save parameters for each instruction, allowing multiple instructions to run alternately in a predetermined order. In summary, the compiler can generate the appropriate timing parameters for each PE and configure them into the context register, so that all the PEs in the PEA can complete their computational tasks in the correct instruction order. PEs of such structures are adopted by many statically scheduled software-defined architectures, such as MorphoSys [14], HReA [27], and Softbrain [10]. In addition to placing context registers inside the PEs, some structures (e.g., HReA) are designed with a context manager at the periphery of the PEA, mainly to control the execution timing of different rows and synchronize the execution of PEs within the same row, and this execution pattern is prevalent in systolic array computations. In addition, this more coarse-grained control reduces the amount of timing information to be saved, but adds an additional constraint to the mapping at compile time by requiring that data must be transferred from the PE in the previous row to the next.

(2) Delay Matching in Static Scheduling

In static scheduling, delay matching is an important factor affecting the hardware throughput. In Fig. 3.10, the buffer in the input and output ports are generally implemented with FIFOs for adding extra delay. As shown in Fig. 3.11, an example is provided to illustrate how FIFOs and computations affect delay matching. Specifically, the data from the top left PE (denoted as PEs) is sent to the bottom right PE (denoted as PEd) through two paths. Assuming that the delay is 1 cycle when the data is forwarded by the switch or PE, the delays on the two paths are 5 cycles and 2 cycles, respectively, while the FIFO depth at the PE input is 2. If II is set to 1, as shown in Fig. 3.11a(2), PEs sends data once per cycle, and PEd can only receive one operand in the 4th cycle and cannot start the operation. Moreover, all buffers on the path are

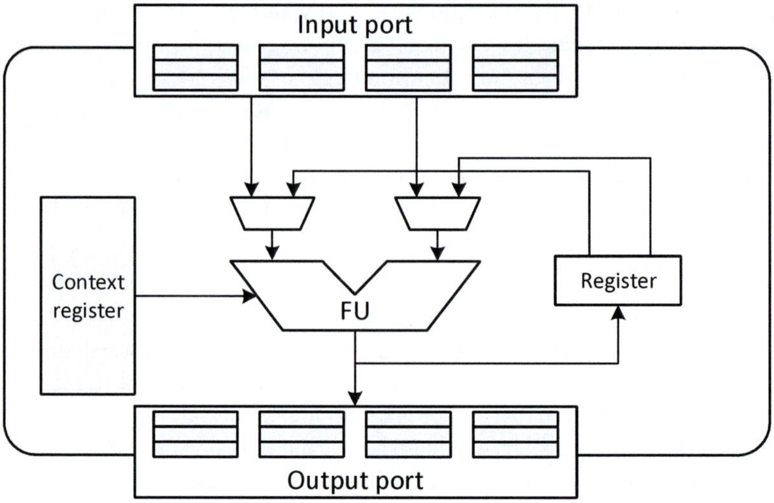

Fig. 3.10 Structural diagram of static-scheduling PEs

full, which leads to data loss on the shorter paths and failure to complete the operation correctly the next time PEs sends data. In this case, as shown in Fig. 3.11a(3), II is set to 3/2, i.e., PEs can ensure correct operations by inserting a no-operation (NOP) instruction after every two operations. The delay mismatch results in an increase in II and a decrease in the throughput. Increasing the FIFO depth can help to balance the path delay. In general, when the difference in path delay between the two input operands of a PE is m and the FIFO depth is greater than m, the path delay can be fully balanced by FIFOs, otherwise the throughput will be affected. Some mapping algorithms can alleviate the delay matching problem, such as long path routing and PE forwarding [34, 35] as shown in Fig. 3.11b. Increasing the FIFO depth is the most direct and effective way to solve the delay mismatch problem but at the cost of higher hardware overhead.

(3) Control-flow Processing in Static Scheduling

Static scheduling is mainly suitable for regular applications that do not contain control flow, whose computational behaviors can be statically predicted. In this case, static scheduling can enable the most efficient computation with simple hardware. Conversely, if the application has uncertain behaviors, the performance that can be achieved with static scheduling will be significantly degraded. For example, for an access operation, if the access latency is 1 cycle in most cases and 4 cycles in rare cases (occurrence of port conflicts), the compiler must conservatively estimate the delay to be 4 cycles for all access operations to ensure that the scheduling scheme functions correctly even in the worst case, otherwise the computational results may be wrong. In this case, even if there are no port conflicts most of the time and the theoretically optimal II should be close to 1 cycle, this conservative assumption of

Fig. 3.11 Impact of delay matching on throughput in static scheduling

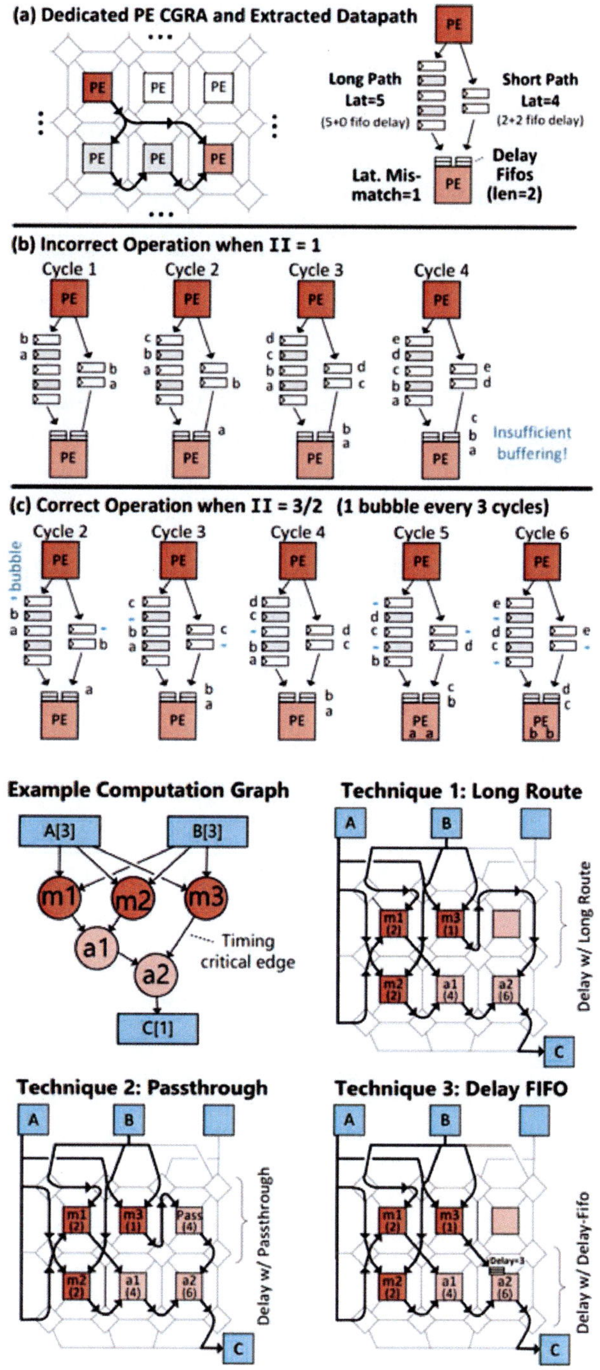

the compiler will cause the II of all PEs to be set to 4 cycles, with all PEs operating only once every 4 cycles, and the hardware utilization and throughput being only 25% of that in the best case. In addition, we have given an example of static scheduling containing branches in Fig. 3.5, where the compiler assumes that all paths will execute in the case of multiple possible execution paths and uses the worst-case II corresponding to the longest path as the overall initialization interval, also leading to a significant reduction in the utilization and throughput.

(4) Summary

According to the preceding descriptions, static scheduling mainly relies on compilation algorithms to solve complex instruction scheduling problems, thus keeping the hardware simple and efficient. Therefore, the research on software-defined architectures with static scheduling, whose PEs have similar simple structures, focuses on scheduling algorithms. Static scheduling is mainly applicable to regular computations. For example, many accelerators [36, 37] use static scheduling for efficient pipelined or systolic array execution patterns in digital signal processing, neural network acceleration, and other domains with predictable behaviors. However, since the compiler cannot obtain dynamic characteristics of the application at runtime, the uncertainty in the application caused can significantly reduce the performance and energy efficiency. In addition, the static scheduling scheme generated by the compilation algorithm heavily depends on the hardware architecture, and changing the FIFO depth or the number of registers may cause the compiler to produce a scheduling scheme with completely different performance.

2. Dynamic Scheduling

Dynamic scheduling is a more flexible strategy than static scheduling. Unlike static scheduling, dynamic scheduling allows the compiler to configure the corresponding data communication paths and PE instructions according to the data dependence of the instructions, with no need to arrange a specific execution order for the instructions in advance. The dataflow-driven instruction execution mechanism ensures that instructions are executed when data is ready and there are no resource conflicts (mainly for input and output channels). For applications with dynamic characteristics (such as graph computing), dynamic scheduling can achieve much higher performance than static scheduling.

Accordingly, the hardware structure of dynamic scheduling is more complex than that of static scheduling, and dynamic scheduling requires extra mechanisms to detect the status of data channels and other hardware resources to ensure that instructions are executed correctly. The difference between static and dynamic dataflow has been introduced previously, and the following will explain the hardware structures with typical architectures as examples.

(1) Static-Dataflow PEs

Static dataflow is the easiest way to implement dynamic scheduling. To implement static dataflow, it is required to add basic structural state detection mechanisms within the PE to dynamically trigger instructions. The structural states include at least the following:

① The input channel is not empty. Since OoO execution is not allowed in static dataflow, inter-thread data will arrive at the input channel sequentially in order. This ensures that multiple sets of input data of a PE must come from the same thread once they reach the input channel. Therefore, the PE in static dataflow only needs to detect whether the input channel contains data, and no tag matching on the input data is needed.
② The output channel is not full. If the instruction in PE A is to send the computational result to PE B, it must be ensured that the output channel is not full, otherwise the computing instruction in PE A will not be executed. For example, if the instruction in PE A is executed faster than that in PE B, the data sent by PE A will gradually pile up in the input channel of PE B and cannot be consumed in time, which will eventually lead to the blocking of PE A as well. If there are instructions sent to PE A from other PEs before PE A sends data, they will also be gradually blocked. This mechanism is usually called backpressure, that is, a consumer (such as PE B) that executes more slowly will block all previous data producers (such as PE A). The presence of backpressure ensures the correct transfer of computational results, but also makes the PE with the lowest throughput in the PEA a bottleneck that limits the average iteration cycle.
③ The FU is not occupied. If some operations of an FU need to be completed in more than one cycle, the occupied cycles of the FU will also block the execution of new instructions.

The DySER architecture was described in detail earlier in the explanation of the design of single-instruction PEs. DySER is also a classic static-dataflow architecture. DySER uses the valid and credit signals to form a set of handshake signals to ensure the states ① and ② mentioned as above. In addition, Plasticine [20], which has different signal definitions, uses a similar hardware design to detect structural states. Note that static-dataflow structures can contain data cache channels, which function primarily to increase the delay to match unbalanced paths, as well as to prevent the occurrence of data blocking. These channels are usually implemented as FIFOs with no tag matching mechanism, so the data is still required to arrive at the input channel in order, otherwise a computation error will occur. Therefore, static and dynamic dataflow cannot simply be differentiated by the input buffer.

(2) Dynamic-Dataflow PEs

The essential difference between dynamic and static dataflow is that dynamic dataflow allows OoO execution of threads and allows data from different threads to reach the input channel of the PE in a random order (within the allowed cache depth). This allows the PEA to exploit more fully fine-grained ILP at the cost of far more complex hardware structures than static dataflow. The key design points of dynamic dataflow are how to ensure the correctness of OoO execution at a small cost and how to solve the problems such as asynchronous access and deadlock caused by the random order of thread data. The following uses SGMF [23] as a typical architecture to introduce the hardware structure of dynamic dataflow.

① Tag Matching

The data token in dynamic dataflow contains tags, which are used to distinguish and match data from different tasks or threads. As a major part of the hardware overhead of dynamic dataflow, the tag matching mechanism has a significant impact on the performance. Figure 3.12 shows the structure of the PE in SGMF, and the focus here is on the input buffer. In the buffer, each data field includes one thread identifier (TID) and three operands (op1 to op3). Every TID is unique for tagging function. The data cache is not implemented using FIFOs, and data is not stored and read in order of arrival. When data arrives, the exact location of the data in the buffer is determined by its TID. For example, the cache depth in the figure is 4, and the location of data is determined by the lower 2 bits of the TID upon arrival. The data with TIDs 0 to 3, regardless of the order of arrival, will be stored independently in the appropriate location. In this way, the number of data with the same TID equals that in each area of the buffer. Upon arrival of all operands corresponding to a certain TID, the whole set of data is read from the buffer at the same time and the instruction can start to execute. This mechanism, called explicit token store (ETS), enables efficient tag matching and data detection logic and forms a first come first service (FCFS) model, which is the basis for dynamic dataflow to implement OoO execution. In fact, this mechanism has been used to introduce the execution process of dynamic dataflow, as shown in Fig. 3.6.

② Memory Interfaces of Dynamic-Dataflow

Due to the OoO execution of computing instructions, the results will also be transferred in a random order, which results in the OoO arrival of data of all PEs in the

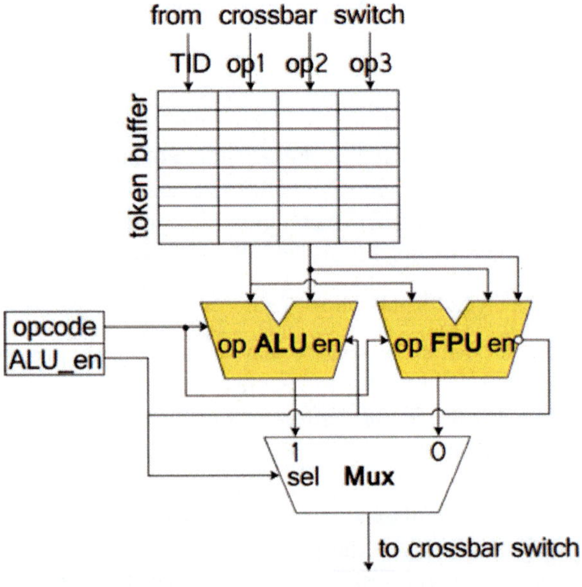

Fig. 3.12 Structural diagram of dynamic-dataflow PEs

PEA. Therefore, the input channels of the PEA must have buffers to support out-of-order accesses. Figure 3.13 shows the load/store (LDST) unit in SGMF, and the input data is also stored in the corresponding area of the buffer with the TID as the address. Note that there is a reservation station at the output of LDST, which stores the TIDs of instructions that have sent memory access requests. In this way, multiple threads can access memory in parallel. For example, if the memory access requests of threads A and B are sent to the cache one after another, and a cache miss occurs when thread A loads data while a cache hit occurs when thread B sends data later, the data loaded by B will be returned before A. By comparing the TIDs of the memory return data with the TIDs of threads A and B stored in the reservation station, the out-of-order data returned by the memory access can be correctly matched with the threads. Note the difference between the reservation station here and the reorder buffer (ROB) in the OoO CPU, both of which are used to temporarily store instructions. However, the ROB is mainly for returning commutating results in the instruction order, while the reservation station is used to improve the parallelism of out-of-order data returns. The reservation station allows memory access requests between threads to be executed in a non-blocking manner, thus effectively improving the parallelism of memory access instructions, but the drawback is that out-of-order accesses will break the data locality and tend to reduce the rate of cache hits.

③ Dynamic-Dataflow Deadlock

Ideally, if the cache depth is infinite, all threads can have unique corresponding areas in the cache, in which case all instruction parallelisms in the application can be fully exploited. In fact, the cache depth is very limited, which limits the degree

Fig. 3.13 Structural diagram of dynamic-dataflow memory access units

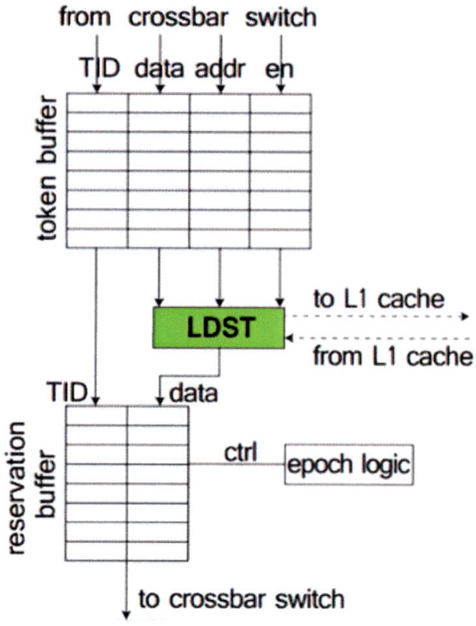

of parallelism. Worse still, the thread disorder will lead to deadlock [38] problems. Figure 3.14 shows a typical example of deadlock occurrence, where the cache depth is 2 and two LD units load two operands of the PE. The deadlock is mainly caused by the data circular waiting due to the thread disorder. In the example, threads 0 and 2 have the same memory location and so do threads 1 and 3. If a cache miss occurs when an operand is loaded by threads 0 and 1, and the data of threads 2 and 3 and the other operand of threads 0 and 1 return first, the buffer is full in this case. However, because the tags of the corresponding data do not match, the PE keeps waiting for the data tags to match before starting the operation, and the data returned from the memory is waiting for the PE to release the corresponding buffer area, thus causing a deadlock. The essential cause of deadlocks is that the length of the thread disorder exceeds the range allowed by the buffer depth. One solution is to add constraints on thread disorder. SGMF groups threads by size of buffer (Epoch). As shown in the figure, the depth is 2, and two adjacent threads are divided into a group. For example, threads 0 and 1 are in a group, and so are threads 2 and 3. Threads within the same group allow OoO execution while those in different groups must be executed sequentially. For example, threads 2 and 3 can enter the buffer only after data of threads 0 and 1 is emptied from the buffer. This mechanism can effectively solve deadlocks, but the size of the group and the buffer have an important impact on parallelism.

④ Summary

The structural design of the dynamic-scheduling PE and dynamic-dataflow architectures are presented with SGMF as an example. The tag matching, memory access and deadlock problems must be considered by all dynamic-dataflow software-defined architectures. TRIPS [22] is another classic dynamic-dataflow architecture that uses the explicit data-graph execution (EDGE) instruction set to store operands and declare the target instruction address for data communication directly in the instruction definition. TRIPS uses frames to store instructions and data from different threads and uses the compiler to avoid deadlocks. Although TRIPS and SGMF differ in architecture because of different instruction sets, many units share similar logic from the

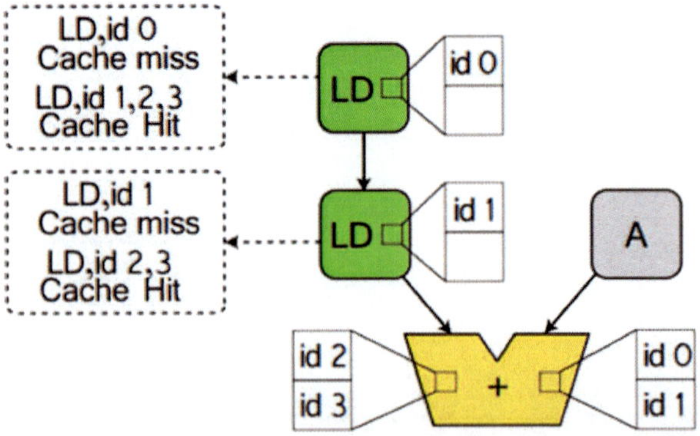

Fig. 3.14 Deadlocks in dynamic dataflow

perspective of dataflow. For example, both frames and reservation station are used to spatially isolate instructions and data from different threads.

In general, a static-dataflow architecture can be regarded as a special case of a dynamic-dataflow architecture with the buffer size of 1. Therefore, the implementation of static dataflow is simpler than that of dynamic dataflow, but static dataflow achieves less computational parallelism. The most critical structural parameter of dynamic dataflow is the size of the data buffer. The larger the data buffer, the higher the computational performance. However, the area and power overhead of the hardware increases significantly with buffer size, often even more than the FU overhead, which will reduce the energy efficiency. Therefore, the choice of the data buffer size requires the designer to make an appropriate trade-off between performance and energy efficiency. In addition, the deadlock problem is an important bottleneck limiting the performance of dynamic-dataflow architectures. It is typical to avoid it at the cost of some parallelism (e.g., SGMF), but this usually prevents the hardware from reaching its full computational potential.

3.1.1.4 Summary

In the above, we focused on the design space of single- and multi-instruction and static- and dynamic-scheduling PEs, and introduced the design considerations of various software-defined architectures with typical architectures as examples. The summary is made below from the perspectives of structural complexity, computational performance, and energy efficiency.

1. Structural Complexity

Multi-instruction and dynamic-scheduling PEs are much more complex than single-instruction and static-scheduling PEs. The fundamental reason is that for the latter case (e.g., DySER) the PEA has only the computing function without control logic. The instruction mapping and scheduling tasks are implemented by the compiler or other peripheral units. The design of dynamic dataflow logic is quite complicated, especially to solve problems such as deadlocks caused by OoO execution in spatially distributed PEAs at a low cost.

2. Computational Performance

For small-scale regular applications, software-defined architectures of different structures are usually able to fully exploit the data parallelism in the application and efficiently utilize hardware source to achieve high throughput. For irregular applications with uncertain characteristics such as control flow, dynamic-scheduling PEs can fully exploit ILP and achieve higher performance than static-scheduling PEs due to their ability to dynamically handle dependence and unfixed delays at runtime. For large-scale applications (with the number of instructions exceeding the PEA size), multi-instruction PEs can be used in a TDM manner, requiring only initial configuration to complete all computational tasks, while single-instruction PEs need to be reconfigured multiple times, introducing additional configuration time, making fully pipelined execution impossible, and significantly reducing performance.

3. Energy Efficiency

Experimental results in Revel [39] show that single-instruction and static-scheduling PEs are relatively energy efficient, although the irregular characteristics of applications will reduce their computational performance and energy efficiency. This is primarily because multi-instruction and dynamic-scheduling PEs typically use data caches and instruction caches, as well as the corresponding data matching and instruction scheduling logic with considerable overhead. Although these additional units can improve flexibility and performance, they are typically underutilized, leading to a waste of resources on non-critical paths. Moreover, the consumption of static power reduces the overall energy efficiency of the architecture.

Based on the above comparison, several rules of software-defined architectures can be summarized.

(1) As the complexity of PEs increases, the architecture becomes more flexible in handling dynamic characteristics and can be applied to a wider range of application fields, but the energy efficiency decreases as well.
(2) The simplification of the hardware structure means the complexity of the compilation algorithm. For example, for single-instruction and static-scheduling PEAs, the compiler needs to complete complex tasks such as control flow processing and instruction scheduling.
(3) There is no optimal architecture, only the most suitable architecture for a particular scenario. The requirements of application characteristics and usage scenarios determine what software-defined architecture should be chosen.

4. Development Status of the Design of PEs

In recent years, software-defined architectures have been used more often as domain-specific accelerators (DSAs). Since DSAs usually have similar computational characteristics and simple control patterns, the hardware only needs to support these specific patterns to achieve efficient acceleration. Therefore, for energy efficiency purposes, many software-defined architectures (e.g., convolutional neural network accelerators [36, 37]) use single-instruction and static-scheduling PEs to complete regular systolic array computations. In some of the software-defined architectures for multi-domain acceleration (e.g., DySER, Plasticine, and MANIC), single-instruction and static-dataflow PEs are widely used as the computing engine. In general, the complex multi-instruction and dynamic-dataflow PEs are less and less because the most important metric now is energy efficiency and current cutting-edge applications often do not contain or contain only simple control flows. In addition, the design space of PEs has been explored thoroughly, and the proposal of new PE structures is no longer a research hotspot in recent years. Mainstream software-defined architectures tend to use existing simple hardware structures and compilation algorithms to optimize various domain-specific scheduling strategies and achieve the high energy efficiency.

3.1.2 On-Chip Memory

The actual computational performance that can be achieved by a software-defined architecture is primarily determined by the throughput of computing and memory access. Ideally, adding more PEs in parallel would enable a proportional increase in the processing capability of a software-defined architecture. However, in many application scenarios, it is difficult for the PE to reach its peak performance because the memory bandwidth cannot meet the computing needs. As a result, a large amount of PEs must stall and wait for data, resulting in a severe reduction in resource utilization rate. Moreover, in recent years, the growth in commutating speed has far outpaced the increase in memory access rate, and the difference between computation and memory reaches several orders of magnitude. Therefore, memory is the bottleneck in most high-performance computing architectures. Furthermore, on-chip memory is usually the largest part of the computing architecture in terms of area and power consumption, which makes the memory system one of the most critical modules in the software-defined architecture.

As shown in Fig. 3.15, a typical software-defined computing architecture has a multi-level on-chip memory structure for alleviating the bottleneck problems.

(1) The PE usually contains a small amount of memory for storing temporary data between instructions, or constants required for the computation. The variables stored here are used by only a few instructions within the PE, with the best localization and the highest reuse rate.

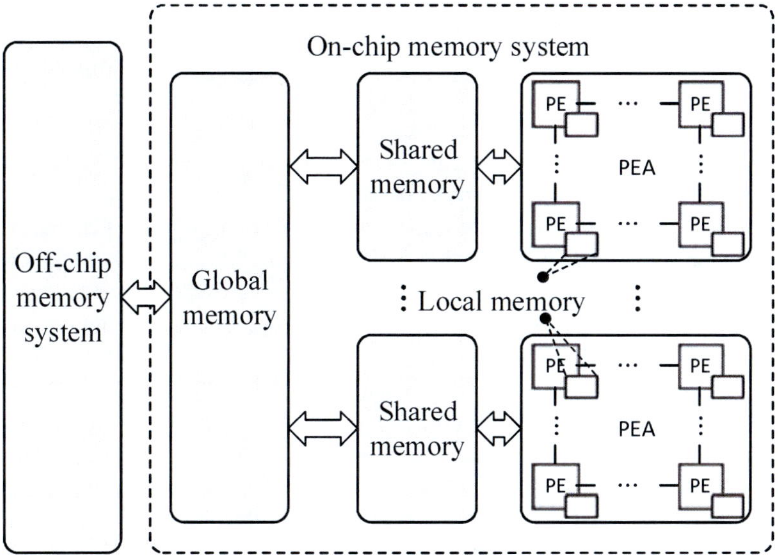

Fig. 3.15 Structural diagram of the on-chip memory of software-defined architectures

(2) At the PEA level, the interconnection network in the PEA has many buffers that can temporarily store and transfer intermediate results between multiple computing instructions. This is different from the use of register files to store temporary data in other architectures such as processors. Using the interconnection network for temporary data transfer has higher data bandwidth and is more efficient than using register files. In addition, the PEA may contain extra memory for storing local data needed in the PEA, thereby reducing access to global memory.
(3) There is usually a global memory outside the PEA. One role is to share data between PEAs and the other role is an interface for communication between PEAs and off-chip memory systems. For energy efficiency and bandwidth reasons, software-defined architectures use the software-managed scratchpad memory (SPM) like GPUs and rarely use more general-purpose cache systems.

The internal design of the PE has been described in detail in the previous section, while the design of the interconnection network will be presented in Sect. 3.1.3. This section details the SPM-based on-chip memory system and briefly introduces the application of software-defined cache systems in software-defined architectures.

3.1.2.1 Software-Managed SPMs

The SPM is a high-speed on-chip memory with independent addressing, which is explicitly managed by software. Compared with the cache, the SPM is not transparent to programming. Specific instructions to be explicitly inserted in software for reading and writing SPM, and the memory access behavior is completely controlled by the programmer. Different from the cache, the SPM directly stores the required data without additional space for tags, as well as the complex design of tag matching, replacement logic and coherence protocols. Compared with the cache, the SPM will be more energy efficient due to its simple structure, and the SPM with higher capacity and bandwidth can be used under the same constraints in area and power. Moreover, without an implicit data replacement strategy, the SPM has relatively fixed and predictable delays, which is especially critical for architectures that rely on static scheduling. However, the SPM has the following drawbacks: (1) Programmers need to understand the hardware structure to properly and efficiently control the behavior of the SPM, which raises the use threshold of the SPM; (2) The SPM does not have a coherence protocol, which makes it difficult for users to ensure coherence by controlling the timing when used in distributed scenarios; (3) The SPM has compatibility problems. For example, when structural parameters such as the capacity of the SPM change, rewriting the code to the new structure is required. Due to these drawbacks, the SPM is not used in GPPs currently, but is widely used in GPUs and domain-specific accelerators. Software-defined architectures are often used in specific domains. Moreover, applications in the same scenario generally have similar access characteristics, and users or compilers can predict memory access behaviors without specific requirements for the generality of memory, so software-defined architectures usually use the SPM as on-chip memory to pursue higher energy efficiency and data bandwidth.

1. Main Metrics of the SPM

The main metrics of the SPM are latency, bandwidth, throughput and capacity. Among them, latency refers to the time required from the memory access request to the data return, including one-way latency and round-trip latency. In processors, latency has a significant impact on performance [40], which may even outweigh the impact of bandwidth. However, in software-defined architectures, since applications are typically executed in a highly pipelined manner, latency can be hidden by pipelining without significant impact on performance if instruction parallelism is sufficient. In contrast, bandwidth and throughput are more critical metrics. Bandwidth is the maximum size of data that can be processed by the memory per unit time, and throughput is the amount of data processed during operation. Since software-defined architectures are widely used for accelerating data-parallel regular applications and the PEAs contain sufficient computing resources, the data bandwidth that the SPM can provide in such scenarios directly determines the highest performance that the architecture can achieve. In addition, the capacity is also one of the important metrics of the SPM. If all the data in the application are loaded into the SPM, the off-chip low-speed memory access can be completely avoided. However, the capacity is closely related to the area and the static power consumption. In the meanwhile, dynamic power consumption and latency of the large-capacity SPM access will be higher. Therefore, a trade-off among multiple metrics must be made to select the appropriate capacity while meeting the needs of the application.

2. Multibank SPMs

As mentioned earlier, as the SPM capacity increases, the corresponding latency and power consumption increases significantly. Multiple independent small-capacity memory blocks are used to form a large-capacity SPM, where each memory block is called a bank. Figure 3.16a shows the structure of an 8-bank SPM, where the controller at the periphery of the bank is responsible for mapping the access requests to the corresponding banks according to the access addresses. In Fig. 3.16b, the data is stored in different banks according to the low bits address, and each bank has independent access ports. It means that if there are 8 memory access requests to different banks at the same time (e.g., data in addresses 0–7), the data can be loaded in a single cycle in parallel as shown in Fig. 3.16b. The number of banks determines the maximum number of instructions that can be processed in parallel, and therefore directly determines the bandwidth of the SPM. In general, the bandwidth of the SPM is linearly related to the number of banks.

3. Coalesced Memory Access and Bank Conflict

The key to fully utilizing the SPM bandwidth is the parallel access requests for data in different banks. As shown in Fig. 3.16b, the parallel memory access requests to

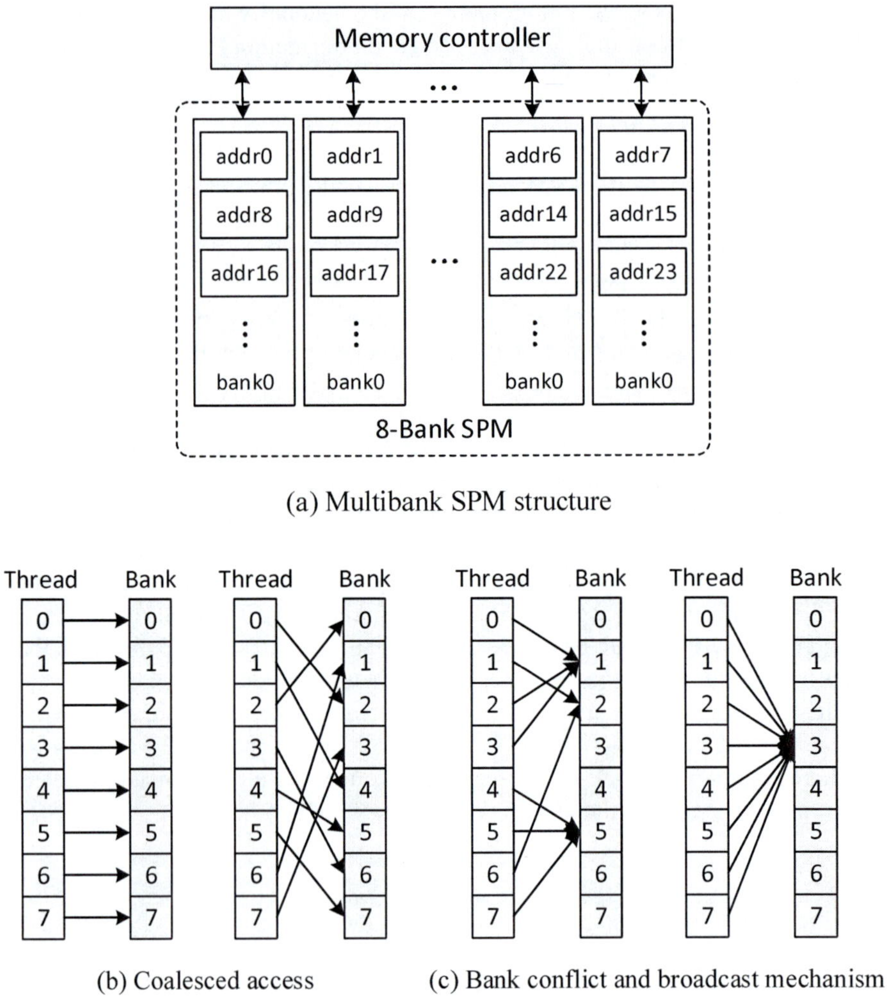

Fig. 3.16 Schematic diagram of the parallel access of the multibank SPM

consecutive blocks and the data is distributed in all banks, which is called coalesced memory access. Coalesced memory access is the most efficient access method, which can obtain all the requested data through a single memory access operation and make the throughput reach the upper limit of the bandwidth of the SPM. In an extreme case, as shown in Fig. 3.16c, all memory access requests in a certain cycle are for the same bank, which is called a bank conflict. In the event of a bank conflict, the memory access instructions originally issued in parallel will be executed sequentially due to the bandwidth limit of a single bank. For each memory access, the data required by a single instruction can be returned, and other instructions need to stall for additional cycles waiting for data, thus taking a total of 8 cycles to complete all requests. In this case, the SPM has the lowest bandwidth utilization, with only one bank working and the rest idle. Frequent band conflicts will affect the performance and waste bandwidth. In addition, bank conflicts will lead to varied memory access latencies. In this example, the eight instructions issued simultaneously to fetch data experience different latencies. For computing architectures that rely on static scheduling, the unpredictable latency can cause the compiler to determine the timing according to the worst-case scenario of the memory access, which can severely reduce performance.

In a static-scheduling software-defined architecture, since instructions work in statically allocated time steps, bank conflicts can be avoided by compilation algorithms or programming optimizations to arrange the order of access instructions. In addition, some software-defined architectures (e.g., HReA) implement a broadcast mechanism in the memory controller. If parallel access requests for the data at the same address cause a bank conflict, the memory controller will send only one access request to the bank and broadcast the result to all instructions. For dynamic-scheduling software-defined architectures, it is difficult to predict the specific access timing of instructions due to the dynamic execution mechanism of dataflow, where multiple access instructions arrive at the SPM port asynchronously. In general, since PEs send access instructions independently, other requests can still use idle banks and maintain a high bandwidth utilization in the case of bank conflicts on a few access instructions, as long as the parallelism of access instructions is sufficient. At the same time, the dataflow mechanism is inherently tolerant of unfixed latencies. Therefore, bank conflicts do not have a significant impact on the performance of dynamic-scheduling software-defined architectures.

4. Centralized Memory and Distributed Memory

The memory systems used by software-defined architectures can be summarized into two main categories, namely centralized memory and distributed memory. Centralized memory means that a global SPM is shared by all PEAs, such as SGMF. Since the data of all PEAs needs to be obtained from the same memory, the global SPM must be able to provide sufficient bandwidth and capacity, so this type of architecture often uses an SPM with a very large number of banks. However, all access instructions are concentrated in the global memory, which increases the probability of bank conflicts; since the global memory must maintain a large capacity, and the capacity of individual banks must also be relatively large, accessing large banks will cause greater dynamic power consumption; the global SPM relies on a large number of banks to provide bandwidth, but it is difficult to match the number of banks with the number of PEs, so the data bandwidth that the global SPM can provide is still the performance bottleneck of the architecture if there are many parallel access requests.

In pursuit of sufficient bandwidth, some software-defined architectures employ distributed memory. As shown in Fig. 3.17a, TIA utilizes distributed small-capacity SPMs in PEs, and each PE can use the local SPM for data storage. Because the capacity of each SPM in the PEA is very small and the PE has a very short path to the memory, access to local data is efficient. At the same time, distributed memory can provide enough bandwidth so that each PE accesses local data without being affected by other access instructions without bank conflicts. However, distributed memory also has certain limitations. First, how to distribute the data evenly to different PEs is a key problem. If the data is not evenly distributed, only a small number of SPMs in the PEs are effectively utilized, while most of them are idle, which will lead to a decrease in utilization. For compilers, the even distribution of data in many applications is always a difficult problem. Second, some of the data needs to be shared by multiple PEs, which requires PEs to add extra instructions to transfer data between SPMs frequently. These extra access and communication instructions will reduce the computational efficiency. Especially when the PEA size is large, the communication overhead is too large and may even exceed the computation time, resulting in performance degradation. A cache system is used in TIA to solve the data sharing problem, but the use of cache leads to a reduction in energy efficiency. Because of these problems, distributed memory structures within PEAs are used by only a few software-defined architectures, most of which have a complex multi-instruction PE structure. The internal SPM can play a significant role only when a single PE is able to perform relatively complex computations on all local data.

Currently, software-defined architectures are designed with diverse memory systems mainly based on application scenario requirements. Although there is still no uniform structural standard, the design of multi-level memory systems has become the main trend. Figure 3.17b shows the architecture of Plasticine [20], where the PEA is composed of the pattern computation unit (PCU) and the pattern memory unit (PMU). In particular, the PCU mainly performs vector computational tasks,

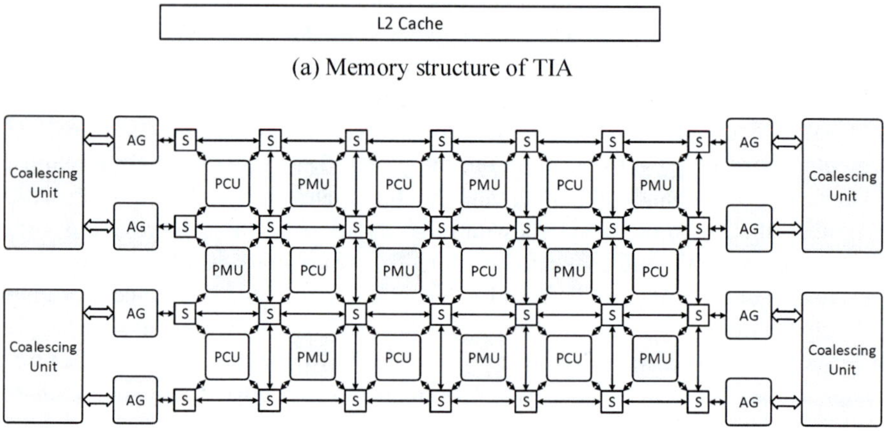

(a) Memory structure of TIA

(b) Memory structure of Plasticine

Fig. 3.17 Distributed memory structure

while the PMU is responsible for data loading and memory management. Focusing on the memory structure, Plasticine has a multi-level memory structure.

(1) Four double data rate (DDR) channels are used to interact with off-chip memories, and each channel contains multiple address generators (AGs) and coalescing units (CUs), as shown on the left side and right sideof Fig. 3.17b. AGs can be configured to generate a sequence of pattern-specific access addresses that are merged into coarse-grained access instructions by the CU to access off-chip memory through the DDR channel. Off-chip accesses are the slowest and least frequent, and have the largest data granularity, so coarse-grained off-chip accesses can improve efficiency after the CU merges sparse access requests.
(2) The PMUs in the PEA are distributed on-chip memory that contain configurable data paths for calculating address sequences for local accesses and also contain a 4-bank SPM for buffering data. The PMU allows high-throughput accesses to scalar data as well as vector data with the same width as the number of banks. PMU mainly takes advantage of the distributed SPM to provide sufficient bandwidth and ensure computational efficiency. Moreover, since PMUs are directly connected to the interconnection network, computational data that needs to be shared can be transferred directly between PMUs without occupying computational resources in PCUs.
(3) The PCU consists of four parallel computational pipelines composed of FUs. There is a FIFO in each pipeline for buffering the scalar and vector data required for computation, and the pipeline contains no other memory units except for the pipeline registers. The temporary data needed for computation is buffered in the PCU, which has the smallest data granularity and is most frequently accessed, so the FIFO is used to store the data.

Different memory levels in Plasticine correspond to varied access costs, characteristics and levels of data granularity, which also reflects the main principle of the design of memory systems in most state-of-the-art software-defined architectures. The prime considerations should be given to the access pattern, data granularity, memory layout of the target application, and data should be stored in partitions based on the preceding information, making the data more frequently used with small access granularity closer to the PEA. Similar to Plasticine, in the mainstream software-defined architectures, each PEA (e.g., 4×4 FUs) is usually configured with one SPM as a data cache, and the SPMs between multiple PEAs form a distributed memory structure. This is more hierarchical than the centralized structure or fully distributed structure in TIA, and can be considered a tradeoff. It has a more balanced hardware overhead and provides higher data bandwidth.

5. Dynamic Management of SPMs

The management of the SPM is entirely dependent on program instruction control, and essentially it does not have the ability to manage data dynamically as the cache does. Some of the work focuses on how to give SPMs certain dynamic characteristics in a lightweight way, so that they can adaptively adjust their resources at runtime according to the real-time demands of the task. This section provides a brief description of some of the work.

In a multitasking computing system, when multiple tasks use hardware resources simultaneously, there are inherent differences in computational load between tasks, and therefore static differences in resource requirements. If the tasks show dynamic characteristics, the computational load at runtime changes over time, which will make dynamic differences in the resource requirements of individual tasks. In the case of differences, evenly distributing hardware resources will lead to reduced computational performance and waste of resources. For example, two tasks T1 and T2 are running simultaneously. T1 is a control-intensive task with low demand for data bandwidth, requiring only two banks, while T2 is a data-parallel task, with the bandwidth as its performance bottleneck, requiring as many banks as possible. In the case of a 16-bank SPM, if the resources are evenly distributed to T1 and T2, each with 8 banks, undoubtedly T1 will waste some banks, while T2 will suffer from performance limitations due to insufficient bandwidth.

For static differences, the compiler can distribute the initial resources of the task more reasonably through static profiling to avoid the problem of load imbalance to a certain extent. For dynamic differences, the compiler cannot obtain the real-time resource requirements of the task, so it usually requires a dynamic mechanism on the hardware to solve the problem. The requirement-aware online hybrid on-chip memory (ROHOM) [41] provides a dynamic mechanism for the bank distribution problem of SPMs. It is designed for systems where caches and SPMs coexist, and uses the miss rate of the cache access to evaluate how many SPM resources the task needs to use. If the miss rate of the cache is low, it means that the task does not need many data blocks and the cache is able to meet the task demand well enough without more SPMs; on the contrary, a high miss rate means that the task needs more data blocks than the capacity of the cache or the task accesses are irregular, so the cache cannot meet the task demand and more SPMs should be used to buffer those frequently missing data blocks to reduce accesses to the cache, thus reducing the dynamic overhead incurred by cache misses.

In addition, the dynamic management of SPMs includes another typical work, online page clustering for multibank scratchpad memory (OCMAS) [42], which dynamically manage the power consumption of a multibank SPM. Depending on how often the data is accessed, the data in the application can be divided into hot data and cold data. If the hot data and cold data can be put into different banks, the utilization of the bank storing hot data in the SPM will be high. The bank storing cold data, which only receives access requests occasionally, can enter a low-power state most of the time, thus reducing the overall power consumption of the SPM. Based on this idea, OCMAS counts data access frequency at the granularity of page table and proposes an algorithm to dynamically set the threshold to distinguish hot data from cold data and store them in different banks respectively, thus using fewer banks to store as much hot data as possible and using more banks to store cold data and enter the low-power state.

Moreover, in many domain-specific accelerators, especially neural network accelerators such as DRQ [43], dynamic data granularity management mechanisms are commonly used to configure the width of the data stored in the SPM and the amount of data in a single access. When the bit width is configured to be small, multiple sets of data can be accessed in one load, increasing the width of the vector and the throughput of subsequent computations; conversely, if the bit width is large, the computational accuracy can be improved, despite the slower computational speed. The granularity configuration mechanism allows users to choose the priority between the performance and the accuracy according to their needs, without hardware changes.

The main advantage of the SPM in terms of power consumption and bandwidth lies in the simplification of behaviors brought by its explicit instruction control. Using the dynamic mechanism to manage resources can make the behaviors of the SPM more complex and increase the design overhead. Therefore, great care should be taken to reduce the overhead incurred by the additional dynamic logic, and lightweight dynamic mechanisms should be used whenever possible, otherwise the overhead is likely to outweigh the performance or power benefits it brings. The current research on the dynamic management mechanisms of the SPM in software-defined architectures is not extensive, mainly because the use of compilers for static analysis has satisfied the needs of most scenarios, and the introduction of complex dynamic management mechanisms usually reduces the computational energy efficiency that is the most critical metric of software-defined architectures.

3.1.2.2 Software-Managed Cache System

The cache technology was originally proposed to address the performance mismatch between computation and memory of processors. Currently, as the difference between computation and memory performance is growing, the cache remains the primary solution to alleviate the problem. The multi-level cache system has been well established, so the basic structure of the cache system will not be repeated here. Currently, one of the research hotspots on the cache is the software-defined cache system, which can dynamically adjust its internal structure and key parameters adaptively with application changes in order to achieve better performance and power consumption than traditional caches. Jenga [44] is used as an example to briefly introduce the research on the software-managed cache system.

Based on the analysis of the memory access behaviors of a variety of programs, the cache hierarchy and capacity required to achieve optimal performance varies. For example, some applications only perform computations on small data sets, and an L1 cache that can store all the data is the best choice, while some applications have larger data sets, and using a larger L2 cache to store more data on chips is a better choice. In addition, due to the parallel feature of multi-core systems, multiple applications with completely different memory requirements may use the cache system at the same time.

In order to dynamically adjust the cache resources to the application requirements, Jenga provides several key solutions: First, Jenga builds a virtual cache system, called the virtual hierarchy, between the software application and the physical memory. The

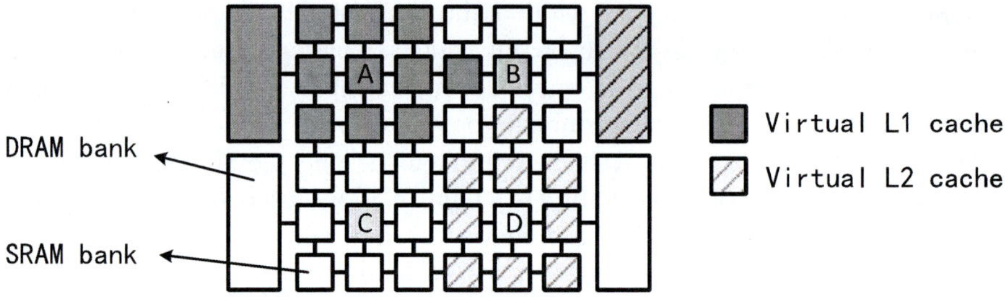

Fig. 3.18 Example of virtual cache system in Jenga

virtual hierarchy mainly aims to (1) virtualize the actual memory system into a multi-level cache system, thus maintaining the transparency of cache programming, and (2) adaptively change the structure of the actual memory system depending on its runtime characteristics, thus meeting the needs of the application. Figure 3.18 shows a typical example of the virtual cache hierarchy in Jenga, where four applications, A to D, are running in parallel and the underlying hardware contains 36 distributed SRAM banks. In this example, different applications require varied caches, where A requires L1 caches with large capacity, applications B and D need large-capacity L2 caches, and application C has the lowest demand for caches. Since the traditional cache system divides the resources equally among the four applications, A, B and D will all have low performance due to insufficient resources, while application C will waste some resources. Jenga, however, relies on a dynamic detection mechanism that incrementally adjusts the operation of each application to achieve optimal performance.

Initially, each application is allocated a separate SRAM bank as its private cache. At runtime, Jenga monitors the cache miss rate of all applications, and when the miss rate is higher than a certain threshold (which can be statically specified or dynamically determined using an algorithm) for an application, Jenga will try to increase the capacity and bandwidth of the L1 or L2 cache for that application by allocating more bank resources to it. As shown in Fig. 3.18, since application C has minimal requirements for caches, and the initial cache has already achieved a high hit rate, Jenga will not allocate additional resources to it. For application A, Jenga will keep adding banks to it until its L1 cache hit rate falls below the threshold. As for applications B and D, adding banks as their L1 caches does not significantly reduce the miss rate due to the very large data sets required for their operations, and Jenga will try to add L2 caches for them. In addition, if the SRAM resources still cannot meet the application requirements, Jenga will allocate some banks of the dynamic random access memory (DRAM) as L2 caches to the application to increase the capacity as much as possible, such as applications A and B. If increasing the capacity of the cache does not reduce the miss rate, Jenga will also try to adjust the cache replacement strategy for the application.

By dynamically detecting the miss rate and incrementally adjusting resources, Jenga is able to flexibly allocate application resources until the miss rate of all applications is below the threshold, at which point application performance is considered to be close to the optimal value. Jenga only requires the hardware to support a simple cache miss rate detection mechanism, so the hardware cost is low. From the software perspective, although the cache structure that Jenga ends up with can be very complex and diverse (as shown in Fig. 3.18), the user does not need to change any code when programming, and accesses to different cache systems by different applications are automatically managed by the cache system. In addition, if an application has changing requirements for caches at runtime, i.e., its access pattern is dynamic, Jenga can track the real-time demand of the application and quickly adjust resources for it according to the changing miss rate.

In recent years, the problems of memory wall and power wall have become increasingly serious. Although caches have always been the research hotspot in the field and their performance has been fully exploited, they have never been able to solve their power consumption problem because additional operations such as tag matching and data replacement are required to complete a data load or write in caches, and these operations cannot be avoided. The software-defined architecture gives particular emphasis on power consumption, and the application scenario also does not need the generality of caches, so caches are rarely used by software-defined architectures.

3.1.2.3 Summary

Currently, due to the growing speed gap between computation and memory, memory is gradually becoming a bottleneck in all kinds of computing architectures, and the design of memory systems is a top priority in architecture designs. Unlike other architectures, software-defined architectures aim to achieve efficient computations in one or several scenarios, rather than general-purpose computations. Because of this, software-defined architectures place no emphasis on the generality of the memory, and energy efficiency and bandwidth are more critical in the design of on-chip memory. In this case, the SPM has become the most widely used on-chip memory in software-defined architectures, while the cache is relatively less used.

In the previous two subsections, we discussed how to reasonably design the hierarchy of the on-chip memory, and focused on the comparative analysis of centralized and distributed memory systems. These two structures can be used as boundaries in the memory design space, between which a compromise with multiple levels can be designed. The most important principle in the design of on-chip momeory is that the characteristics of the application scenario must be analyzed in detail, including access patterns, data reusability, data capacity, and bandwidth requirements. Often, different data blocks will have different needs, at which point we need to differentiate between different data blocks and place data with higher reusability and smaller granularity closer to the PEA to reduce the need for accesses to global memory. In fact, the memory systems in software-defined architectures are becoming increasingly diverse and complex. For example, in Plasticine, memory and PEs are fully coupled in a PEA, and even the memory units can communicate and perform simple computations. Although the memory systems in different software-defined architectures may seem to vary, they are mostly designed to allow the memory system to provide sufficient data bandwidth and avoid performance loss caused by computing resources waiting for data through mechanisms such as paralleled memory and computation, and distributed memory.

3.1.3 External Interfaces

In the above section, the software-defined architecture is able to achieve high performance and energy efficiency for two main reasons. One is the efficient computation model combining temporal and spatial computation. The other one is the domain-specific design of PEAs, control logic, and memory systems, all of which are designed at the cost of generality. So far, software-defined architectures are primarily positioned as domain-specific accelerators and are rarely applied to general-purpose computations alone. Similar to other accelerators, software-defined architectures typically act as coprocessors in real systems, coupled with GPPs (e.g., DySER) to collaborate on the computational tasks of the application. At runtime, the processor is responsible for executing the program control instructions, sending the computational task to the PEA when a compute- or data-intensive code region is encountered, and receiving the results and storing them in the corresponding location after the computation is completed. The communication between the software-defined architecture and the processor is called the external interface. There are also various types of external interfaces depending on how the software-defined architecture collaborates with the processor. The following details several typical ways in which software-defined architectures can interface with other architectures for communication.

3.1.3.1 Tight Coupling Versus Loose Coupling

Software-defined architectures can be loosely or tightly coupled with GPPs [45]. In a loose-coupling implementation, the software-defined architecture is interconnected with the CPU as an accelerator, and communication is mainly through direct memory access (DMA). In this way, software-defined architectures and CPUs can complete computations and memory accesses independently, and the parallel operation pattern leads to higher parallelism and performance. For example, software-defined architectures and CPUs can use different areas of the cache at the same time, thus making full use of memory bandwidth. Figure 3.19 shows how MorphoSys is loosely coupled with the CPU as an accelerator. The TinyRISC CPU is responsible for executing control-intensive part and loading configuration and data for the software-defined architecture. The loose coupling mechanism has the following problems: (1) DMA allows limited bandwidth for data transfer, and frequent data communication is likely to become a performance bottleneck in the system; (2) the

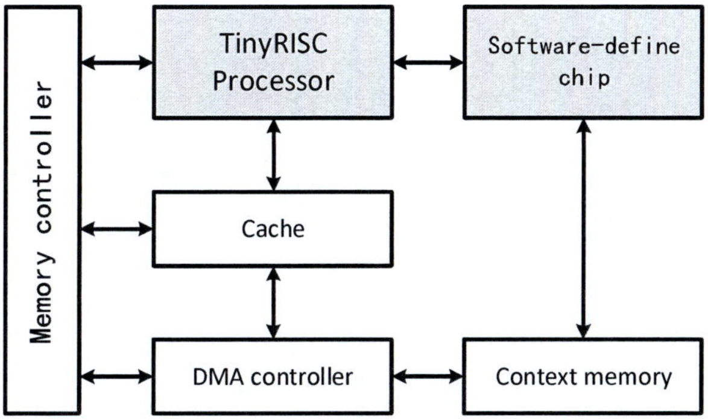

Fig. 3.19 Loose-coupling implementation of MorphoSys

compiler must divide the application into two separate parts and map them to the corresponding architectures, otherwise data synchronization between the CPU and the software-defined architecture will serialize the operation of the program and seriously degrade performance.

In a tight-coupling implementation, the software-defined architecture and the CPU are on the same chip, sharing all on-chip resources, and the computation of both alternates, with only one architecture performing computational tasks at each moment. Figure 3.20 shows a tight-coupling implementation of ADRES, which has two operation patterns, namely CPU computation and PEA computation. In CPU computation, the PEs and register files in the top line of the PEA are configured as the VLIW, and the rest of the chip is turned off by clock gating, power gating, etc. to save power. In PEA computation, all units are activated to complete compute-intensive tasks. In this tight coupling mechanism, there is no need for explicit data transfer between the CPU and the PEA via DMA or bus, etc. The CPU only needs to write data to the register file and the PEA can get the correct data from the corresponding location after configuration switching. Moreover, the CPU and PEA work alternately, partially maintaining the serial semantics of the program code and eliminating the need for data synchronization, so the control logic is simpler. The main disadvantages are that the CPU and the PEA cannot work independently of each other, losing potential task-level parallelism, and that many PEA resources are not fully utilized in CPU computation.

3.1.3.2 Shared Memory Versus Message Passing

In systems consisting of GPPs, software-defined architectures, and other accelerators such as ASICs, it is necessary to divide parallel programs into multiple computational tasks that are allocated to each computing architecture for execution to take full advantage of the processors in different computing domains. Usually, data communication is required between multiple tasks to share certain data blocks, so the system must provide efficient interfaces between processors to meet the communication needs. In order to hide the underlying hardware implementation details of the system, hardware communication modules are often abstracted as communication

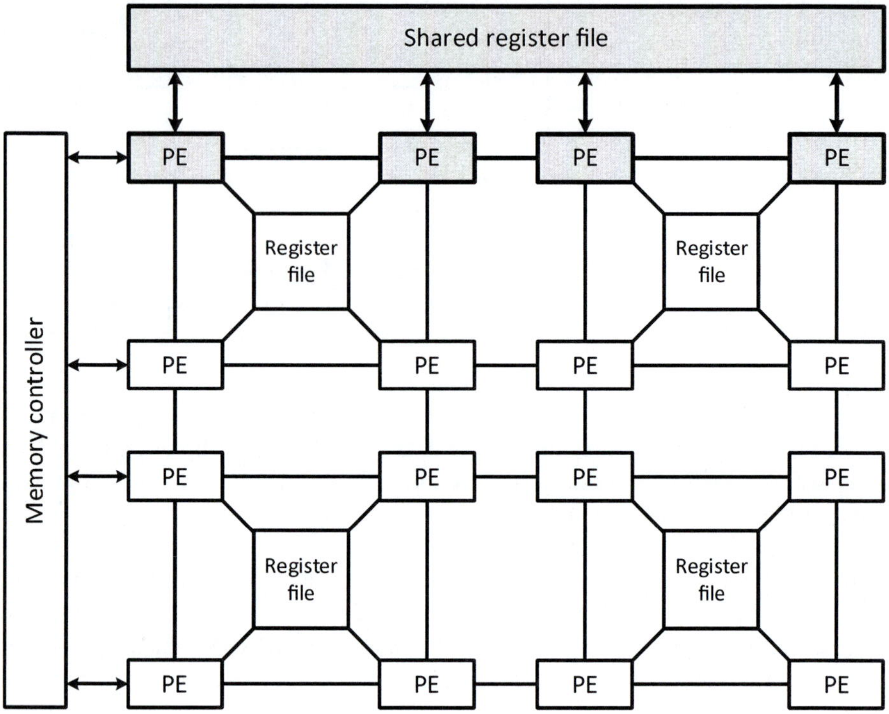

Fig. 3.20 Tight-coupling implementation of ADRES

interfaces in high-level parallel programming languages. Commonly used external interface models can be divided into shared memory and message passing [46]. The shared memory model is closer to the underlying hardware implementation, while the message passing model is mainly implemented through system calls, library calls, etc.

1. Shared Memory Model

The shared memory model virtualizes task-level parallelism in a multi-core computing system into thread-level parallelism in a uniprocessor. In a uniprocessor system, multiple threads multiplex processor and memory resources in a TDM manner, with a private address space for each thread and a shared address space between different threads. Data transfer between multiple threads is mainly done by reading data from and writing data to the shared address space, and modifications to the shared data by a thread can always be observed by other threads. In a multiprocessor system, multiple threads may be executed in parallel on multiple processors, but the compiler will ensure that the address space of the threads remains similar to that on a uniprocessor, i.e., a shared address space is allocated for multiple processors. Thus, parallel programs based on the shared memory model are very similar to serial programs in that the programmer can use memory access instructions to complete data transfers between threads. Note that although the shared memory model in software abstracts the communication of parallel tasks as reading data from and writing data to a shared memory area, in practice the underlying process of program operation is related to the specific architecture. For example, in a multi-core system interconnected by Network on chips (NoCs), the compiler may accomplish the same execution semantics as shared memory by properly arrange the order of communication instructions. It is not necessary to have a shared memory area on the hardware as long as the communication process of the parallel program is consistent with what is declared in the program. Therefore, the communication interface is the way the programming model abstracts the underlying hardware and does not necessarily correspond to the actual hardware. However, in general, the communication interface in the programming model often corresponds to the actual communication model of the hardware, which enables the programmer to understand the actual execution process of the hardware when writing the code and provide more reasonable optimization for the parallel program.

(1) Memory Consistency and Cache Coherence

In order to improve the memory access performance of each processor in a multiprocessor system, each processor usually contains a local cache so that most of the access requests can be responded to in the local cache with low latency. Figure 3.21 shows the structural diagram of a typical acceleration system using the shared memory model, where three processors are interconnected via a bus and share the memory space. The processors can be CPUs, GPUs, software-defined architectures, or other accelerators. Each processor has a local cache that stores frequently accessed shared data, thus reducing the latency of accessing such data. The cache is generally connected to the memory controller via a bus for data transfer.

The shared memory model is very programmer-friendly and only requires reading and writing arrays to complete the communication. However, in hardware, memory consistency and cache coherence [47] in multiple caches are always unavoidable problems in the shared memory model and may cause considerable performance overhead. The cause of these two problems lies in the distributed cache structure in multiprocessor systems. As shown in Fig. 3.21, although there is only one version of the shared data in the shared memory, each processor may keep a copy of such data in a separate private cache.

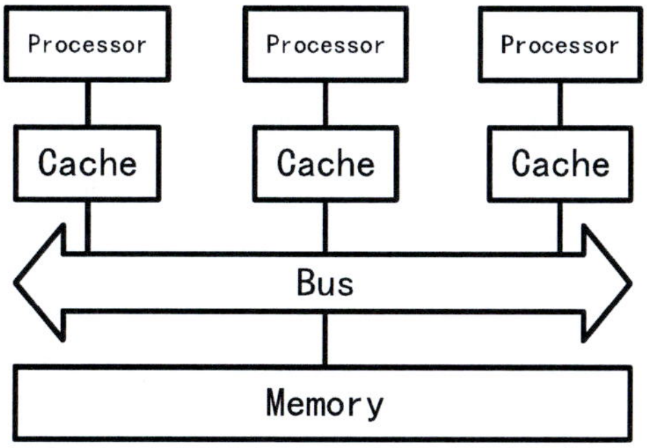

Fig. 3.21 Structural diagram of the multiprocessor system based on the shared memory model

The consistency problem of shared data means that different processors may observe different values when reading shared data from the same address. For example, suppose the shared variable x is initially stored in memory only, with a value of 0. Processors 1 and 2 read the variable x in turn and save x = 0 in their respective local caches. Then, processor 1 performs a write operation to the variable x to change its value to 1. If the caches in this system all use a writeback mechanism, i.e., a write operation to a variable in the cache does not immediately result in a change to memory, processor 2 will read the value 0 in another read operation to the variable x. However, according to the serial semantics of the program, the correct value of the variable x should be 1. Therefore, the shared variable x becomes inconsistent.

The coherence problem of shared data means that when shared data at different addresses is modified, different processors may observe these modifications in a different order. For example, suppose processor 1 sequentially performs write operations on the shared variables x and y to change their values from 0 to 1. Processor 2 reads the value of y first and reads the value of x if the value of y is 1. In theory, the value of x that processor 2 reads is 1, because processor 1 change the value of x before y. However, it is quite possible that, although processor 1 performs the write operation to x first, the change in the value of y in memory occurs before the change in the value of x due to reasons such as the OoO of the write cache or instructions. In this case, processor 2 will read the value of x as 0. At this point, the shared variables x and y no longer remain coherence with each other.

In order to address the consistency problem, multiprocessor systems are required to comply with the cache coherence protocol in hardware design to meet the following two basic principles: ① if the reads and writes to an address are divided into different phases, and all reads between two writes and the next write are considered to be in the same phase, in each phase, any device can read the value of the last write of the previous phase; ② There is at most one device in the system to write to an address at a time, and no other devices can read or write at this time, but multiple devices are allowed to read the same address simultaneously. In this way, any processor in a multiprocessor system can always observe the latest value of the shared variable, even though there may be multiple copies of the shared variable in distributed private caches. Cache coherence requires specifying the memory model during system design, that is, specifying the rules of memory ordering, and programmers need to take into account the uncertainty of program execution results caused by the memory execution order when designing programs.

(2) Interface Protocols Based on Shared Memory

The Coherent Accelerator Processor Interface (CAPI) [48] is the standard interface for accelerators promoted by IBM POWER8 and later chips. In Fig. 3.22, by adding the coherent accelerator processor proxy (CAPP) on the POWER CPU and the POWER service layer (PSL) on the accelerator, the POWER CPU can share memory space with the accelerator through the PCIe channel, and the accelerator can directly play a role like the CPU core and access memory directly through the interconnect.

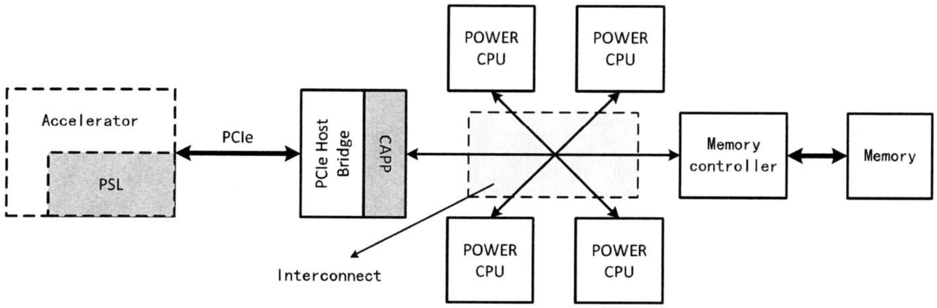

Fig. 3.22 CAPI interface protocol framework

In traditional heterogeneous computing, the CPU is connected to the accelerator through a PCIe channel, but the accelerator can only interact with the CPU as an I/O device, which requires a large number of calls of drivers to complete the address mapping between the CPU and the FPGA. Each call requires the execution of thousands of additional instructions, which greatly increases latency and results in performance waste. The CAPI allows accelerators to access shared memory directly using valid physical addresses. This simplifies accelerator programming and reduces the latency for the accelerator to access memory.

2. Message Passing Model

The biggest advantage of the shared memory model is the simplicity of its programming model, which requires fewer additional modifications compared to serial programs to implement parallel programs. However, with increasing number of processor cores and size of acceleration systems, the cache coherence and consistency problems of the shared memory model become increasingly serious, and the high complexity and poor scalability of the interface protocol implementation prevent the performance of the shared memory model from meeting the needs of applications in many scenarios. In contrast, the message passing model has better scalability and can avoid hardware overhead incurred by consistency and coherence protocols. As a result, the message passing model has been more and more widely used in complex computing systems with large-scale clusters and multiple processors, despite more complex programming.

In the message passing model, the memory space of each computing device is independent of each other. The programmer first needs to divide the application into multiple independent tasks, each of which uses a local data set on a different computing device. Since there is no shared memory space, data synchronization between tasks cannot be accomplished through shared variables. Therefore, the programmer needs to use the communication primitives provided by the programming model to specify information such as the data and communication time that tasks need to communicate.

Another typical application of the message passing model is to establish a static connection between processors, i.e., to specify the data consumer-producer relationship between processors that need to communicate. When the connection is established, the data generated by the producer device is immediately consumed by the consumer device. At this point, communication between the two devices is done by dataflow transfer directly through a fixed channel without additional instructions.

Considering the typical scenarios of the message passing model, for example, in large-scale clusters where the communication overhead between different devices is extremely large, programmers need to minimize the data dependence and balance the load between different tasks. In addition, memory in the message passing model is not transparent, so the programmer must manually allocate the memory space for the data required by each task and specify the transmission channel and communication path for the data, either for the allocation of local data sets or for communication and synchronization between different processes. Therefore, the message passing model requires a higher programming threshold, and programmers must have some knowledge of the hardware structure of the acceleration platform to write efficient parallel programs. However, the advantages of the message passing model are also evident. The manual adjustment by the programmer allows computer systems to scale to larger sizes more efficiently without wasting excessive hardware resources to implement cache coherence protocols. Like the shared memory model, the message passing model is an abstraction at the software level and is not directly related to the specific implementation of the underlying hardware, and the implementation of the communication primitives on different platforms is done by the compiler.

Currently, many acceleration systems use both the shared memory model and the message passing model, i.e., the message passing model is used as an interface between processors to ensure system scalability, while the shared memory model is used between multiple cores within a processor to lower the programming threshold.

The message passing interface (MPI) is a cross-platform and language-independent interface protocol standard. It is not a specific programming language, nor is it bound to any specific hardware or operating system, so it is highly portable. Although MPI is based on message passing, it can also be implemented on computers with a hardware architecture that employs shared memory. In parallel programs using MPI, each process has a unique process identifier and its own address space, and they cannot share data directly. MPI provides various interfaces such as Send/Recv for process communication. Since the programmer explicitly specifies the details of data distribution and communication within each process, the compiler or the underlying hardware does not need to impose additional constraints on data accesses, thus avoiding the problem of excessive data synchronization overhead in the shared memory model.

3.1.3.3 Summary

Software-defined architectures often play the role of domain-specific accelerators in computer systems, which need to interact with GPPs such as CPUs through external interfaces to jointly complete computational tasks. This section first describes the ways that the software-defined architecture is coupled to the GPP, and then introduces two external interface models, shared memory and message passing. Currently, in small or medium scale of multi-core CPUs, the shared memory model is often used to simplify programming and to exploit the parallelism of multi-core processors at a lower threshold. And in special scenarios such as large computer clusters, the message passing model is often used to meet the needs of system scalability and avoid the waste of system parallelism caused by the consistency and coherence of shared data. Since the shared memory model and the message passing model are software-level abstractions of the external interface, they can be used simultaneously to better meet the needs of different developers and applications when the hardware allows.

3.1.4 On-Chip Interconnects

In the CPU, data transfer between instructions is mainly done by reading from and writing to the register file. In software-defined architectures, on the other hand, computing instructions are mapped to different PEs in the PEA, and communication between instructions is done through on-chip interconnects. The producer–consumer relationship between instructions is statically analyzed by the compiler, and the communication path can be statically specified, i.e., the sender's data is sent to the receiver through the path pre-specified by a static interconnect, or the path can be dynamically determined, i.e., a dynamic interconnect allows the packets to reach the destination after the router selects a forwarding path. The on-chip interconnect serves as the carrier for communication within the software-defined architecture, and its parameters such as bandwidth and latency have a significant impact on performance. In communication-intensive applications, data communication can take even longer than computation, and then the on-chip interconnect becomes the main bottleneck in the overall system.

3.1.4.1 Topology of On-Chip Interconnects

There are various implementations of on-chip interconnects [49]. The first is the bus (Fig. 3.23a), which is an information transmission channel multiplexed by multiple devices in a TDM manner. The bus implementation is the simplest, but as the number of devices increases, the bus bandwidth allocated to each device decreases and scalability is poor. The second is the point-to-point interconnect (Fig. 3.23b), which establishes a connection between each pair of devices that may be communicating. This kind of interconnect allows the highest bandwidth, but it has poor scalability, with the interconnect area squared as the number of devices increases. The third is the crossbar switch (Fig. 3.23c), which differs from the point-to-point interconnect in that each device is connected to the switch through only one link, rather than having a link connection with every other device. However, it has poor scalability,

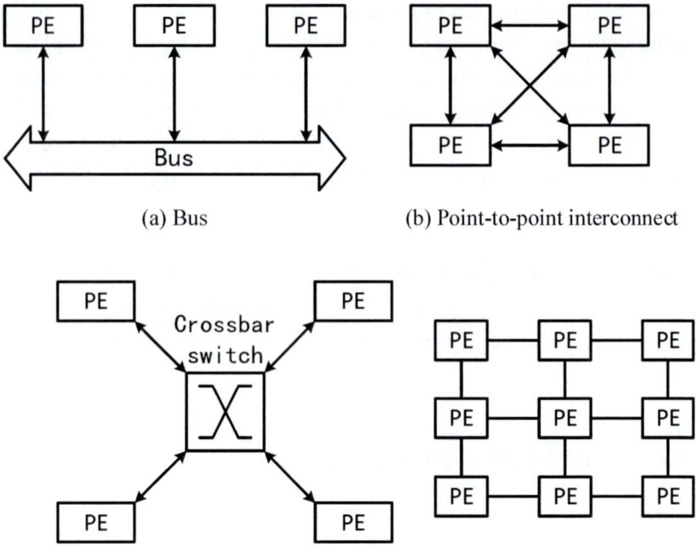

(a) Bus (b) Point-to-point interconnect

(c) Crossbar switch (d) NoC

Fig. 3.23 Implementations of on-chip interconnects

rapidly increasing area, power consumption, and latency with the number of devices, and poor support for broadcast communication, making it difficult to achieve one-to-many or many-to-one communication. The fourth is the NoC, which allows each device to be connected to the adjacent one through switches or routers. The NoC can use topologies like mesh and torus and achieve a better balance between scalability, bandwidth, and area.

For on-chip interconnects of software-defined architectures, it is highly desirable to establish a flexible communication network between various physical units to increase communication bandwidth and computational parallelism. Software-defined architectures are typically interconnected using NoCs and optimized for topologies to achieve a balance between bandwidth, area and power consumption. For example, Fig. 3.24 shows the on-chip interconnect structure of SGMF. SGMF contains a total of 108 physical units: 32 memory access units, 32 control units, 32 compute units and 12 special compute units. Figure 3.24b shows the connections between the physical units and crossbar switches. In particular, each unit is directly connected to the nearest four crossbar switches as well as the nearest four units. Figure 3.24c shows the connections between the crossbar switches, each of which is directly connected to the two nearest crossbar switches and the two crossbar switches with the distance of two hops.

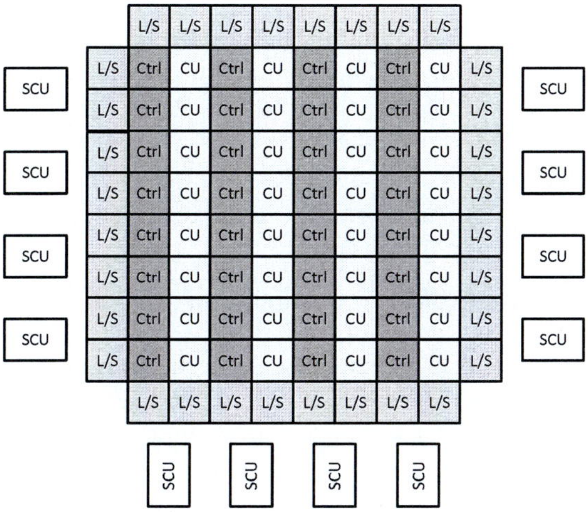

(a) Distribution of physical units of SGMF

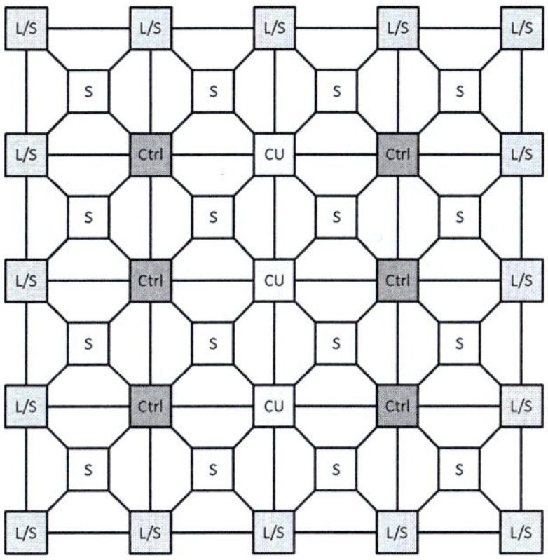

(b) Connections between the physical units and crossbar switches of SGMF

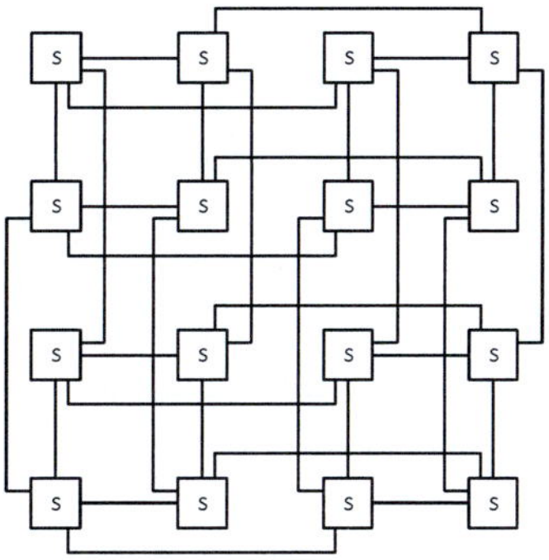

(c) Connections between the crossbar switches of SGMF

Fig. 3.24 NoC topology in SGMF

3.1.4.2 Characteristics of On-Chip Interconnects for Software-Defined Architectures

Although on-chip interconnects for software-defined architectures have a similar design space to multi-core CPUs, the target applications have the following special communication characteristics due to the domain-specific usage scenarios of software-defined architectures [50]:

1. Multi-granularity Communication

To increase data parallelism and reduce the cost of controlling and configuring software-defined architectures, many PEs in the array often adopt vector computation, and both the input and output of these units create heavy demand for vector communication, requiring higher interconnect bandwidth. At the same time, however, there are still a lot of scalar communications in the PEA, because the PEA also requires the transfer of control and configuration information, and there are also many acyclic computations and some reduction operations. Vector and scalar communications pose a challenge for the full utilization of the interconnect bandwidth. Simply using a wider interconnect to accomplish both vector and scalar communications would tend to cause a waste of bandwidth.

2. Heavy Demand for Broadcast Communication Patterns

In the simplest case, the producer sends data to the consumer, maintaining a one-to-one communication pattern between them. However, in software-defined architectures, computation results are often sent to multiple recipients, creating a one-to-many communication. For example, the predicate computed by a branch statement needs to be sent to multiple instructions within the branch. Broadcast communication can cause severe blocking of interconnects, which can be optimized by exploiting the overlap of multiple paths to avoid unnecessary communication and improve the performance of software-defined architectures.

3. Mass Communication Between PEs and the Memory

In a multi-core CPU, each core is connected to a private cache so that most of the data can be found in the private cache, avoiding remote communication from one core to other cores, so multi-core CPUs tend to have smaller bandwidth requirements for NoCs and are more sensitive to network latency. In software-defined architectures, PEs usually do not have private caches, but there are many SPMs distributed in the PEA, as shared memory. Each read or write operation of the PE will need to pass through the interconnection network, consuming the bandwidth of the interconnection network.

4. Statically Determined Communication Patterns

In the software-defined architecture, the dataflow graph of the program can be determined at compile time. By mapping the dataflow graph to physical units, the communication patterns between PEs and memory units are fixed, and the compiler can determine exactly which link in the on-chip interconnect is the critical link with the highest communication load. Therefore, software-defined architectures often employ static networks to enable efficient communication between physical units.

3.1.4.3 Dynamic Network Versus Static Network

In traditional multi-core processor architectures, the on-chip interconnect is often dynamic. Data is transferred dynamically in the form of packets, which record the information such as the source and destination. In particular, when a packet arrives at the router, the router performs computations and selects the output of the crossbar switch to be used based on the source and destination. Then, the packet competes with other packets coming from the same input or requesting for the same output port, which is determined by arbitration. After the port access is successfully obtained, the packet is able to go out of this output port to the next router.

In software-defined architectures, static networks can be used to implement on-chip interconnects after dataflow graphs are mapped to physical units because the communication paths between physical units can be statically determined at compile time. In a static network, a fixed connection is established between the ports of crossbar switches within the router at configuration time. At runtime, packets from different ports are output directly from the corresponding port of the static connection to the next router. As a result, packets no longer need to compute the routing path or go to switch arbitration, which can reduce routing delays, as well as the additional power consumption and area caused by these two steps, thus often achieving higher bandwidth within the same hardware constraints.

When designing on-chip interconnects for software-defined architectures, the design considerations of both static and dynamic networks are analyzed [50] to select the optimal combination for different applications.

1. Static Networks

(1) Flow Control

Since data in a static network does not go through processes such as dynamic routing computation during transmission, flow control cannot be performed at each forwarding as it is in a dynamic network. In some fully static software-defined architectures, the interconnection network can contain no flow control mechanisms. In architectures designed for applications with regular computations, all communication steps can be analyzed by the compiler, so that only statically configured communication sequences and reasonable resource allocation are needed to avoid data channel conflicts. However, in dataflow-based dynamic-scheduling architectures, flow control is a must to achieve dataflow computation. In static networks, there are two main flow control schemes that can be used. The first one is to directly establish a credit-based flow control mechanism between the source and destination, i.e., to use the credit signal at the source to record whether there is a free buffer at the destination, and to allow sending data only when there is a free buffer, which is used by DySER. The second is to use ping-pong buffer in the array, where one buffer is responsible for receiving the data sent by the producer and the other buffer is responsible for sending the data to the consumer. The two buffers work in parallel to avoid network congestion. The first option only requires adding registers with backpressure function at the input and output of the PE, so the hardware cost is relatively low, but the communication performance may be degraded due to the long path for credit exchange between the source and the destination. The second option requires adding buffers in the communication path, which is relatively costly. In software-defined architectures, static-dataflow PEAs typically use the first option to reduce cost, and dynamic-dataflow PEAs tend to use the second option to improve performance.

(2) Bandwidth

By increasing the number of ports of crossbar switches in a static network, more flexible connections can be established between PEs in the PEA, and more data can be transferred in parallel, thus increasing the communication bandwidth. However, as the number of ports increases, the area of the crossbar switch is squared, and its power consumption and latency also increase significantly. The selection of bandwidth needs to be based on application requirements. When performance is the key metric and communication bandwidth limits the overall performance, increasing the number of switch ports can effectively address system bottlenecks.

(3) Customized Sub-networks

Since there are often both vector and scalar communications in software-defined architectures, two sub-networks with different data widths and configurations are used to handle vector and scalar communications. The on-chip interconnect can be more flexible and the interconnect bandwidth can be better utilized while the total network bandwidth remains unchanged.

2. Dynamic Networks

(1) Routing Algorithm

The router uses a specific routing algorithm to dynamically select a communication path for each incoming packet. The routing algorithm has the following typical implementations. Most commonly, a routing table is used. A routing table is maintained on each routing node, which records the output directions that need to be selected from the current node to other nodes, and packets are routed by looking up on the routing table. The routing table is simple to use and can employ algorithms to obtain a relatively optimized path and make full use of the network bandwidth, but the disadvantage is that for larger networks, the routing table takes up a large area. The second option is to pre-encode the routing path within the packet, which is transferred along the specified path. This method takes up too much space within the packet and is therefore less commonly used. The third one is adaptive routing. In particular, to reach the destination, the packet usually has multiple directions available and selects a relatively non-blocking one for transmission based on the blocking conditions that may be encountered in each direction. Although this implementation allows packets to avoid blocked areas with optimal performance, the operation is complicated. Specifically, the information transmission and computation associated with network congestion is often complex, and it tends to increase the routing delay of critical paths. All things considered, the routing table is the best choice for software-defined architectures. This is because the array size is generally not too large, and there are not many communication paths, which can be statically specified in routing table configurations. In addition, the use of node routing tables enables optimization of the network structure to improve performance.

(2) Virtual Channel Allocation

The essence of packet transmission in the network is to use the buffers on the physical channel for location changes. By dividing the buffers on a physical channel between different types of packets, a single physical channel can perform the function of multiple virtual channels. Virtual channels are critical to solving problems such as deadlocks and blocking. However, the introduction of virtual channels needs to be accompanied by a solution to the virtual channel allocation problem. Virtual channel allocation is usually implemented by using an arbiter or a FIFO. The FIFO stores the IDs of idle virtual channels, among which the front-end idle virtual channel of the FIFO is taken for use each time. This method is more area and power efficient than using an arbiter and can reduce the critical path latency.

(3) Input Buffer Size

Theoretically, as the number of buffers increases, the throughput of the network increases accordingly, but the marginal effect of this approach is diminishing. The network area and power consumption increase significantly. When dynamic networks adopt credit-based flow control, the depth of the buffer is typically 2–4, and further deepening the buffer brings limited performance gains. If mechanisms such as the ping-pong buffer are used, especially for dynamic-dataflow PEAs, the depth of buffer is likely to be much greater, and the degree of parallelism is higher, thus imposing higher requirements for the communication bandwidth.

(4) Broadcast Communication

For software-defined architectures, due to the tight timing constraints of the network and limited hardware resources, the common practice for broadcast communication is to copy the packet and use a tree communication path generated by the compiler, to reduce the packet forwarding.

(5) Deadlock Avoidance Scheme

Deadlock means that when the dependence on routing resources between multiple packets forms a dependence ring, every packet is blocked because it waits for other packets to release their resources and also cannot release the resources it occupies, so all the packets are blocked permanently. There has been considerable work in the NoC domain [51, 52] on how to avoid routing deadlocks as well as protocol deadlocks. For software-defined architectures, to ensure the point-to-point communication timing of packets under the limited communication resource budget, a possible solution is to use virtual channel differentiation to separate different information flows to avoid deadlocks.

3.1.4.4 Summary

The software-defined architecture enables data interaction between instructions in temporal computation by transferring the generated data through on-chip interconnects. The bandwidth of the on-chip interconnect is of great significance for exploiting the computing potential of software-defined architectures. To achieve a better trade-off between bandwidth, area, power consumption, latency, etc., software-defined architectures often use NoCs to implement on-chip interconnects. Compared to NoC applications such as multi-core CPUs, software-defined architectures have some special communication requirements, such as multi-granularity communication and broadcast communication, which pose a great challenge for the full utilization of the bandwidth of on-chip interconnects. In addition, since the communication characteristics of applications in a software-defined architecture can be statically determined at compile time, static networks or a combination of static and dynamic networks can be used. Specifically, the design space of both networks is taken into account for the optimal combination, thus achieving better flexibility while increasing the bandwidth of on-chip interconnects.

3.1.5 Configuration System

One of the key features of software-defined architectures is that hardware functions are defined by software, rather than only for a single application scenario as in the case of ASICs, which is achieved by relying on a configuration system on the hardware. Figure 3.25 shows the structure of a typical configuration system. The configuration system reads configuration information from memory through the configuration bus, and then sends the configuration information, after certain processing, to the PE, interconnect switches and other modules in the corresponding PEAs in order to rewrite the internal configuration information registers, thus changing the function of the circuit. Function configuration is an additional phase in the operational flow of the software-defined architecture, which requires additional time to switch functions. How to handle the relationship between configurations and computations has a significant impact on performance, hardware utilization, and other metrics. In the following, reconfiguring of circuit structures and changes in circuit functions are defined as configuration.

Configurations in software-defined architectures come in many forms. In particular, they can be classified into static configuration and dynamic configuration, depending on whether configuration switching is allowed at runtime. In addition, depending on whether the software-defined architecture allows reconfiguring some resources, they can be divided into full reconfiguration and partial reconfiguration. The selection of the configuration system is closely related to the structure of the PE because each PE has varied size of configuration contexts and reconfiguration frequency. A detailed introduction to these concepts and the optimization of the configuration system is described.

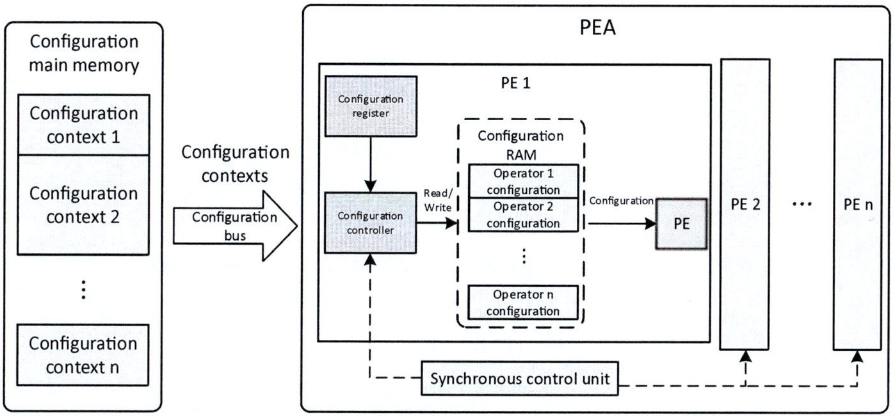

Fig. 3.25 Structure of a typical configuration system

3.1.5.1 Full Reconfiguration Versus Partial Reconfiguration

1. Full Reconfiguration

Full reconfiguration means that all elements in the array are configured synchronously and start the computation after the configuration is complete. In full reconfiguration, the computation and configuration are completely separated without overlap, so the implementation is the simplest. The impact of the configuration on the operation flow was briefly mentioned in Sect. 3.1.1 when introducing the single-instruction PE. For a fully reconfigured array, the size of the applications that can be mapped to the array at a time is limited by the resources available on the array. For a small application, all instructions can be mapped to the array at a time. The application can run in a fully pipelined manner after the mapping, and this type of computation is most efficient. However, the application size is usually larger than the number of computing elements of the PEA. Then, it is necessary to divide the application into multiple kernels, and each kernel should be able to map to the array. When the execution of a kernel completes, the entire PEA is reconfigured to switch to the execution of next kernel. The dot product of vectors with a length of 16 is taken as an example, as shown in Fig. 3.26a. The vector dot product cannot be mapped at a time due to the array size limitation. Here, the vector dot-product calculation can be performed in two steps: (1) calculate the dot product of vectors with a length of 4 four times to obtain four intermediate results; (2) switch array configuration to addition reduction to perform an adder tree calculation on the four intermediate results and obtain the final result.

If the application size is much larger than the array size, the array will need frequent configuration switches between the kernels and this results in overheads in two perspectives. On the one hand, since all PEs need to stop running each time they are reconfigured, and configuration and computation are performed sequentially, the total overhead incurred by frequent configurations will not be negligible. On the

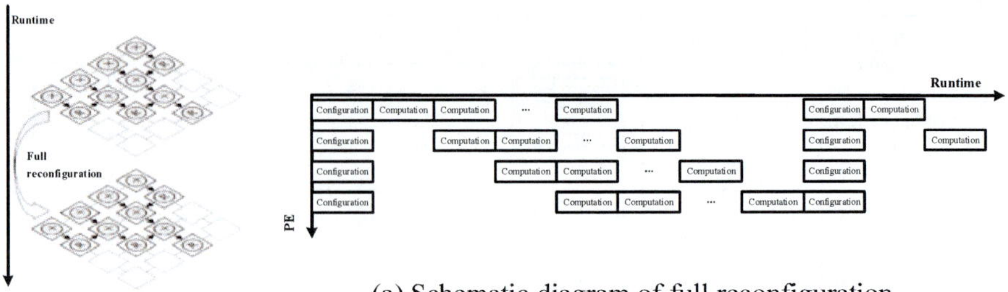

(a) Schematic diagram of full reconfiguration

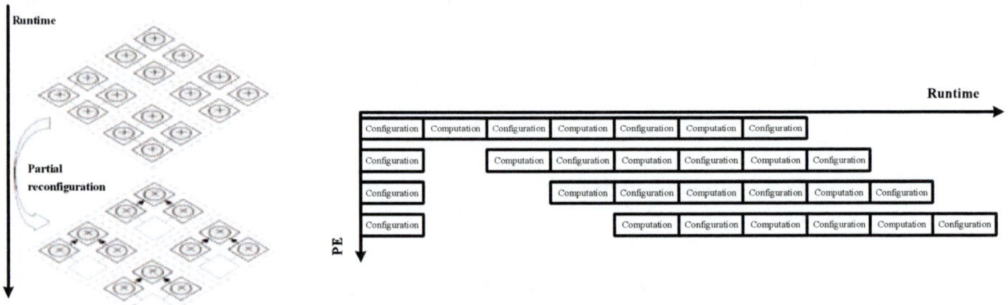

(b) Schematic diagram of partial reconfiguration

Fig. 3.26 Operational flow of full reconfiguration and partial reconfiguration

other hand, before the start of the reconfiguration, the intermediate results calculated by each core must be stored in memory, so the data needs to be read from the memory after the reconfiguration. The additional memory read and write overhead can cause significant performance degradation and power consumption increase when the scale of the data is large. Therefore, full reconfiguration is mainly designed for smaller-scale applications. If large-scale applications are to be computed, reasonable task allocation needs to be considered as much as possible to reduce the amount of intermediate data that needs to be buffered and to reduce the reconfiguration frequency. Otherwise, the cost of reconfiguration will severely impact the performance and energy efficiency, and the advantages of array computation cannot be fully utilized.

2. Partial Reconfiguration

The main reason for the overhead in full reconfiguration is the complete separation of the computation and configuration, which causes complete stagnation of the array during configuration. In contrast, partial reconfiguration divides the array into multiple subarrays, and the reconfiguration of each subarray is performed independently. Thus, while some subarrays are being reconfigured, other subarrays can still perform computations, thus enabling the overlap of computation and configuration to hide the configuration time. Moreover, in partial reconfiguration, when using subarrays to perform computations, the operators in the application are compressed to the subarrays, allowing higher resource utilization of the subarrays and sometimes improved computational performance. As shown Fig. 3.26b, the array is divided into four 2 × 2 subarrays, each of which independently completes the calculation of the vector dot product, with a length of 4. One vector dot-product calculation contains a total of four multiplications and three additions, a total of seven operators, so the subarray needs two configurations to complete one dot-product calculation. Assuming that both multiplication and addition can be completed in 1 clock cycle, the subarray requires a total of 1 cycle to complete multiplication and 2 cycles to complete addition.

In the case of four parallel subarrays, regardless of the configuration time, the dot product of two 16-element vectors can be completed in 3 cycles, i.e., throughput up to 32/3 data per cycle. The array can only receive 4 sets of data for dot product in each cycle at most under full reconfiguration. From this perspective, the mode with subarrays running in parallel and configured independently achieves higher performance than the fully reconfigured array. Of course, this is just a rough performance comparison with no consideration about the time consumption of configuration switching. However, as can be seen in Fig. 3.26, the subarray operation pattern has higher resource utilization with fewer idle PEs, and does not require PEs to perform data routing operations, thus potentially achieving higher performance.

Since the partially reconfigured software-defined architecture divides the hardware resources into several parts with independent computing capabilities, it is particularly suitable for multi-tasking processing, whose multiple tasks run independently of each other on different subarrays. Overall, the configuration frequency of subarrays in partial reconfiguration is higher than that in full reconfiguration. However, subarray configuration requires less time and can be enabled in parallel with computation, thus having less impact on performance. Currently, some software-defined architectures allow dynamically changing the array's operation pattern. In other words, a single task can be accelerated by using all resources of the entire array, and the subarray operation pattern can be utilized when many tasks need to be executed, to better support multitasking. For example, the compilation algorithm proposed by Pager et al. [53] enables configuration transformation, in particular, transforming the configuration originally for fixed-size arrays (e.g., 4×4) into that for subarrays (e.g., 2×2), so that the two operation patterns can be switched automatically without recompiling at runtime. The PPA [54]'s hardware contains the similar configuration transformation circuit to enable scaling of the array size in hardware.

In fact, the division of partially reconfigured subarrays can have various granularities. For example, both BilRC [55] and XPP-III [56] use a more fine-grained configuration. In particular, configuration is done only for a single PE every time. The configuration data spreads along the interconnects among PEs until the reconfiguration of the entire array is completed. While a particular PE is being configured, other PEs in the path that are already configured can start working. Software-defined architectures such as Chimaera [57, 58] and PipeRench [12, 13] employ line-by-line configuration techniques, which enable the arrays to be configured sequentially in lines. Usually a line in the array can be viewed as a stage of the pipeline, and when a line completes its computational task and outputs the results to subsequent lines, this line can start reconfiguration while the subsequent pipeline stages are still performing computation. In line configuration pattern, configurations of the entire array are propagated along the lines, enabling computation in lines to form pipelines and reduce the reconfiguration cost.

2. Static Reconfiguration Versus Dynamic Reconfiguration

1) Static Reconfiguration

Static reconfiguration is a configuration pattern in which all configuration tasks are completed before the start of the computation and no functional reconfiguration is performed after the start of the computation due to the relatively costly configuration time. The most typical architecture with static reconfiguration characteristics is the FPGA. Since both LUTs and interconnects in FPGAs need to be configured at the bit level, this results in a huge amount of configurations (typically, hundreds of megabits) and long circuit reconfiguration time (typically, tens to hundreds of milliseconds). Therefore, the FPGA needs to load all the configurations before the circuit can start working. To reconfigure the circuit function, the current computational task must first be interrupted. Although some studies [59] divide the FPGA into multiple parts, each of which is controlled independently to support runtime configuration, the reconfiguration overhead is still considerable and even exceeds the time required to complete the computation in many applications due to its overly fine-grained bit-level configuration. For these reasons, FPGAs typically have only static reconfiguration characteristics. For statically configured architectures, the application scenario is severely limited by the available hardware resource. For example, when the application is too large, the circuit cannot be integrated into the FPGA.

For other software-defined architectures, such as the multi-instruction and dynamic-scheduling PEs introduced in Sect. 3.1.1 (e.g., TIA), although they are configured at the word level, each PE contains multiple instructions, and all instructions need to be configured for the entire array, which still takes longer. Also, due to the instruction caches, it is possible to map the complete application onto the array at a time. Consequently, these architectures are also typical statically configured.

2) Dynamic Reconfiguration

In contract to static reconfiguration, dynamic reconfiguration is an operation pattern in which the software-defined architecture performs functional reconfiguration during computation. Dynamic reconfiguration enables small arrays to perform large-scale computational tasks through multiple configuration switches. Dynamic reconfiguration is valid on the premise that the array configuration's size is small and the array's function can be reconfigured in a short period of time. At least, the cost of reconfiguration should be much less than that of computation in the entire running process, otherwise, dynamic reconfiguration will seriously degrade the performance. A typical software-defined architecture with dynamic reconfiguration characteristics is the CGRA. Compared with FPGAs, the PEs and interconnection networks in CGRAs are configured in words, and the array size is much smaller than that of FPGAs, so the configuration overhead can be effectively reduced. In general, the CGRA is able to complete a functional reconfiguration in tens of nanoseconds, which is basically negligible compared to computational tasks.

Software-defined architectures with single-instruction PEs typically support dynamic reconfiguration, such as the aforementioned DySER and Softbrain, mainly because simple array structures also require less configuration information. Of course, they all provide corresponding mechanisms to further reduce the configuration cost. For example, the fast configuration mechanism in DySER and the asynchronous execution of tasks in different arrays triggered by stream in Softbrain take advantage of partial reconfiguration to hide the configuration time by parallelizing configuration with computation. In addition, HReA [27] employs a configuration compression technique to recode the bit stream to reduce the amount of configurations, and is designed with a decoding circuit in hardware, which can effectively improve the loading efficiency of configurations.

3. Summary

The configuration system is one of the key modules in a software-defined architecture, which has a significant impact on the architecture's overall operational flow. This section introduces the main concepts of partial reconfiguration, full reconfiguration, static reconfiguration, and dynamic reconfiguration with corresponding basic characteristics. In general, for small applications, it is simplest to use statically and fully reconfigured software-defined architectures, which can take full advantage of the high performance and efficiency of spatial computation and pipelined computing in arrays. Considering larger applications, dynamic, partial reconfiguration is a better design choice, because functional reconfigurations can be performed at a much smaller configuration cost until the entire application is complete. Due to limited space, this section does not describe the specific hardware design of the configuration system, and interested readers can refer to related materials [60].

3.1.6 Summary

This section focuses on the design of PEs, on-chip memory, external interfaces, on-chip interconnects, and configuration systems for software-defined architectures. Rather than presenting too many circuit-layer implementation details, this section provides and analyzes multiple design options in as much detail as possible in the design of each module, aiming to build a relatively complete, multi-dimensional architectural design space. In fact, as mentioned several times in this section, the specific design choices for each dimension should be determined according to the requirements of the application scenario. This is one of the most dominant trends in software-defined architectures today, that is, to design different architectures for different application domains and take full advantage of their energy efficiency, rather than sacrificing energy efficiency for the higher generalization, or losing flexibility due to excessive customization as ASICs. In addition, new software-defined architectures prefer to use simple hardware structures and leave the complex logic functions to compilation algorithms for higher hardware energy efficiency.

The overall design space for software-defined architectures has been explored relatively well over the past two decades or so, but the above only presents design choices for individual dimensions, without an in-depth discussion considering other dimensions. The following sections will describe how to combine the different dimensions rationally to build a truly efficient computing architecture. In addition, agile development tools are now available in academia and industry, which can rapidly generate hardware architectures, support automatic design parameter exploration, and build complete software-defined architecture systems from applications efficiently. Section 3.2 will present the typical development frameworks.

3.2 Development Frameworks of Software-Defined Architectures

Section 3.1 describes the complete design space for software-defined architectures, and gives several typical design choices in each dimension. However, it is insufficient to consider only an individual dimension, when designing an efficient software-defined architecture in practice. This is primarily because the different dimensions are not actually independent of each other. For example, different types of PEs have different requirements for the memory system, and bring distinct designs of the configuration system. Since multiple dimensions are often interdependent and interact with each other, considering only single dimensions does not necessarily lead to the most appropriate solution, which requires a comprehensive consideration of the relationship between multiple dimensions. Moreover, in the same dimension, the various schemes introduced in Sect. 3.1 can be considered as the boundaries of the entire design space, and the design choices presented are often several boundary schemes in the same dimension with distinct structural characteristics and different preferred metrics. For example, multi-instruction and dynamic-scheduling PEs are more applicable to applications, but have complex structures, high hardware overhead, and relatively low computational energy efficiency; single-instruction and static-scheduling PEs have simple structures, low hardware overhead, and higher computational energy efficiency, but are usually only designed for applications with regular computation pattern. In this case, it is often possible to reasonably combine different structures, make a compromise design between several schemes, retain the advantages of several structures, and make up for the disadvantages of each other, thus obtaining a more balanced design objective. This hybrid design strategy is also one of the prevalent trends in software-defined architectures.

From the above analysis, we can see that the software-defined architecture has a huge design space with many design options for each dimension, and multiple dimensions are interrelated with each other. Therefore, it is a very complex task to choose the most suitable architecture design as required. Section 3.2.1 first describes how the cutting-edge research in the domain deals with the relationship among computation, control, and data access, which are the three key factors that have the greatest impact on performance. The hybrid array structure and hybrid interconnect structure are then used as examples to introduce how to effectively balance and explore several design options in single dimension.

In addition, there is still no unified design scheme for software-defined architectures, which means that the architecture needs to be redesigned for each new domain. When it is necessary to validate the architecture of your own design, you often have to write behavior-level simulators and RTL-level code, which consumes a lot of engineering time. To shorten the development cycle, there is already research in academia proposing agile hardware development frameworks that provide a complete tool chain to automatically generate the underlying hardware architecture from a high-level behavioral description language, and allow rapid exploration and optimization of architectural parameters. Section 3.2.2 introduces two typical development frameworks that can help users to quickly validate ideas and save time cost for design. It is necessary to mastering these automated tools as an essential skill for every architecture designer, along with the rapid development of these agile hardware frameworks.

3.2.1 Architectural DSE

This section discusses how to build software-defined architecture systems properly, using a typical architecture as an example. There are two main design approaches for handling the relationship among multiple design dimensions. The first one is decoupling design, which divides the array's functions into computation, access and decoupling in a more fine-grained manner. The independent functional modules are designated to these functions separately. Since each module only needs to support a limited number of functions, which can be achieved using only simple hardware. By decoupling, hardware circuits can be composed of multiple simple modules that work together through specific interfaces, rather than making the array too complex by placing all functions inside it. The other design approach is tight coupling, such as near data processing (NDP) and processing in memory (PIM), which takes the memory as the main body and puts the computation as close as possible to the memory unit, or even embedded in the memory array, in order to solve the memory bottleneck problem. For the selection of a single dimension, the focus is on the hybrid design, which explores a compromise among several complementary options and finally selects the design node closest to the requirements.

1. Decoupling of Computation, Control from Access

The instructions contained in the application can be classified into computing instructions, control instructions and access instructions. From the software perspective, these instructions are highly related and executed sequentially until the entire computational task is completed correctly. However, from the hardware perspective, different types of instructions have varied requirements for hardware resources. In particular, computing instructions need to complete general arithmetic operations on integers or floating-point numbers, such as addition, subtraction, multiplication and division; control instructions mainly require conditional computation, such as comparson; access instructions include instructions related to address computation in addition to loading and storage, while only multiplication and addition of integers are required to complete address computation. If all instructions are enabled in the compute array, each FU must support all arithmetic functions. In practice, instructions require only a small portion of the computational resources. For example, PEs that are mapped with control instructions will only be used for conditional computation, and all arithmetic computation resources within the PE will be idle. Resource utilization can be effectively improved if separate modules can be designed for each instruction according to its characteristics, with each module providing only the computational resources required for that instruction.

In addition, another advantage of instruction decoupling is the higher degree of parallelism. In many applications, describing the application in high-level language with sequential execution semantics imposes additional dependency and reduces program parallelism. After instruction decoupling, the data dependency and control dependency of each part of the instruction can be analyzed more directly. The sparse vector-vector product (SPVV) is taken as an example, and the algorithm is shown in Fig. 3.27. The source code shown in Fig. 3.27a is written in high-level language, and the interior of the loop body contains access instructions to the idx and val arrays of the two sparse vectors, conditional comparisons of the idx values, as well as multiplication and addition arithmetic computation. If we just look at the structure of the code, there is a loop-level dependency in the code, i.e., the next loop cannot start until i1 and i2 have been computed and updated in the previous loop. In addition, there are dependences between instructions inside the loop body, and multiple branch statements imply intensive control dependences; accesses to arrays contain data dependences on i1 and i2, and complex dependences will severely reduce ILP. However, if the complete operational process is analyzed by decoupling of computation, access and control instructions, each part of the instruction block is viewed independently, and more parallelism can often be exploited based on the analysis of the dependences between instruction blocks. Figure 3.27b shows an example of decoupling operation, where each vertex in the figure stands for the execution of a single instruction, and the edge represents the fulfillment of dependences through data transfer. Since each module can be executed relatively independently, it waits for data to arrive only when synchronization is required, which can effectively increase the parallelism of computation. The following two subsections describe the instruction decoupling mechanism in more detail.

1) Decoupling of Computation from Access

Decoupling of computation from access (such as the decoupled access execution (DAE) architecture) allows computation and access to be executed in parallel, increasing the parallelism of both computation and access instructions. As shown in Fig. 3.27, there are four memory loading modules that load all the data in array idx and array val from vector r1 and vector r2 respectively, while the PE receives only valid data for multiplication and addition. The conditional PE (cmp) receives and compares the data of idx1 and idx2, and controls the access and computation behaviors depending on the comparison result. If the condition is true, the PE starts to compute, otherwise the data is ignored and the PE does not need to compute;

```
class row:
  int idx[];
  float val[];
  int cnt;

float sparse_dotp(row r1, r2)
  int i1=0, i2=0;
  float total=0;
  while (i1<r1.cnt && i2<r2.cnt)
    if (r1.idx[i1]==r2.idx[i2])
      total+=r1.val[i1]*r2.val[i2];
      i1++; i2++;
    elif (r1.idx[i1]<r2.idx[i2])
      i1++;
    else
      i2++;
```

(a) Sparse matrix-vector multiplication

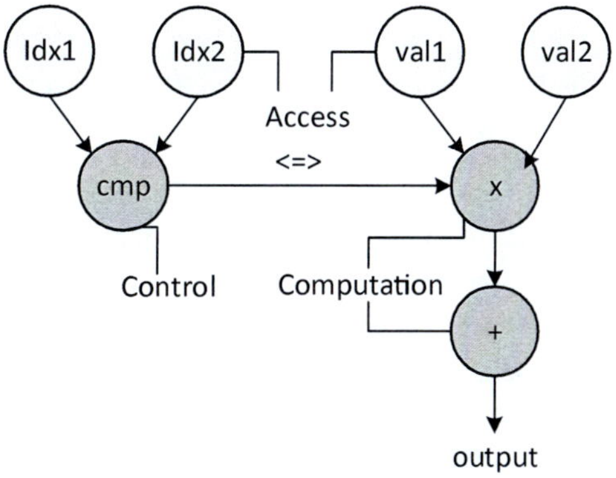

(b) Decoupling operation example

Fig. 3.27 Decoupling of computation, access, and control (SPVV as an example)

the loading unit decides to keep or discard the data loaded in arrays val1 and val2 depending on the computation result.

This operation pattern allows computing instructions and access instructions to be run on different modules. The computing instruction only needs to wait for receiving data and starts the corresponding operation, while the memory instruction only needs to load the configured data sequentially. The collaboration between these two instructions is ensured by the conditional instruction. In hardware, the addresses designated by the access instructions are all contiguous with progressive increasement. Therefore, the computation required by access instruction can be done with a simple adder without other resources in the compute array. Hence, more PEs in the array are allowed to be used for arithmetic operations. For example, the access address needs to be computed by some PEs in a 4×4 array, thus only a few PEs can perform effective multiply–add operations. After decoupling, all PEs can be used for multiply–add operations. More loop unrolling operations can be performed on the original loop with sufficient memory bandwidth, thus significantly improving computational performance.

The decoupling operation mechanism in this example also has some limitations. Without the use of compute arrays, the access module can only generate several fixed-pattern access sequences (e.g., sequential linear access) using its simple integer arithmetic circuits. Consequently, randomized access instructions are aggregated into coarse-grained, pattern-specific access instruction sequences to fully utilize the address computational efficiency in the access module. As shown in Fig. 3.27, the decoupling execution actually loads all the data that may be needed. It performs more loading instructions on the val array, and discards some of them according to the conditions. However, the sequential execution only loads the needed data based on the condition-based determination results. Decoupling of computation from access here employs redundant loading to extend the access instruction, which transforms deterministic data access to accesses of all the data in the val array. In this instance, the loading task can be done with a simple and efficient access unit. In general, in many domains of software-defined architectures, the vast majority of data accesses in applications are regular with similar and simple characteristics, so the decoupling of instructions from accesses can be widely applied to many acceleration domains.

At present, more and more software-defined architectures adopt the decoupling design. Typical architectures with computation-access decoupling include Plasticine [20] and Softbrain [10]. Softbrain is a typical architecture with multi-level memory decoupling. Figure 3.28 shows the architectural diagram of Softbrain, where SD is the task scheduling unit to handle dependences between streams. SD distributes the computational tasks, which satisfy the dependences, to other modules for computation. Softbrain uses the CGRA as its computing core, which contains only a single-instruction and static-scheduling compute array (similar to DySER) internally. The CGRA is responsible only for effectively completing computational tasks, but does not execute any access-related instructions or contain any memory units. Softbrain contains multiple memory engines, including the MSE, SSE and RSE, which are responsible for automatically generating access addresses based on the configuration and accomplishing different levels of access tasks. In particular, the MSE sends the generated addresses to the main memory for accessing off-chip data; the SSE contains SPMs internally to cache on-chip data, and generates access requests mainly to the internal SPM; the RSE temporarily stores the intermediate computation results of the array in current iteration for future use. All these units contain configurable AGs with support of the integer adders and multipliers for address computation. By configuring specific parameters, the AG can automatically compute the corresponding access addresses and access instructions. As shown in Fig. 3.28b, Softbrain can support linear, stepwise, overlapping, and repetitive access patterns after the configuration of start address, step, single step size, and step count. This is fully adequate for the applications they are intended for.

The compiler assigns the access sequences, which configures the AG, to the memory engines according to their corresponding levels when Softbrain is used to accelerate certain applications. All the computing instructions are mapped to the CGRA. The memory engine accesses the memory independently, obtains the data and sends it to the vector interface of the CGRA during the process. The desired computation is lanched on the array once the data in the vector interface is ready. The computation results are stored in the corresponding memory after going through the output vector interface. The whole process keeps running iteratively until the computational task is completed. It can be seen that decoupling of computation from access in the hardware architecture allows the compute array and access engine to be more dedicated to specific tasks, resulting in a more streamlined module structure with higher utilization. In software-defined architectures, streamlined architecture often provides higher computational performance and energy efficiency. In addition, the hardware design of the overall architecture is also more modular. Consequently, in the case of new application requirements and computation models, only the parameters supported by the AG need to be modified without changing other modules, thus reducing the time cost of architecture design.

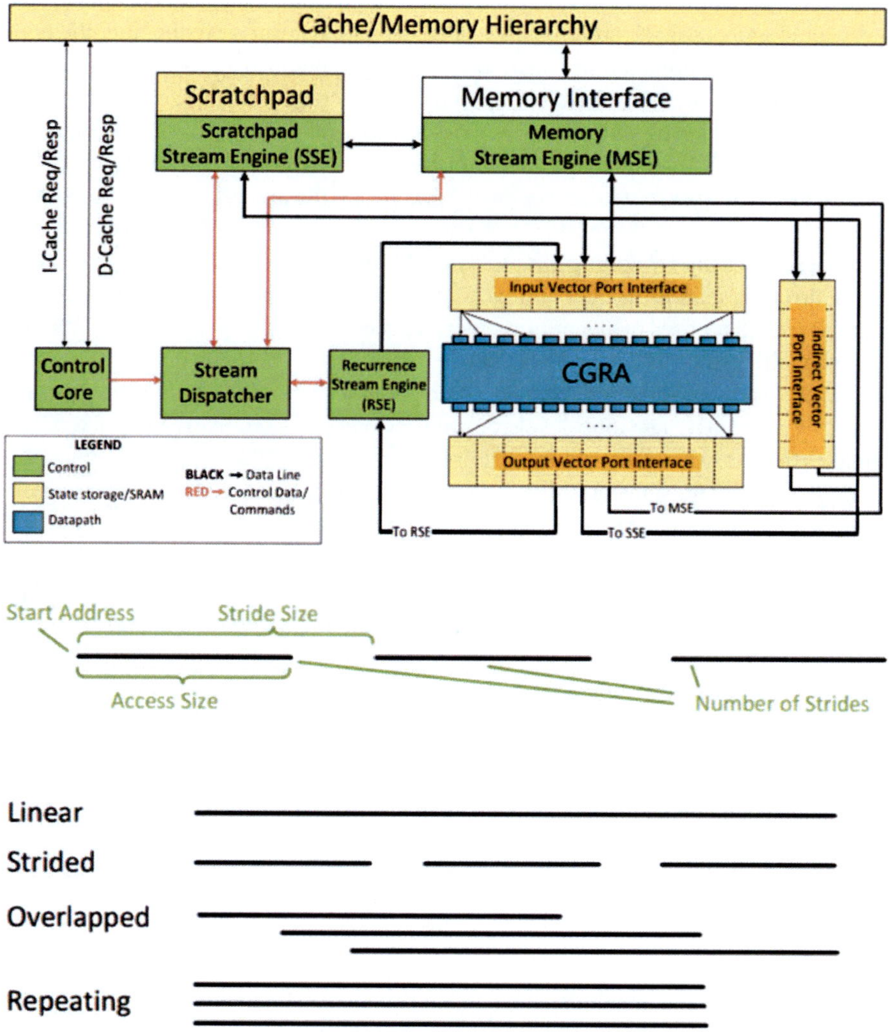

Fig. 3.28 Architectural diagram of decoupling of computation from access (Softbrain as an example)

2) Decoupling of Computation and Control

The preceding describes the decoupling of computation from access, but in fact computation and control instructions can also be decoupled, which can further simplify the PE's functions. As introduced in Sect. 3.1.1, the computational efficiency is reduced if the control flow is supported in single-instruction and static-scheduling compute arrays. The advantages of simple arrays might be fully exploited if the control instructions can be offloaded to the dedicated units outside of the array. This is feasible since only a limited number of control flows exist in target applications of software-defined architectures. It would be more efficient to use a sequentially executed processor for other control-intensive applications. Meanwhile, the control flows in certain applications also conform to a specific pattern for simple design. Based on these observations, it is possible to decouple partial control flows from the compute array. Nevertheless, the fine-grained control instructions inside the loop body, e.g., SPVV, require frequent interaction with computing instructions. It is more efficient to execute these control flows inside the array. For example, Softbrain takes such control flows in SPVV as a stream-join class [61]. The processing can be accelerated easily by adding simple control logic to the PE.

Current research on decoupling computation from control in software-defined architectures focuses on decoupling loop-level control flow to support different loop-level parallel patterns. Loop-level control flow affects the coarse-grained TLP. Therefore, the array cannot process such control flow efficiently, making the parallelism drop significantly. The loop-level control flow pattern is relatively more fixed, which makes the hardware design simpler. For example, Fig. 3.29 shows Plasticine's decoupling of control flow. In particular, it uses three control structures to support sequential operation, coarse-grained pipeline, and streaming control, respectively.

Similarly, ParallelXL [62] employs a continuation passing mechanism to dynamically generate or eliminate tasks at runtime using special instructions, which supports a variety of flexible task-level control flows such as fork-join, data parallelism, and nested recursion. DHDL [63] provides primitives such as the counter, pipeline, serial, parallel, and meta-pipeline for describing task-level control flow in the programming model to facilitate the generation of decoupled acceleration architectures in hardware.

2. NDP: Tight Coupling of Computation and Memory

Contrary to the aforementioned design idea of decoupling, another research direction in software-defined architectures lies in tightening the coupling between different modules, placing the modules that are originally closely connected and communicate frequently as closely as possible to reduce the frequency and cost of data transfer among them. The most typical application of this idea is the tight coupling of computation and memory, including NDP and PIM. PIM focuses on circuit-level studies, which will be presented in Sect. 3.3. This section briefly discusses NDP.

To achieve tight coupling between computation and memory for efficient data movement, NDP can be employed to decentralize computation near the memory or place memory closer to computation. The latter can be realized by the existing multi-level caches or replacing the on-chip SRAM with the embedded DRAM, while the former is still a direction to be explored. This is because if the computation is decentralized to the memory, the whole system has multiple computation areas and is no longer a traditional von Neumann architecture, which revolutionize the design of the system architecture.

Currently, NDP is primarily used to enhance the performance of domain-specific accelerators. The most popular approach in industry connects 3D-stacked memory chips to computing chips with interconnection vias on silicon. The latest GPU accelerator of NVIDIA's GPU (A100-80 GB) uses HBM2e memory. Several DRAMs are stacked together and interconnected with computing chips on a single silicon chip in

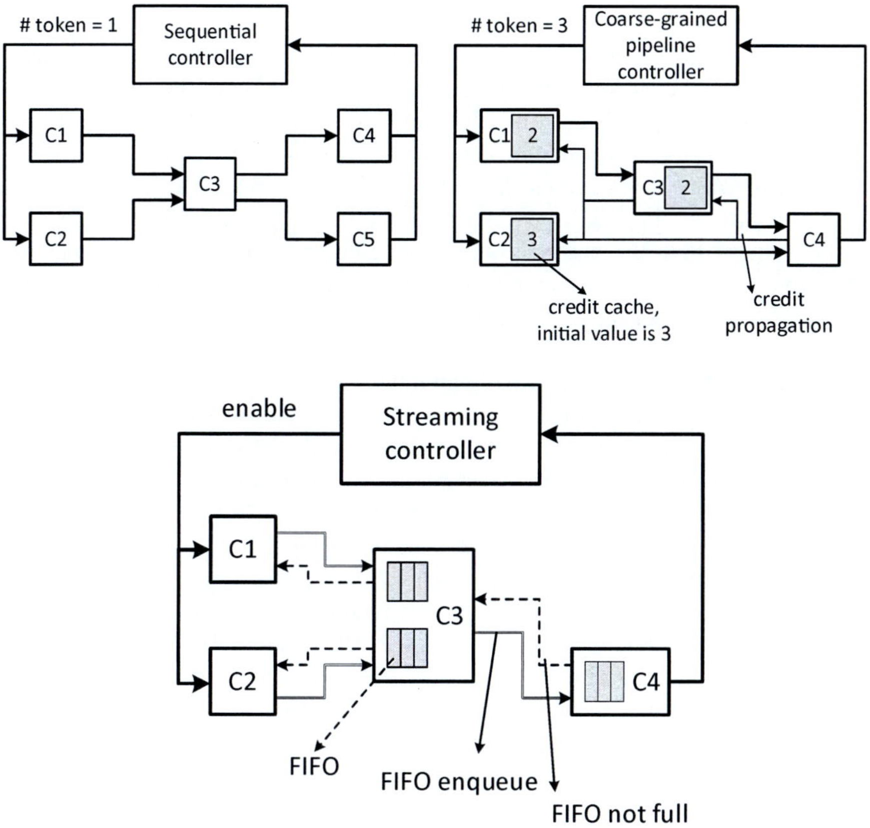

Fig. 3.29 Control pattern in Plasticine

HBN2e, whose bit-width is up to 1024 bits with the bandwidth of several hundred GBps. For intellectual property reasons, another process, The hybrid memory cube (HMC) is preferred in academia to explore the possible advantages and design space of NDP due to the IP issues of HBM. HMC is jointly developed by Micron and Samsung in 2011. Similar to HBM, HMC uses the through-silicon via (TSV) technology to connect multiple DRAM chips together in three dimensions. The difference is that HMC requires additional logic chip to implement memory control and I/O conversion at the bottom of the DRAM stack. In addition, the HMC external interface is no longer parallel, but uses a high-speed serial interface with support of direct communication among multiple HMC stacks. Due to the built-in NDP characteristics of the HMC architecture, many research efforts exploring the possibility of embedding different computing cores, such as CPUs, GPUs or other customized logic, in the logic chip. There is also work that explores how to combine the HMC with SDCs for application acceleration.

GPUs and programmable devices might not be suitable for HRL [64], considering the strict limitations on area and power consumption of the logic layer in the HMC.

The logic layer also needs to be functionally reconfigurable for different applications, whereas customized circuits can hardly provide reconfiguration capabilities. The SDC can be a better solution to this problem. In addition, the area cost will be high if fine-grained PEs such as LUTs in FPGAs are used. Otherwise, the power consumption caused by interconnection will be high with complex control flow, if the coarse-grained PEs are used. As shown in Fig. 3.30, HRL uses an SDC that combines fine-grained units with coarse-grained units on the HMC's logic layer of to achieve both the power efficiency of fine-grained architecture and the area efficiency of coarse-grained architecture. The performance of HRL is up to 92% of a dedicated accelerator for graph computing or deep neural networks. HRL is not only an architectural DSE for combining SDCs with NDP, but also uses the mixed-grained array architecture that will be highlighted later.

In addition to HMC, the research on NDA [65] tries to embedding the NDP functions into commercial standard memory modules. As shown in Fig. 3.31, three methods are proposed in NDA to enable data export through the TSV to DDR3's data path, respectively, at different data levels of the DDR. The TSV stacks DDR3 chips with SDC-based accelerators. A standardized DIMM module with computing capabilities is built by integrating eight such stacking systems on a printed circuit board (PCB). The difference between this module and the normal DRAM DIMM is that a SDC accelerator is stacked on each DDR3 chip. This SDC accelerator is composed of four PEAs in NDA.

The DIMM module uses the same standard DRAM interface for data interaction with the control processor. It also needs modified ISA to enable the computation on the accelerator. NDP provides substantial gains in energy efficiency and performance in many applications compared to that with only CPUs and an SDCs as the coprocessors. In addition, the design space of this SDC accelerator is fully explored in NDP considering the impact of the operating frequency, instance number, and different connection sturucture on the performance of different applications.

3. Hybrid Array Structure

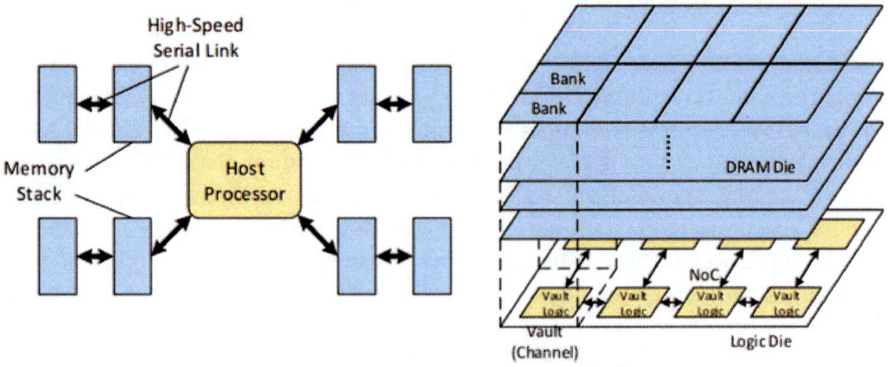

Fig. 3.30 HMC memory system used by HRL

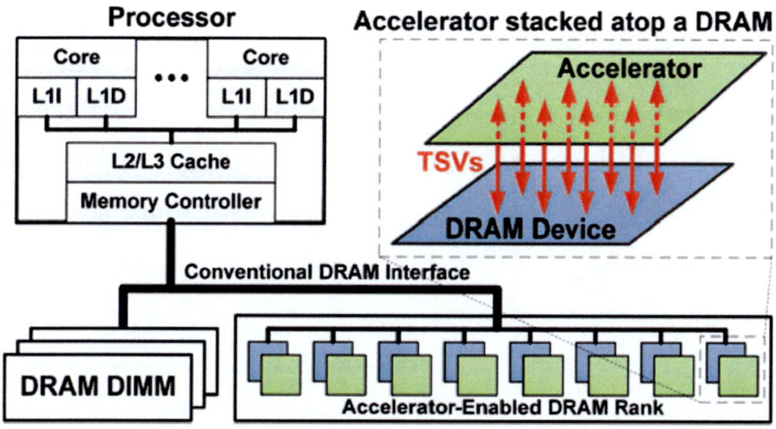

Fig. 3.31 Schematic diagram of the NDA architecture

A hybrid design method is an effective scheme when design choices are made for a particular architecture on multiple dimensions. Suppose there are two complementary designs A and B as the design boundaries, another design scheme C might be obtained based on the trade-off between these two borders. Ideally, scheme C should be dynamically configurable, i.e., it can be configured to work as scheme A or B, or schemes between them. This hybrid design method can effectively combine multiple existing solutions, taking full advantages of each single solution. This section describes the mixed-grained array structure combining static scheduling and dynamic scheduling.

1) Mixed-Grained Versus Variable-Grained Arrays

As highlighted previously, many software-defined architectures support configurable data granularity in the array, considering the impact of data and compute granularity on the architecture. First, different algorithms have different requirements for data granularity. For example, in the cryptographic algorithms such as AES, 8-bit data computation is sufficient. Hence, the 32-bit PEs can support four operations simultaneously. In the deep neural networks, 8-, 16-, or 32-bit fixed-point numbers are commonly used, and some algorithms even use1-bit data to represent the weights [66]. Dense matrix operations require higher bit-width as the vector's element to improve the throughput. The data granularity in these algorithms directly affects the computational accuracy, performance and energy consumption. Therefore, the hardware with a mixed-grained array structure allows the same array to be used in multiple scenarios.

There are many software-defined architectures with mixed-grained arrays, such as DGRA [61], whose interconnection and PE structures are shown in Fig. 3.32. DGRA arrays support 16-, 32- or 64-bit data communication and instruction computation. In fact, the DGRA array consists of four independent sets of 16-bit interconnection and PEs. In each set, the interconnection network has its own control logic, and each PE supports 16-bit computation independently. Several sets can be collaborated simultaneously to enable computation with larger bit-width if required. For example, four

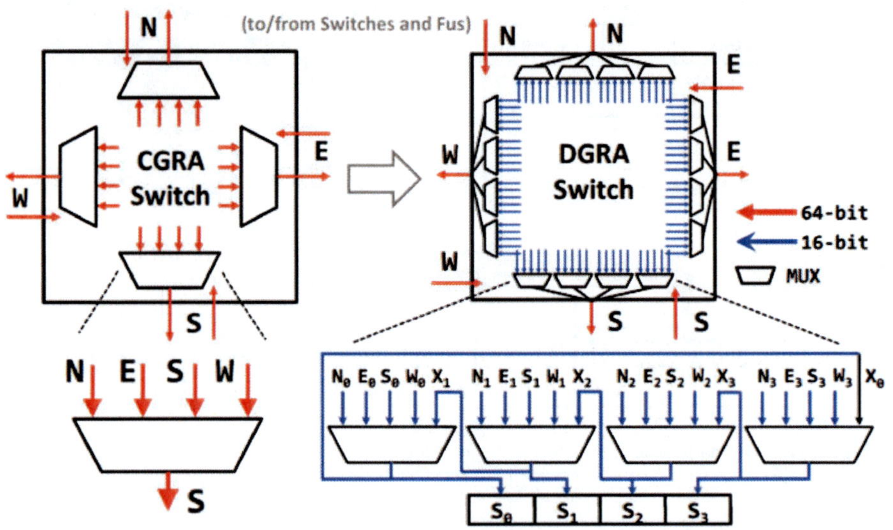

Fig. 3.32 Structural diagram of mixed-grained interconnects and PEs in DGRA

sets of switching elements in the interconnection network are configured identically when the data width is 64-bit. They simultaneously deliver different bits of that data to the same destination. When the data arrives at the PE, Four groups of FUs calculate the corresponding part once the data arrived. Finally, the results are concatennated as 64-bit data to other units.

The combination of low bit-width arrays into large bit-width arrays in the DGRA is highly flexible to support variable bit-width computation simultaneously. For example, different algorithms in an application have different requirements for bit width. It is only necessary to configure the hardware resources according to the corresponding bit width. Consequently, mixed-grained computation can be performed directly without switching configurations, although this will pose higher requirements on compilation.

The hardware overhead of DGRA mainly comes from the control cost associated with controlling each subarray independently, e.g., each sub-interconnection network requires independent data addressing control (e.g., mux). Each sub-FU also requires a separate operand cache.

Various mixed-grained arrays in software-defined architectures are used to satisfy different scenarios. For example, HReA is designed for the general-purpose acceleration domain, where many applications contain complex control flows. It supports 1-bit and 32-bit data communication, while 1-bit interconnection network applies to control signal transfer without occupying the data communication network. Evolver [36] is designed mainly to accelerate deep neural networks with support of 8-, 16-, or 32-bit fixed-point computation and on-chip reinforcement learning. The computing precision in Evolver can be adjusted at runtime according to the performance and power requirements of the usage scenario. Morpheus [67] is an SoC that consists of multi-grained heterogeneous computing blocks, such as FPGAs for fine-grained computing, CGRAs for 32-bit coarse-grained data, and PiCoGAs for 4-bit medium-grained data.

2) Hybrid Array of Static and Dynamic Scheduling

Section 3.1.1 has discussed the design principles of static- and dynamic-scheduling PEs in detail. In fact, the two scheduling mechanisms are complementary on some occasions. For example, static scheduling has higher computational performance and energy efficiency for regular applications while dynamic scheduling has higher computational performance for irregular applications. Static scheduling has lower hardware overhead, while dynamic scheduling requires more area and power consumption. Applications generally contain various levels of control flow, such as control of loop iterations memory access control. In addition, the majority of instructions inside the loop body are computing instructions. In this case, using only a single type of PEs cannot meet the needs of the overall application. It is better to use both types of Pes in the array, and distribute the application tasks to them according to their advantages. For instance, allocating the computationally intensive part to the statically scheduled subarray, while the part containing the control flow is mapped to the dynamically scheduled PE. This allows a good combination of the advantages of both PEs while avoiding the disadvantages of each as much as possible.

In recent years, some research work in the software-defined architecture has explored this hybrid array with static and dynamic scheduling. The array structure of Revel is shown in Fig. 3.33, where sPE is the statically scheduled PE in Softbrain and dPE is the dynamically scheduled PE in TIA. Since Revel is mainly designed for compute-intensive applications with limited control flow, the number of sPEs is greater than that of dPEs. During application division, Revel maps all computations inside the loop body to the sPE, and other instructions including initialization instructions, loop control instructions, and access control instructions, to the dPE. The communication between dPE and sPE is fulfilled by a switching interconnection network in hardware implementation, whose timing correctness is ensured by the compiler. The dPE handles the dependences between loops during the process. Once the execution condition of a loop is satisfied, a token is sent to the sPE for computation in the loop. These two PEs run in parallel and work together through data communication, enabling flexible and efficient acceleration of different parts of the application.

Currently, software-defined architectures combining static- and dynamic-scheduling Pes are not widely used., There are several possible reasons for this. From the software perspective, the hybrid PE poses extreme requirements for compilation. On the one hand, the compiler needs to divide the application reasonably to take full advantages of heterogeneous arrays. There is still no efficient algorithm for subgraph division yet. On the other hand, if the correctness of the communication between two PEs is ensured by the compiler, the space for the compiler to explore instruction scheduling will be greatly increased. It is also more difficult to find the optimal scheduling scheme. From the hardware perspective, the hybrid array structure also makes the design space too large. For example, how to allocate the number

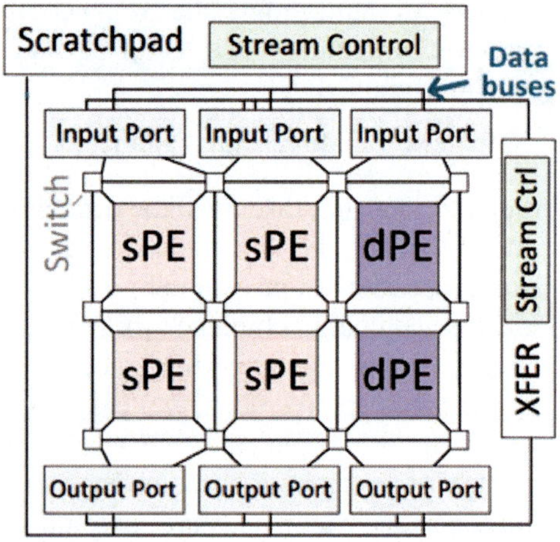

Fig. 3.33 Hybrid array structure with static and dynamic scheduling in Revel

of the two kinds of PEs and how to design the interface between them to ensure correct data communication at a small cost, remain a critical and challenging task.

4. Hybrid Interconnect Structure

The main feature of software-defined architectures is the pipelined execution in spatial domain, which requires frequent data communication. This places higher demands on the design of interconnection networks. On the one hand, interconnection networks should be flexible. If the interconnection network is inflexible, which means the data transfer path is determined at compile time, relatively lower utilization will be caused due to unbalanced usage of network resources with some switches occupied all the time while others stay idle. On the other hand, the network must provide high-bandwidth and high-throughput communication. The data width in software-defined architectures keeps growing. The network bandwidth must increase accordingly, otherwise communication will become the bottleneck in the entire architecture.

Interconnection networks can be classified into two categories: static and dynamic. Static networks are based on circuit switches, which are configured statically. The data communication path in static networks remains unchanged at runtime. Dynamic networks are NoCs based on package switch. They are consist of dynamically addressed routers. The main feature of static networks is that they have the simplest structure, and higher area and energy efficiency (average area and power required per data transfer) of the hardware implementations. Therefore, higher energy efficiency and data bandwidth can be obtained to transfer large bit-width data (e.g., vectors) in static networks under the same hardware budget. The disadvantage of static networks is mainly that they are highly related to application characteristics. and rely on the compilation algorithm to reasonably distribute the communication paths in a balanced manner across the network. Communication imbalance problems might happen within the static network for many irregular applications, where some resources are not fully utilized. In addition, switch units in static networks cannot be shared by multiple communication paths. The compiler must map all data transfers in the application to different switches to avoid conflicts. In this case, the resources provided by the static network must exceed the maximum demand of the application to ensure that the application is mapped correctly, thus causing a decrease in resource utilization. The characteristics of dynamic networks are very different from those of static networks. In particular, the router structure in dynamic networks is far more complex, because it requires dynamic addressing, flow control, and deadlock avoidance logic. Therefore, dynamic networks have greater hardware overhead and lower area and energy efficiency for data communication. The most important advantage of dynamic networks is their flexibility. In particular, the communication path can be queried dynamically at runtime by using the router algorithms based on real-time flow information rather than being statically specified by the compiler. Moreover, idle routers can afford some communication tasks, which makes dynamic networks' utilization higher. Meanwhile, the router's internal data cache can be shared by multiple packets at the same time. Therefore, even if the size of the network is smaller than the demand for data communication in the application, the TDM ensures that the application run correctly after mapping, but at the cost of frequent network congestion and even deadlocks.

The literature [50] provides an in-depth exploration of the interconnection network design space for software-defined architectures and proposes a hybrid interconnect architecture. Similar to the motivation of hybrid array architecture design, static and dynamic networks have complementary features in many aspects. Combing both networks in an array can yield better trade-off. This work places both static and dynamic networks in the array, and makes them work in parallel with no direct communication interface. For application partitioning, the compiler first assigns the communication paths with frequent communication, high bandwidth demand and large bit-width to the static network, while the remaining scalar communication with infrequent communication but irregular data paths to the dynamic network after thorough analysis of all data paths. In this way, the static network serves as a large-bandwidth, high-efficiency primary interconnection network to meet the communication requirements of the application, while the dynamic network shares the rest of the communication tasks to avoid non-critical data paths occupying the static network. Both networks are efficiently used. Experiments show that the hybrid network achieves higher area efficiency and power efficiency than the static network while achieving the same performance.

Table 3.3 compares the important metrics of different network architectures. Some results refer to the experimental results in the literature [50]. Note that the impact of interconnection networks on the overall software computing architecture is related to a variety of factors. Overall, the communication characteristics of the application, and the structure of the PEs are the most important factors to determine the choice of interconnection network. For example, dynamically scheduled PEs make more use of dynamic networks, while statically scheduled PEs prefer to use static networks. Hybrid networks make it difficult to ensure correct timing of computation and communication. Moreover, even for the same network, the configuration of its internal parameters (e.g., number of virtual channels, flow control strategies) also has a very significant impact on the overall performance. In the experiments of literature

Table 3.3 Comparison of static, dynamic, and hybrid interconnection networks

Indicators	Static network	Dynamic network	Hybrid network
Bandwidth	High	Low	Medium
Latency	Low	High	Medium
Utilization	Low	High	Medium
Flexibility	Low	High	High
Area/Bit	Medium	Low	High
Energy/Bit	Medium	Low	High
Scalability[a]	Medium	Low	High

[a]Scalability refers to the performance gains that the network can provide as the array structure increases

[50], the same application using dynamic networks shows a 8× performance gap between the best and worst configurations. In conclusion, interconnection networks, as the important components to connect multiple modules in a software-defined architecture, must be designed considering the impact of the application and other hardware modules on the communication. Moreover, the most appropriate design scheme can be finally determined only after the appropriate metrics are satisfied based on comparison of multiple feasible schemes.

3.2.2 *Examples of Agile Hardware Development*

As the continued development of IC process technology, the transistor number on a single chip grows exponentially. Thereby, the complexity of the on-chip architecture gradually increases, making the development cost for architectures higher and the development cycle ever longer. In addition, due to the slowdown of Moore's Law, the single-core performance of GPPs is approaching its bottleneck, which makes domain-specific dedicated accelerators gradually become a promising direction. In this context, the development cost of software-defined architectures is becoming one of the biggest obstacles to their own development. First, as a complex computing system, the software-defined architecture has a large multi-dimensional design space. To design an efficient architecture, a variety of alternatives must be provided for comparison to obtain the expected one, which will consume a lot of development time. Second, the application-domain algorithms are changing rapidly in recent years., As a typical domain-specific accelerator, the software-defined architecture needs to quickly update its own structure to adapt to the emerging domain's expectations.

The traditional hardware development is a typical waterfall model, that is, the hardware development cycle is divided into a number of top-down, interlocking basic activities, such as problem definition, requirements analysis, writing code, and running tests. The basic processes are independent of each other and executed in a chronological order. The project results can only be seen at the end of the entire project lifecycle. Figure 3.34 shows a typical development process of the software-defined architecture, where the designer first writes a function-level simulator in a high-level language (e.g., C++) for functional validation and DSE. Once the functional simulation is complete, it is also necessary to write RTL-level code for the entire system using HDLs (e.g. Chisel, Verilog, SystemVerilog, and VHDL), perform behavior-level simulation and testing, and generally synthesize the RTL design into the FPGA for validation.

After the test is completed, hardware engineers need to use EDA tools to synthesize, place and route the RTL design, and confirm that the post-layout simulation timing meets the requirements. The overall design can be taped out only after all the work is done. The overall workflow is lengthy, with many steps, and designers need to master multiple languages and tools. Moreover, this development model is inefficient with a long iteration cycle. If there is a problem at a certain step or if the architecture requirements change subsequently, the whole process often needs to be repeated, which takes a lot of time. Obviously, the key feature of the waterfall model is that it cannot adapt to the rapid changes in user requirements. The long development process of traditional architecture cannot adapt to the fast-changing application scenarios. Therefore, both industry and academia are focusing on the research of agile hardware development tools for lower design cost and shorter development cycle.

Agile hardware development refers to a development model that uses automated software tools to generate corresponding low-level hardware circuits directly from the high-level architectural designs. Agile hardware development consists of agile high-level design, agile validation, agile physical design, and agile product development. Ideally, designers only need to describe the architecture at the functional level using high-level languages (e.g., Python) or DSLs (e.g., Halide), then the architectural DSE can be completed by simply changing input parameters. Corresponding tools are taken to automatically generate synthesizable RTL-level (e.g., Chisel) or even lower-level code and configurations. Meanwhile, there are corresponding simulation validation tools at each design level to facilitate rapid validation by designers. This development model allows designers to save a lot of development time and focus more on architecture design, avoiding a lot of engineering time spent on tedious validations. At the same time, designers only need to be proficient in user-friendly, high-level language for design and validation without deep understanding with the entire flow, thus significantly lowering the development threshold. In particular, with the availability of a large number of open-source tools, individuals are also able to develop complete architectures in a short period of time, which will greatly facilitate research efforts in software-defined architectures.

Although agile hardware development is one of the hottest research topics in the field of architecture design, open-source and complete development platforms are still in their infancy. This section introduces two more mature agile development platforms for software-defined architectures, namely the Agile Hardware (AHA) project and the Decoupled Spatial Architecture Generator (DSAGEN) project at Stanford University.

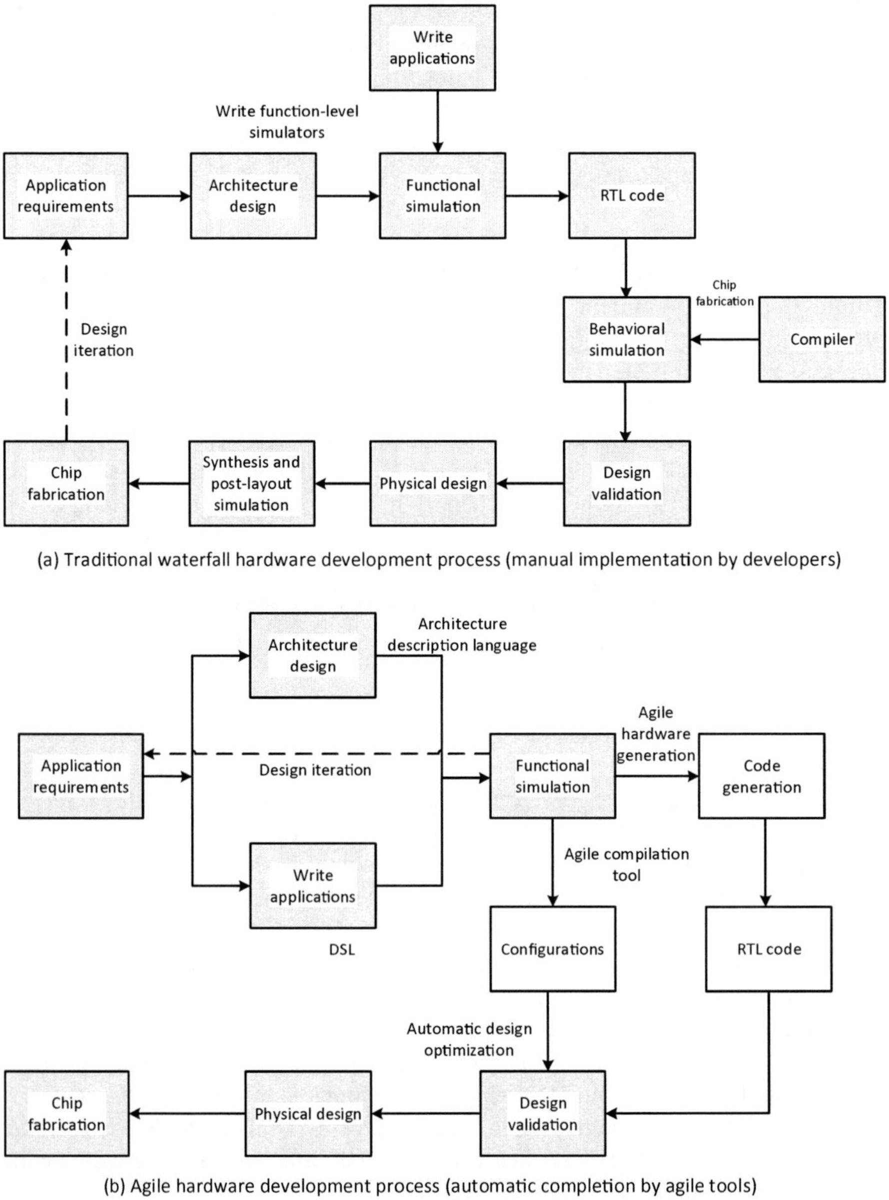

Fig. 3.34 Comparison of traditional hardware development process and agile hardware development process

1. AHA Project

In software design, individuals can develop large-scale applications and products in a short period using various software tools and frameworks due to the well-developed ecosystem. However, hardware development often requires large teams to work for months or years, which seriously hinders the development of hardware and weakens the interest of people in hardware development. The AHA project is dedicated to lowering the development threshold for collaborative hardware/software systems through agile development, also making hardware development easy, fast, and attractive. To achieve this goal, the AHA project has created open-source tool chains that can be used for rapid hardware design, application development, and simulation validation. Currently, AHA has been able to quickly build mature ARM and CGRA SoCs using these tool chains.

Today's computer aided design (CAD) tools have been very powerful, especially HLS tools, which have been able to allow developers to design complex systems. However, traditional CAD tools seem inefficient in the design of software-defined architectures, where hardware architectures and application algorithms are constantly changing, as well as high requirements on performance and power consumption. Agile hardware development for software-defined architectures requires software tools to have the following key features:

(1) Provide a lightweight architecture description language that allows users to efficiently carry out the high-level DSE.
(2) Provide a high-level algorithm description language that allows users to easily describe application algorithms.
(3) Provide hardware automatic generation tools that can quickly generate the underlying hardware description based on the high-level architecture described by the user.
(4) Provide compilation and mapping tools that can map the application algorithm to the user-described hardware architecture and generate the corresponding configuration.
(5) Provide multi-level simulation and validation tools to shorten the user's test cycle.

AHA provides the appropriate tools for the above features and builds a complete environment for agile hardware development [68]. As shown in Fig. 3.35, in AHA's development process, users need to describe the hardware architecture using DSLs, including PEak, Lake, and Canal, which have high-level abstraction and directly correspond to the modules of the software-defined architecture. Meanwhile, AHA provides tools such as Halide [69, 70] and Aetherling [71] for describing applications. The algorithms and hardware scheduling logic are included in the code to facilitate relevant optimizations by the compiler. Once the hardware architecture is determined, the compiler uses the key parameters of the architecture to efficiently complete algorithm mapping and instruction scheduling, then generates the corresponding configuration bitstream. Thereafter, if the hardware architecture changes in subsequent iterations, the compiler tool will automatically track the changes in each

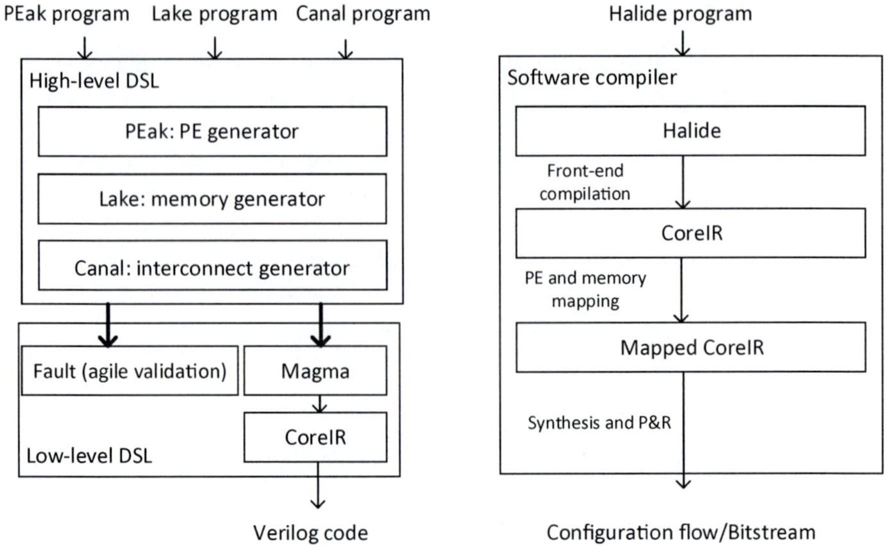

Fig. 3.35 AHA agile hardware development framework

module, analyze the impact of these changes on the application mapping results, and incrementally change the previous mapping results without the need to recompile the application. Based on these features, AHA is able to adapt to the needs of iterative development processes and efficiently support architecture design and iterations. The following sections describe each module in detail.

1) Agile Architecture Description

As shown in Fig. 3.36, AHA uses three DSLs to describe the software-defined architecture: PEak for PEs, Lake for memory, and Canal for interconnection networks. These three languages can be coded independently, with each compiler to generate the corresponding IRs. The output can be functionally validated by the corresponding validation tools. Besides, RTL-level code can be generated by the hardware generation tools.

(1) PEak

PEak is a Python-based DSL for describing the functional model of a PE. A PEak object is a high-level abstract description of a PE, which declares the instruction set, structural state, and the behavior of each instruction. Each PEak object has multiple usages. In particular, it can be used purely as a functional model for simulation or an RTL generator with the corresponding tools (SMT [72], Magma) or a simulation model in other levels of testing.

Figure 3.37 shows an example of building a PEak object by Python, where each PE class init function defines the building blocks of the PE, such as registers, input and output channels, and structural states, while the call function defines the specific

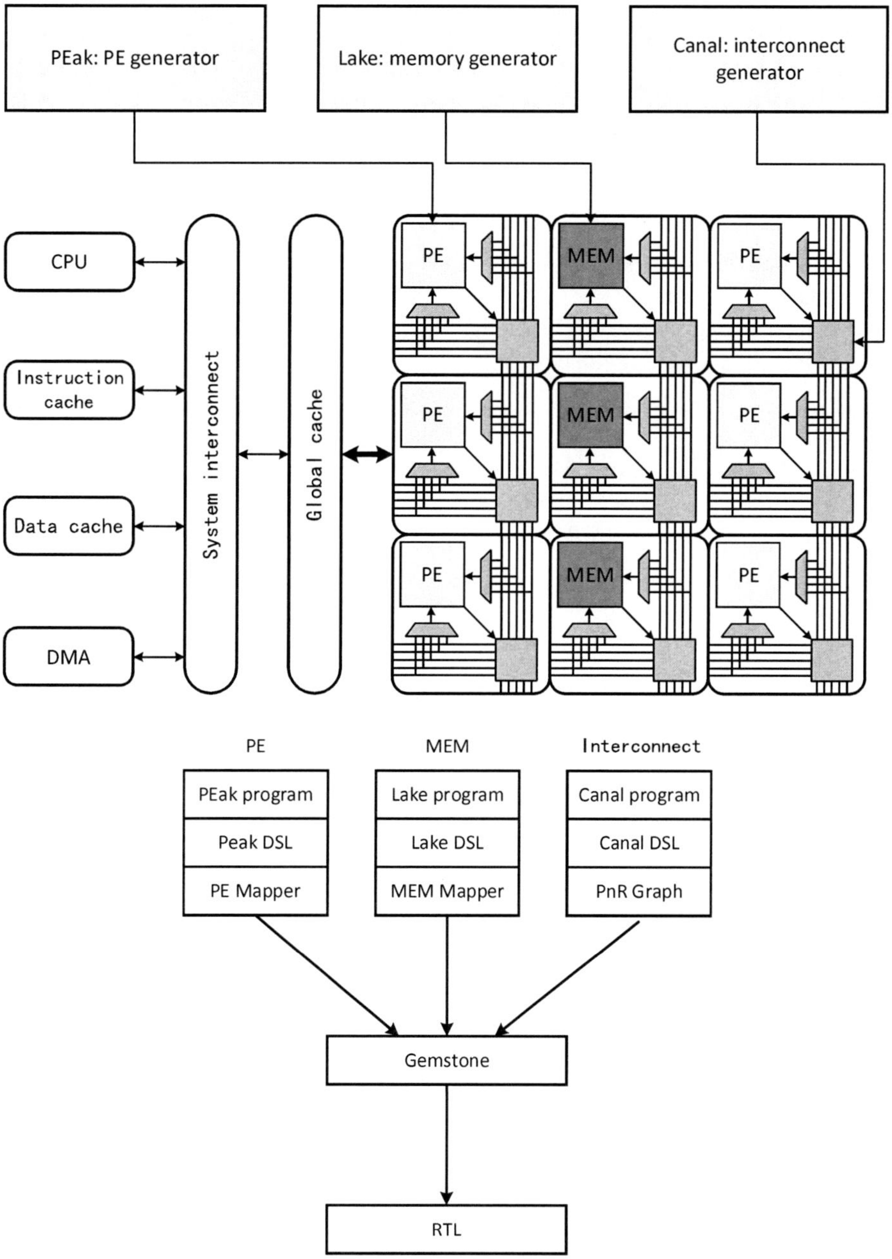

Fig. 3.36 AHA agile hardware generation: PEak, Lake, and Canal

behavior of the PE. The PE contains a data register and a bit register. It can perform functions such as complementary operation, bitwise AND, multiplication, addition, and register value export depending on the configuration. It only requires a direct declaration of the instruction set, the input and output interfaces, and the logic operations to describing the PE by Peak, without considering complex issues such as timing. Compared to HDLs such as Chisel or Verilog, PEak is semantically clearer and easier to read. Furthermore, only a small amount of code is needed to quickly build PEak objects that can be used for simulation and hardware generation. If complex PE structures are required, such as extending instruction sets and variable data bit-width, they can be modified iteratively based on simple structures, thus significantly shortening the design cycle.

(2) Lake

Unlike PEak, Lake builds memory models from a relatively low-level, hardware-centric description, since memory systems in software-defined architectures have multiple levels and multiple structures. Taking a relatively low-level description

```
class Opcode(Enum):
    Add = 0
    And = 1
class Instruction(Product):
    op = Opcode
    invert_A = Bit
    scale_B = Bit
    reg_out = Bit
# Data is Bitvector
Data = Unsigned[16]
```

PE instruction set declaration

```
pe = PE()
inst = Instruction(
  Opcode.Add,
  Bit(0), # invert_A
  Bit(1), # scale_B
  Bit(0)) # reg_out
out,flag = pe(
  inst,
  Opcode.Add,
  Data(2), # A
  Data(3), # B
  Data(5), # C
  Data(0)) # c_in
assert out==Data(17)
assert flag==Bit(0)
```

PE Python execution

```
class PE(peak)
  def __init__(self):
        self.o_reg = Register(Data)
        self.f_reg = Register(Bit)

   def __init__(self,
                inst: Instruction,
                A:Data,
                B:Data,
                C:Data,
                c_in:Bit
               ) -> (Data,Bit):
      if inst.invert_A:
          A = ~A
      if inst.scale_B:
          B = B*C
      if inst.op == Opcede.Add:
          # adc = add with carry
          res,flag = A.adc(B,c_in)
          else: # inst.op == Opcode.And
              res = A & B
              FLAG = (res == 0)
      if inst.reg_out:
         res = self.o_reg(res)
         FLAG = self.f_reg(flag)
      return res,flag
```

PE function declaration

Fig. 3.37 Example of PEak object

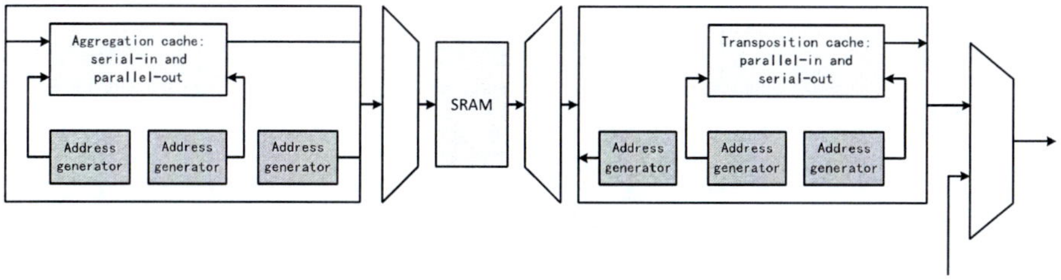

Fig. 3.38 Classical memory system built by Lake

enables hardware designers to carry out more precise DSE and to select the most appropriate memory system. The Lake memory module contains one or more memory units, sub-modules for serial-to-parallel conversion or address generation, and the interconnect structure between these units or modules. Figure 3.38 shows a classical memory system built by Lake, which contains three memory units and multiple address generators. Each module has specific parameters such as port number, port width, and delay information. The address generators have additional parameters for declaring access sequences of specific patterns. These parameters will be analyzed by the Lake compiler for behavioral simulation and compatible generation of hardware structures. Meanwhile, users can quickly generate various types of memory systems by just changing the parameters, or they can enable automatic parameter space exploration through programming, greatly simplifying the memory system design process.

(3) Canal

The Canal tool can generate interconnects to connect the PE and memory system after they are defined by the user. The Canal program uses a directed graph to represent the interconnection structure, where vertices represent the recipients of data and directed edges represent the connection relationships. A vertex can have multiple input edges, representing that the vertex can receive multiple kinds of data, which in hardware generally corresponds to a multiplexer. In Canal, different network topologies can be declared by defining appropriate properties for the vertices. For example, defining a two-dimensional coordinate attribute in a vertex represents a grid-like topology, and defining a class attribute means that the vertex is a port of an array or a pipeline register. Similarly, vertices or edges in Canal have configurable parameters for specifying information such as the bit width and delay to facilitate design parameter exploration. The workflow of Canal is shown in Fig. 3.39.

Canal generates interconnect information according to the designated number and parameters of Pes and memories, including the P&R map, configuration flow or bitstreams for switch units, and RTL-level code for interconnects.

2) Agile Algorithm Description

The AHA project focuses on cutting-edge applications in image processing, machine learning, and other fields, which are described by the Halide [69, 70] language. Halide

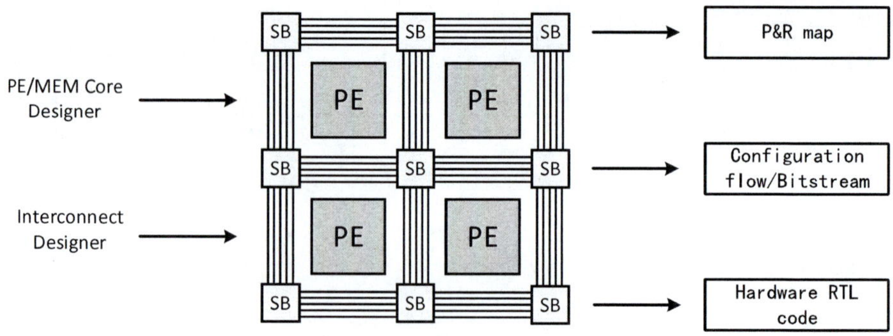

Fig. 3.39 Example interconnection network generated by Canal

is a DSL with DLP, which divides the application into two parts: the algorithm, which is described by C++ functions, representing the computational process, and the scheduling part, which is mainly hardware-oriented, specifying which computations in the algorithm need to be accelerated, the level of data in memory, and loop parallelism. By explicitly specifying such scheduling information, the compiler makes more reasonable optimizations for data distribution and chunking, loop processing, and etc. Figure 3.40 shows an example of compiling and mapping a 3 × 3 convolution operation by Halide. Halide's compilation and mapping of the application consists of the following basic processes.

(1) The compiler first compiles the user's input program into an IR within Halide. This IR does not consider any hardware information, but simply holds the computing core instructions inside the loop. The memory operations are represented by accessing an infinitely large multi-dimensional array.
(2) The compiler converts the Halide IR into an unmapped CoreIR (uCoreIR) [73]. In this process, the compiler transforms the computational statements into the bit-vector computational primitives of CoreIR and extracts all access

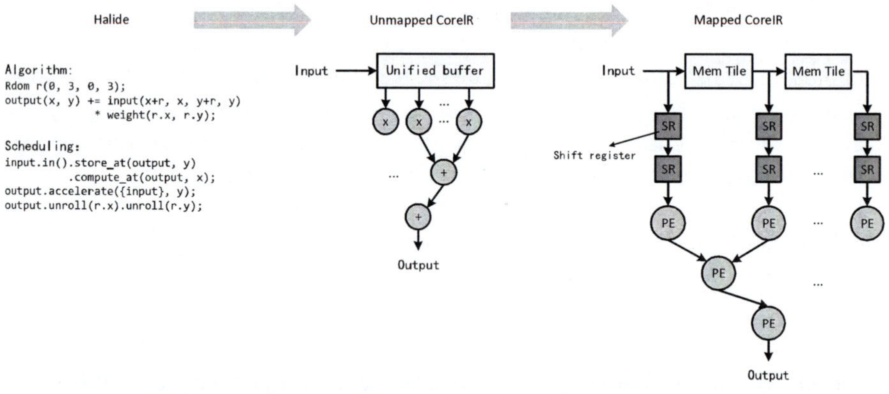

Fig. 3.40 Halide's compilation and mapping of a 3 × 3 convolution operation

instructions into operation instructions for the streaming memory (called unified buffer). The uCoreIR is represented by the dataflow graph consisting of the computing core and the streaming memory. It still only has information about the application without any considerations on specific architecture.

(3) The Halide compiler feeds the uCoreIR into a mapper for initial mapping. The mapper maps the uCoreIR's computing instructions and streaming access instructions to the corresponding PEs and operations on Lake-generated memory respectively with the guidance of the parameters in the hardware module. The compiler makes many normal optimizations on uCoreIR, such as constant folding, subexpression elimination, and invalid vertex removal. The generated dataflow graph here consists of hardware PEs and memory modules, instead of pure instructions. This dataflow graph is called mapped CoreIR (mCoreIR).

(4) The Halide compiler maps the mCoreIR's vertices to the actual hardware array after the detailed hardware architecture with parameters is generated by Peak, Lake and Canal tools. The P&R among the vertices and the corresponding configurations are also accomplished here.

With the help of the Halide compiler, users only need to write the algorithm in C++ and specify the desired scheduling scheme, then the compiler will automatically generate the corresponding configuration according to the architecture without user's intervention. Moreover, if the user changes the hardware architecture after compilation, e.g., the PE's architecture changes, the PEak tool will rewrite the hardware information for Halide. Meanwhile, the compiler will automatically map the updated computing instructions to Pes, without recompiling the whole application. Therefore, using the high-level architecture description language and algorithm description language makes AHA particularly suitable for iterative software-defined architecture development.

3) Agile Hardware Generation

AHA uses several tools for automatic hardware generation, which are briefly described here.

(1) CoreIR, as the IR of hardware, also refers to the corresponding hardware compiler framework. CoreIR actually depicts the interconnections among hardware modules in a directed graph. It also describes the hardware architecture and the logical functions in each module. Although CoreIR does not represent any actual hardware, it can generate functionally equivalent hardware. Therefore, CoreIR can be used as middleware between the high-level description language and the hardware architecture language. The RTL-level code can be compiled and generated by CoreIR.

In Fig. 3.35, the CoreIR of the target architecture is the compiled result of the high-level architecture described by PEak, Lake, and Canal. This CoreIR will be further processed by the Halide and GemStone to generate P&R information and the RTL code separately. In addition, a set of core hardware primitives are provided in CoreIR to enable function enhancement, which allows users or compilers to perform specific IR-level optimizations based on these primitives.

(2) Magma is a Python-based HDL with its abstraction level between Verilog and PEak, similar to Chisel [74]. As a typical meta-programming language, the program written in Magma can be compiled to generate a Verilog file. It is also acknowledged as a generator in the hardware field. The basic abstraction block in Magma is a circuit, similar to a module in Verilog, but it is far more readable than Verilog due to the high-level language features of Python. Besides, all variables in Magma is strongly defined with, specified bit-width and limited operations. Therefore, all programs in Magma are synthesizable.

Magma is taken as the middleware between PEak and Verilog in AHA. The language written by PEak is compiled to generate Magma language files, which are then compiled by Magma's compiler to generate CoreIR representations for Verilog code generation.

(3) Gemstone is a Python-based framework for hardware generators, which can be used to easily design configurable and reusable hardware modules. In fact, Gemstone can be considered as a compiler among multiple generators. For example, the PEak to Magma language conversion mentioned earlier is done by Gemstone. Gemstone is a multi-level compilation framework that can convert a higher-level generator language to a lower-level generator language if there are multi-level generator languages. Therefore, Gemstone can be used in a phased manner. For example, the higher-level hardware framework can be used to finish only functional testing and quickly build a complete system at the beginning of a project, then Gemstone can be used to gradually compile and generate lower-level hardware descriptions for more accurate timing simulations, as well as area and power estimations, and eventually generate Verilog design for synthesis and chip design.

4) Agile Validation Tools

The design tools such as gem5 [75] and GPGPUSim can be used for behavior-level simulation and validation in GPPS or general-purpose graph processing units (GPGPUs). However, there are still no mature high-level simulation tools due to the absence of unified hardware architecture in SDC. In the traditional development process, the testing of hardware architectures usually requires simulation using EDA tools after RTL coding, while behavior-level simulation is fulfilled by the corresponding simulator, written by the designer for a long time.

The Fault tool is provided in AHA project for high-level architecture validation. Fault is a package in Python that can be used to test architectures written in the Magma language. The main goal of Fault is to build a unified interface for execution-based simulation and formal verification using the constrained random verification (CRV), which takes simulation efficiency, portability, and performance as the primary metrics. One of the key features of Fault is its portability. For example, Fault can be used for functional verification after writing a PE using PEak. It can still be used for simulation after PEak is compiled to generate a Magma file. The higher level of abstraction, the more efficient the simulation will be. This approach allows the test code to be reused at multiple abstraction levels. The compiler will ensure the correction of the generated hardware circuits. In particular, if the high-level simulation results are right, the RTL-level simulation results are consistently correct as well. Consequently, Fault can replace the time-consuming RTL-level simulation with the more efficient behavior-level simulation under certain circumstances.

2. DSAGEN Project

Like the AHA project, DSAGEN [76] was originally proposed to address the tedious and lengthy software and hardware development process in the current DSA field. It also provides a tool chain for agile software/hardware iterative development. Unlike the AHA project, the main goal of DSAGEN is to explore development using hardware primitives in a given decoupled design space, rather than providing a high-level architecture description language. DSAGEN poses higher priority on the decoupled design of each software-defined modules. Its development framework provides a variety of decoupled primitives in the software/hardware interfaces. Therefore, the development efficiency of the architecture is greatly improved, since developers use the high-level language to select appropriate modules for a complete computing architecture. Compared to the AHA project, the design flexibility of DSAGEN is reduced due to the limited choice of supported modules. However, the modular hardware design and well-defined design space bring much convenience to users. First, the modular design with limited hardware primitives alleviates the burden on the compiler, which can compile the input program into decoupled dataflow and instruction flow more efficiently. Second, the well-defined design space helps the development tools to automate the DSE, i.e., to iteratively optimize the hardware structural diagram generated by the compiler for the target function. Third, the software/hardware iteration cycle can be effectively shortened and the design efficiency is significantly improved due to the shrinking design space.

1) Overall Development Process of DSAGEN

The basic modules provided in DSAGEN cover a wide range of design options in multiple dimensions of software-defined architectures. Developers can quickly design a variety of domain-specific accelerator architectures under the DSAGEN development framework using high-level languages.

Figure 3.41 shows the overall development framework of DSAGEN, including software compilation, hardware synthesis, hardware DSE, and hardware generation. DSAGEN uses ADG as the middleware between software and hardware, which consists of Pes, memory units, and interconnect units to describe specific architecture instances. Similar to CoreIR, ADG can either be synthesized directly into hardware or used to iteratively generate the optimal architecture description graph in DSE. As shown in Fig. 3.42, the automatic development process of DSAGEN is as follows by taking the cyclic program segment as an example.

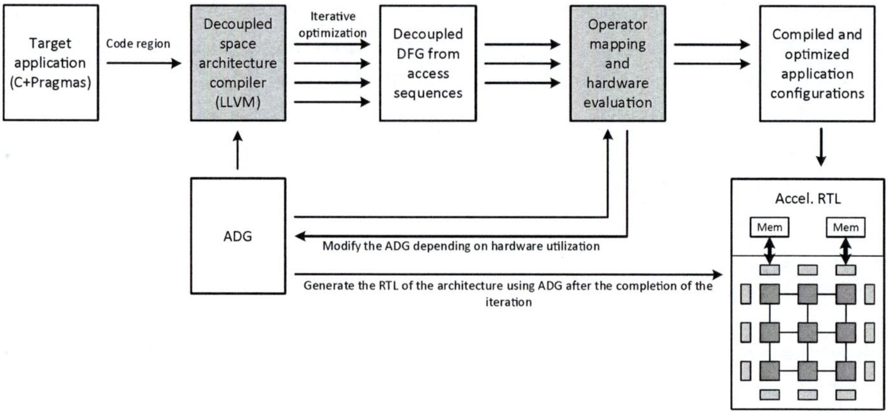

Fig. 3.41 Development process of DSAGEN agile design

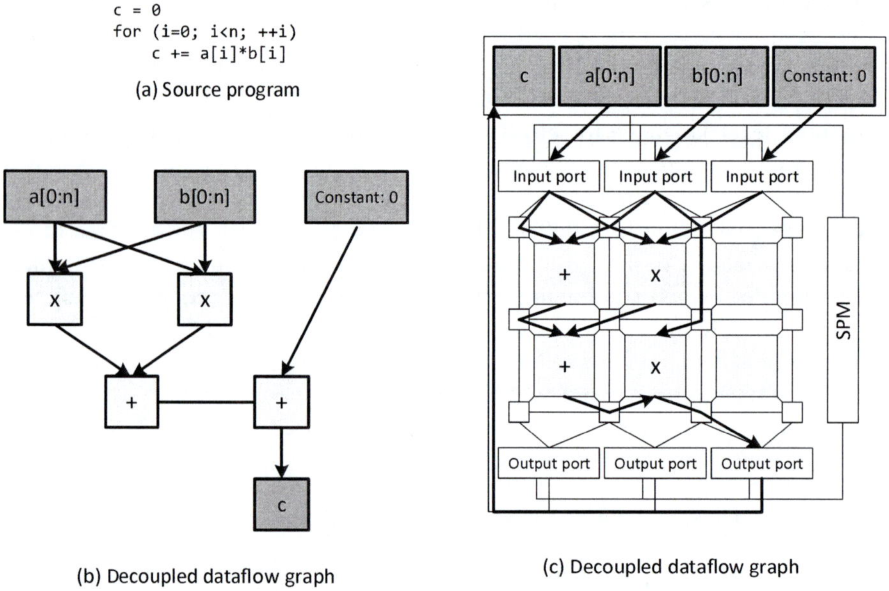

Fig. 3.42 Example of DSAGEN agile development

(1) Decoupled space architecture compilation. The input program in Fig. 3.42a is compiled to generate decoupled dataflow from instruction flow, with memory accesses represented by coarse-grained access dataflow and operators for computing instructions represented by dataflow graphs.
(2) Hardware mapping. The dataflow graph in Fig. 3.42b is mapped to the given topology in Fig. 3.42c. In particular, arrays are mapped to corresponding areas in memory. Data read and write operations are mapped to memory access units

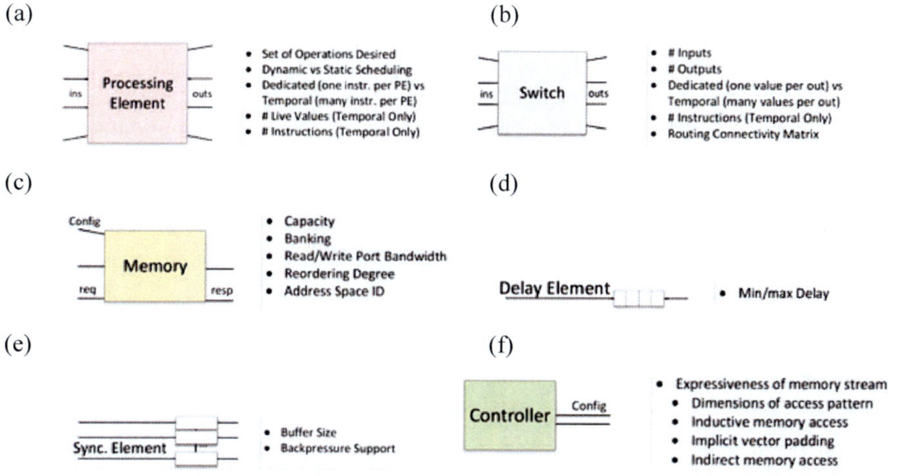

Fig. 3.43 DSAGEN hardware primitives and related parameters

for implementation. Computing instructions are mapped to PEAs, while data paths between instructions are mapped to interconnection networks.

(3) DSE. Figure 3.42 shows that the DSAGEN development framework generates the optimal ADG based on the objective function with the corresponding training algorithm.

2) Hardware Design Space

The DSAGEN development framework defines the overall design space of the architecture, which is divided into five dimensions, namely compute, control, interconnect, memory, and interface. It provides a variety of parametric and modular design options for each dimension.

Figure 3.43 shows the complete design space of DSAGEN. In general, its design primitives are essentially the same as those described in Sect. 3.1:

(1) The computation primitives are used to describe the PEs. DSAGEN supports single-instruction, multi-instruction, static-scheduling, and dynamic-scheduling PEs with specified parameters such as registers and data bit-width. These options allow users to combine multiple types of PEs to meet the performance requirements of multiple scenarios.
(2) DSAGEN uses switches to build interconnection networks. The switch module supports the specification of the internal cache size, and multiplexing the data to output ports in a TDM manner. It can also contain instructions, such as the ability to selectively eliminate some data. After specifying the switch type, the compiler will generate a communication connection matrix to specify the interconnection between switches.
(3) DSAGEN can automatically generate memory models based on SPMs, and specify key parameters such as capacity, number of banks, and port bandwidth.

(4) The latency module is mainly applied to statically scheduled PEAs for balancing path latency. For details, see Fig. 3.11.
(5) The interface[3] primitive in DSAGEN is mainly used to synchronize the input data of the array. For example, an array needs N sets of input data to start the operation, and the interface module can detect the data readiness and ensure the array computation timing.
(6) The control primitive mainly targets the access address flow control. This control module is able to generate multiple patterns of address sequences through configuration.

By setting the appropriate parameters, the hardware module provided by DSAGEN can satisfy most design requirements in software-defined architectures. More importantly, DSAGEN significantly shortens the development cycle of heterogeneous architectures. For example, if you want to use different types of PEs or different numbers of ports for memory modules, you only need to specify the required parameters for each module individually, without repeating the complete design process. Moreover, the design primitives themselves are independent of each other with well-defined parameters. Therefore, the designer can easily perform multi-level optimizations for each module separately. Meanwhile, the compiler can search for the optimal parameters in the parameter space.

3) Agile Hardware Design Compilation

Since the underlying hardware generated by DSAGEN and the target application domain change with the input parameters, the programming model and the compiler design are the main focus of the DSAGEN project. Ideally, despite changes to the underlying architecture parameters, the compiler should still be able to map the application to the array without user involvement and to generate the optimal configuration scheme for the new architecture.

As shown in Fig. 3.44, DSAGEN extends the C language using annotations for the programming model. The codes declared by dsa config need to be accelerated by the software-defined architecture, which corresponds to the array's initial configuration at runtime. There are no data dependences within the loop if it is marked by das decouple. The compiler will analyze the loop's access sequence and use the memory controller to implement the access request, corresponding to the continuous access requests of arrays a and b in the example. The dsa offload is applied to the innermost loop, declaring that the internal computing instructions of the loop will be mapped to the PEA, corresponding to the multiplication and addition operations in this example. The compiler will automatically perform the optimization such as loop unrolling.

The compilation of DSAGEN can be divided into four steps: decoupling access from computation, data dependence conversion, modular compilation, and instruction scheduling & code generation.

---

[3] Note that the interface primitives introduced in Sect. 3.1.3 of this book mainly refer to the interfaces between software-defined architectures and other architectures, which are different from the interface definitions here.

Fig. 3.44 Example of the DSAGEN programming model

```
#pragma dsa config
{ #pragma dsa decouple
  for (i=0; i<n; ++i) {
    #pragma dsa offload
    for (j=0; j<n; ++j)
      c[i*n+j] = a[i*n+j]*b[j];
  }
}
```

(1) Decoupling of access from computation: The compiler analyzes the access sequence generated by the loop body with the dsa decouple annotation and use the SCEV module in LLVM to aggregate the access requests into access patterns supported by the controller. Therefore, the access instructions are decoupled from the computation with corresponding configurations generated in the controller.
(2) Data dependence conversion: The compiler converts all control dependences of the CDFG into data dependences in order to implement the control flow using either the dataflow in the array or the predicate mechanism.
(3) Modular compilation: The compiler compiles each module and checks whether the module meets the requirements specified by ADG, such as verifying whether the interconnection and communication timing between dynamic-scheduling PEs and static-scheduling PEs are correct. If all modules are compiled completely, and the module interconnection structure is correct, the compiler will generate the corresponding ADG, which represents the complete computing architecture.
(4) Instruction scheduling & code generation: The compiler maps the access sequence annotated by dsa decouple to the controller, and maps the dataflow graph annotated by dsa offload and after the dependence conversion to the PEA, after the ADG is provided. In addition, the compiler needs to determine the order of instruction execution and generate configurations for the switches. The relevant compilation algorithms will be described in detail in Chap. 4.

Throughout the compilation process, the compiler generates multiple configuration schemes based on user requirements, with some configurations for optimal performance or power savings. In addition, if the underlying hardware changes for the same application, the compiler only needs to generate a new ADG for the hardware description again, and then map the application algorithm to the new hardware without user's manual modification.

4) Agile DSE

DSAGEN uses ADG to describe the hardware architecture, which is a concise abstraction of the hardware structure. The structure of the generated ADG could be modified by the compiler automatically based on analysis of the original ADG and the application algorithm, which is equivalent to changing the hardware structure. A iterative hardware-software co-optimization strategy is used in DSAGEN, i.e., the architectural ADG and the algorithmic DFG are optimized simultaneously in an iteration, and the overall flow is as follows:

(1) The compiler first attempts to add or remove some components in the initial ADG within the constraints of pre-defined power consumption and area. For example, if current hardware power consumption is less than the pre-defined power consumption, the compiler will try to add PEs or memory ports for higher hardware computing capability. A new ADG will be generated for each modification after this kind of stochastic attempt.
(2) The compiler will try to map all the DFGs and access sequences of the application into the new ADG after loop unrolling. The ADGs that are not mapped complete will be discarded, keeping the remaining ADGs that can complete the computation task.
(3) The compiler evaluates the performance of all the remaining ADGs in step (2), selects the best-performing ADG as the result of this iteration. The iteration should be continued if the updated ADG has better performance than the initial ADG.

Since the design space in DSAGEN is explored based on several given modules, the compiler can quickly obtain the optimal parameters of the architecture. DSAGEN evaluates the ADG's performance by IPC, which can be calculated by the number of instructions, critical path length, etc., while the number of instructions is estimated using the smaller value between the maximum IPC and memory bandwidth in fully pipelined execution. To evaluate the ADG's area and power consumption, DSAGEN uses a regression model to quickly predict each module's resource consumption based on its parameters. Although the predictions are not exact (error of about 10%), they are sufficient for a quick comparison of the relative overhead of different ADGs.

To sum up, DSAGEN basically implements the fully automated exploration of module parameters and hardware structures. Developers only need to set power and area constraints based on application scenarios and give the initial hardware architecture, then the compiler can continuously iterate to search for a better architecture and achieve optimal performance within the constraints. After the architecture optimization is completed, the compiler will also optimize the configuration for the final ADG to make the shortest path for configuration propagation and accelerate the configuration speed. Finally, DSAGEN will generate the corresponding Verilog code based on the optimized ADG and generate a binary configuration bitstream for each module. Due to the highly modular and parameterized architecture, the hardware generation task is simpler in the DSAGEN framework than that in the AHA project.

3.2.3 *Summary*

This section started with a detailed DSE of software-defined architectures, focusing on the decoupled design among the three dimensions of computation, control and memory, with a brief introduction about the concept and examples of NDP. Due to the diverse architectures of software-defined architectures and constantly emerging design schemes, only the most typical architectures are taken as examples to introduce each design scheme in this section. Then, it introduces the concept of agile hardware development, together with the agile development process and corresponding tools in the AHA and DSAGEN projects. Due to limited space, specific details about the tools are available to the reader on the project homepages.

In current design of software-defined architectures, dealing with the relationship between compute and memory remains a research hotspot. The design approach of decoupling compute from memory has been adopted by more and more architectures, and will become the mainstream design framework for new architectures in the future. In addition, with the emergence of novel memory devices, many research efforts have started to explore ways to bring PEAs closer to or even integrate them inside the memory, aiming at narrowing the performance gap between compute and memory and alleviating the memory wall problem. Currently, the process development of novel memory devices is still immature and the overall research is still in the early stages.

In addition, agile hardware architectures are rapidly becoming a research hotspot in the field of software-defined architectures' development process. Agile development platform is necessary to promote the development of software-defined architectures. The POSH and IDEA projects are two projects funded by DARPA's ERI in 2018 to provide continuous R&D support in this direction. The agile hardware development tools have evolved rapidly in recent years. The AHA project uses self-developed tool chains to create a variety of software-defined architectures, while the DSAGEN project use self-developed framework to generate corresponding acceleration architectures for domains such as deep learning and graph computing, whose performance is nearly 80% of ASICs. However, due to the complexity of software-defined architecture design and the lengthy process of chip design, there is still a long way to go before building a mature development ecosystem in hardware development like software engineering.

3.3 Design Space of Software-Defined Circuits

Compared to hardware architecture, the design of software-defined circuits aims to reconfigure the hardware at the circuit level as another software-defined design dimension, thus further increasing the design's flexibility. The definition of the circuit by software includes both the control of key signals such as supply voltage and operating clock, and the configuration of circuit implementations for specific FUs, such as

PEs. Redefining the circuit's function by software can meet various requirements of multiple applications for rapid trade-offs in hardware performance, energy efficiency and computational accuracy without the need re-design of the circuit.

3.3.1 Exploration of Tunable Circuits

As the two basic signals driving the circuit, the supply voltage and the operating clock have a crucial impact on the circuit's performance and energy efficiency. Reducing the supply voltage or clock frequency is the most convenient and direct way to obtain a low-power circuit. Meanwhile, these two signals are also closely related. When the supply voltage is reduced, the movement of electrons will be slowed down accordingly, and the performance of the circuit will be degraded, thus making the system run slower with lower performance. To ensure the reliability of the system, it is often necessary to reduce the operating frequency. When the supply voltage is gradually approaching the threshold voltage, the state of the triode may probabilistically not flip, resulting in incorrect output results, which may eventually have a negative impact on system with strict reliability requirements. If the circuit error shows certain characteristics, its impact on multimedia processing, machine learning and other fault-tolerant applications is negligible, so the circuit supply voltage and clock frequency of these systems can also be slackened accordingly. Besides, the threshold voltage is also affected by the circuit aging, temperature, process deviations, etc., the circuit's performance (maximum clock frequency) will change even under the constant supply voltage. If the software can modify the supply voltage or clock frequency to counteract the impact of these factors, the robustness can be guaranteed with the best optimization on hardware implementation. In summary, adapting the circuit's supply voltage and clock frequency by software allows flexible trade-offs between power consumption and performance, also indirectly manipulates the probability and magnitude of the system error, which can further improve the performance and energy efficiency while meeting the accuracy requirements of fault-tolerant applications to achieving multi-dimensional optimization.

1. Flexible Design and Integration of Multi-voltage & Near-threshold Circuits

The flexible design of multi-voltage circuits not only supports a wide range of performance and power consumption options, but also allows the selection of the minimum supply voltage while meeting the basic performance requirements based of the using scenario, which leads to minimum power consumption. Besides, the system can also selectively turn off the voltage or clock of the inactive circuits, thus optimizing the overall energy efficiency of the system [77]. To maximize the adjustable range of supply voltages and minimize the power consumption, more and more research is focusing on circuit design at near-threshold or sub-threshold voltages [78]. Moreover, since certain errors can be afforded in fault-tolerant applications, this can somewhat reduce the requirements for wide supply voltage circuit designs [79]. In addition, changing the supply voltage and clock frequency of the circuit is used as a compensation method to counteract the degradation effect produced by external factors such as circuit aging, thus ensuring a stable and sustainable system operation [80].

In fact, software-based control of the supply voltage and clock frequency has been used for a long time to optimize the overall energy efficiency. As early as 2002, IBM has designed a PowerPC processor with scalable voltage and frequency, which aims to maximize the energy efficiency for different application requirements by software-based control [78]. In this processor, the software dynamically controls the on-chip voltage manager and the phase locked loop (PLL) frequency divider through the SoC. When the supply voltage changes, the system operating clock will change accordingly to ensure the robustness of the entire system. Meanwhile, the SoC will send a deep sleep instruction when the software detects that some circuits are inactive for a certain period of time. The inactive circuits will keep in deep sleep state with power turned off, after their states are saved on external non-volatile memory, thus reducing the overall power consumption of the system. To ensure that the circuit can operate properly at very low voltages, the processor makes necessary modifications to the conventional circuit design, such as increasing width ratio between the pMOS and nMOS, etc. Noting that these design modifications might cause certain negative impacts on the circuit latency or area at normal supply voltages.

With the rapid development of integrated circuits, their demand for on-chip memory is increasing. Therefore, how to reduce the static power consumption of memory becomes an urgent issue. A flexible multi-voltage design can quickly alleviate this challenge without changing the process. In a DARPA project, MIT designed an SRAM powered by sub-threshold voltages [78] with dynamically adaptive voltages to solve this issue. This SRAM supports an ultra-wide dynamically adaptive voltage from 250 mV to 1.2 V, which is managed by a reconfigurable auxiliary circuit. Moreover, the SRAM adopts an 8 T design [81] in order to support sub-threshold supply voltages. As shown in Fig. 3.45, to reduce the read delay caused at low voltage, it adds two nMOS triodes to the conventional 6 T SRAM as a read buffer at the sub-threshold voltage. In addition, to ensure accurate read and write functions as well as desirable power consumption and delay at different voltages, this SRAM optimizes the size and aspect ratio of the triode, and integrates three reconfigurable writing methods and two sense amplifiers. The measurement results show that the minimum static power consumption of this SRAM is only 1/50 of its maximum value in permitted voltage range.

Expanding the range of the circuit supply voltage not only requires circuit redesign with additional auxiliary circuits, but also leads to degradation of the circuit performance and energy efficiency at normal voltages. However, the circuit without any modifications may produce wrong results when operating at low voltages due to the triode's threshold characteristics. Therefore, witnessing the tolerance to certain errors in fault-tolerant applications, conventional circuits can be applied to fault-tolerant systems by software-based voltage control without excessive additional hardware consumption. In fault-tolerant applications such as multimedia processing, data mining or machine learning, errors in the control module are forbidden, while some errors in the dataflow module can be offset by the processing algorithm or will

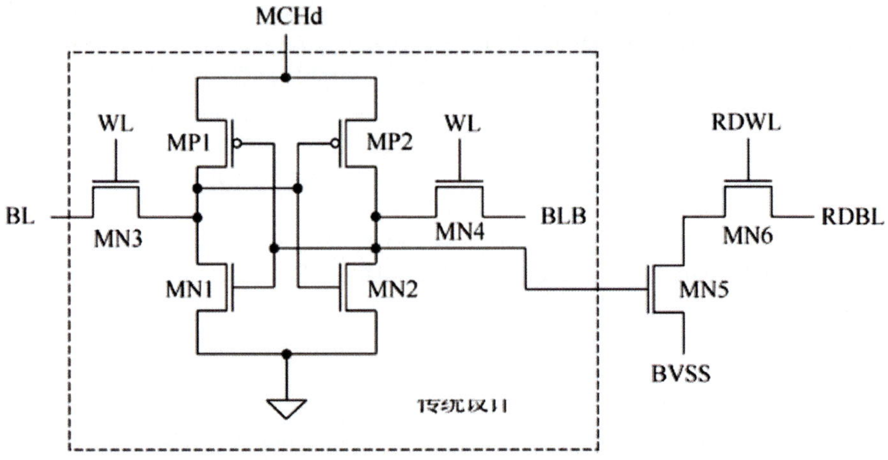

Fig. 3.45 8 T SRAM unit circuit

not have a significant impact on the final result. Therefore, for different application requirements, the circuit's supply voltage can be dynamically adjusted by software to manipulate the error location and range, thus finally optimizing the performance and energy efficiency of the circuit while meeting the accuracy requirements. The study of voltage scaling of circuits using the fault tolerance characteristics of applications has not yet verified on silicon, but the circuit synthesis method has received some attention [79]. With the support of Intel, the University of Texas at Austin has proposed a circuit synthesis method for fault-tolerant applications. This method simulates the algorithm and data to obtain its specific fault tolerance characteristics, then selects some PEs for rounding operation to reduce the area, power consumption, and delay within the given quality (error) constraint. The supply voltage is reduced finally according to the delay reduction to avoid timing violation. In addition, the design uses a heuristic optimization algorithm to obtain the optimal profile on accuracy and energy efficiency. Experimental results show that the circuit synthesis method can reduce the average energy consumption of hardware in fault-tolerant applications by more than 70%. Note that the above synthesis method ultimately yields the RTL implementation of the circuit, and different circuit synthesis results will be obtained for different applications. Therefore, the hardware obtained by this method can only be used for specific applications with limited flexibility. Therefore, the flexibility of hardware might be greatly improved by using the software-defined circuit and integrating the synthesis method into the upper-layer software to control the input and output data and supply voltage dynamically.

The voltage and clock of software-defined circuits can also be used as a solution to the circuit degradation problem. Studies have shown that the threshold voltage is affected by circuit aging, noise, temperature and process deviation, etc. As the circuit ages or the temperature rises, the threshold voltage rises, resulting in performance degradation of the circuit. To ensure the circuit robustness, conventional methods generally set the maximum operating frequency to the allowable maximum value in the worst-case, known as guardband. Therefore, the circuit performance has been severely limited. When the voltage and clock can be reconfigured, the software can adapt the voltage and clock according to the aging state, the operating environment and process deviations, so that each circuit can achieve the maximum performance with minimum power consumption while ensuring robustness. To reduce the guardband and maximize the average performance and energy efficiency, Intel designed a TCP/IP processor with dynamically adaptive frequency, voltage, and substrate bias voltage [80]. Apart from the TCP/IP core, the processor also integrates a dynamically adaptive bias controller, a supply voltage drop sensor, a thermal sensor, a substrate bias voltage generator, and a dynamic clock unit consisting of three PLLs. As the core control unit, the dynamically adaptive bias controller receives messages from each sensor, and then drives the dynamic clock and voltage units according to the changes in voltage drop and temperature, so that the processor obtains the optimal circuit configuration in a given operating environment. Moreover, the processor adds a clock gating function by using the control unit and clock regulation module to further reduce the power consumption. In addition, the processor can effectively solve a series of problems caused by the aging effect, both by reducing the operating clock frequency to ensure system reliability, and by increasing the pMOS substrate bias voltage to compensate for the increase in threshold voltage.

2. Flexible Design of Multi-clock Domain (GALS) System

With the development of multi-core technology and dense integration of multiple FUs on a single chip, the traditional global synchronization architecture is facing unprecedented challenges, because the global clock tree generation becomes more challenging, while the forced global synchronization method also affects the overall performance. Therefore, the global synchronization system has gradually failed to meet the performance and energy efficiency requirements of emerging applications. Although asynchronous systems have certain advantages in terms of power consumption, performance and immunity to interference, fully asynchronous systems also face several problems, such as the implementation of clock re-synchronization or handshake protocol required for global communication. Moreover, the multi-core technology and dense integration not only increases the circuit complexity, but also greatly increases its resource usage and routing delay. Therefore, flexible multi-clock systems, i.e., globally-asynchronous locally-synchronous (GALS) clocking technology, is gradually becoming an effective method to optimize the overall system performance and energy efficiency. The GALS [82] approach was proposed by Chapiro of Stanford University in 1984. It aims to integrate the advantages of both synchronous and asynchronous structures into a single system. In GALS-based systems, the interconnection and communication between local functional circuit modules are implemented asynchronously without theglobal clock, while the circuits within the same module operate synchronously, i.e., different functional modules are driven by different internal clocks, providing a convenient way to deal with multi-clock domain problems.

In the GALS system, the choice of clock domain partitioning and granularity have a crucial impact on the system performance and energy efficiency. The finer the partitioning granularity, the simpler the system clock tree structure and the smaller the circuit consumption, but the finer partitioning granularity will increase the hardware cost of the local clock generator and global asynchronous communication. When the partitioning granularity is larger, the number of local clock generators, and the complexity and hardware cost of system global communication will be reduced, while the larger granularity will increase the complexity and hardware cost of the system clock tree, and degrade system performance. Therefore, appropriate granularity of partitioned clock domain is the premise to ensuring the performance and energy efficiency of the GALS system. Meanwhile, in the general-purpose multiprocessor system, the effect of GALS will also be subject to constraints of the application data and algorithm. Therefore, combining GALS with SDC technology not only increases the flexibility of GALS systems, but also optimizes the performance and energy efficiency of application-oriented GALS systems. The SDC technology enables redefining the partitioning granularity of the clock domain by software, designing clock tree control circuits for the GALS system, and adding voltage gating circuits for each local clock generator according to the application requirements. Moreover, it allows changing the partitioning granularity of the clock domain, and rearranging the local clock and global clock tree, thus balancing the hardware cost of global asynchronous communication among the local clock generator, global clock tree, and modules, and finally getting an application-oriented, optimal clock distribution.

3.3.2 Exploration of Analog Computing

The mainframe computers before 1960s were analog computers with very complex operating procedures that required manual connection. But such computers made man's first moon landing, atom bomb, and nuclear reactors to reality. However, analog computers had almost been replaced by digital or mixed-signal computers for decades due to the high requirements on noise and difficulty in modular abstraction, before the widespread adoption of integrated circuits,. As the Moore's Law is eventually reaching its limit, other evaluation criteria, such as energy efficiency, power consumption, and area are increasingly considered. Industry and academia are no longer obsessed with the design of GPPs, but are exploring the possibilities of domain-specific accelerators and heterogeneous computing. Looking back to the history at this point, analog computing is found to make up for many shortcomings of existing digital circuits. Specifically, the data in digital circuits needs is represented in binary format. Theoretically, the bit width can be increased to achieve arbitrary precision. But our world is analog and often does not require very high-precision operations using digital computation. In addition, when the system interacts with the real world by using a digital system, the analog signal collected by the sensor needs to be converted into a digital signal by an analog-to-digital converter (ADC) before it can be processed. Moreover, it often needs to be converted back into an analog signal by a digital-to-analog converter (DAC) after computation. These two conversions are not simple and generally consume a lot of energy. Meanwhile, high-speed ADCs have always been a very cutting-edge topic in the field of circuit research. Furthermore, clock is necessary to control the digital circuits. Although power gating techniques can be used to shut down the inactive hardware modules, the charging and discharging of the parasitic capacitor on clock path still consumes a lot of energy. In contrast, the process of the analog computing is event-driven without the involvement of clocks, thus reducing the energy dissipation on the clock. Finally, the most prominent advantage of analog computing is that some computation models that are very inefficient in digital circuits, such as multiplication and division, exponential and logarithmic operations, as well as integration and derivation, can be implemented easily based on the device functions of analog circuits.

1. Current state of Analog Computing

1) Applications of Analog Computing

Large-scale analog computers played a significant role in solving ordinary and partial differential equations in the twentieth century. But due to the fast-rising of digital computing, there was no time to explore the possibility of its applications on integrated circuits. In recent years, a number of studies have implemented analog computing on integrated circuits for analog or mixed-signal computing, mainly also for the acceleration of solving ordinary and partial differential equations. Most of these work use a two-dimensional spatial computing architecture similar to that of SDCs, where a number of different computing blocks are distributed in two dimensions to form an array, and the different computing blocks are connected by switch-based interconnection network. This architecture is also known as field programmable analog array (FPAA) [83]. Figure 3.46 shows a typical two-level analog computing architecture. The top level consists of multiple logic blocks that can be interconnected to form certain functions. Each logic block contains many different analog computing modules, which can also be dynamically connected. The major difference between FPAAs and SDCs or FPGAs is that the most fine-grained computing modules of FPAAs are based on analog signals, or at least a mixture of analog and digital signals. In general, the PEs of FPAAs for accelerating ordinary differential equations typically contain classical integrators, and multipliers or also known as variable gain amplifiers (VGAs), and may contain nonlinear functions and additive and subtractive computing blocks to implement specific functions [84]. The integrator builds on the properties of the capacitor, i.e., the integral of the current across the capacitor is proportional to the change in voltage. VGA is a classical design topic for analog circuits with many design schemes. Nonlinear functions can be implemented by the transfer curve of the transistor, or SRAM operating in continuous time with certain reconfigurable capability. Addition and subtraction can be achieved by using only the current signal as a representation and controlling the polarity of two converging currents. In addition, simple exponential and logarithmic functions can be implemented using the exponential current

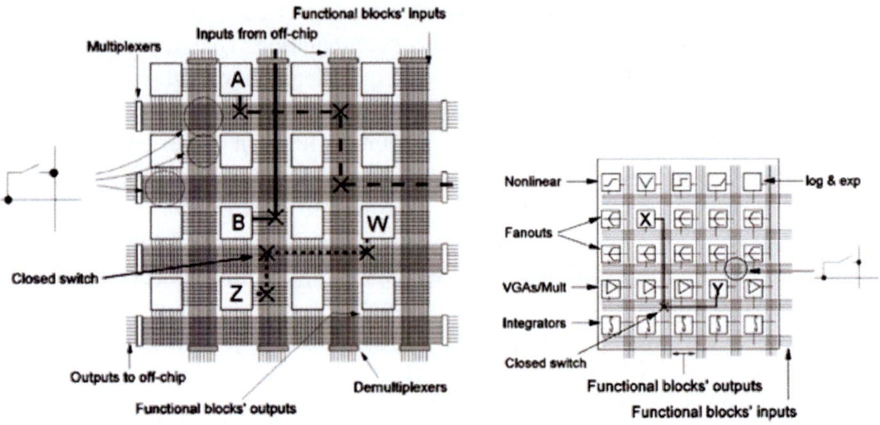

Fig. 3.46 Schematic diagram of a multi-level modular analog computing architecture

characteristics of the diode. All these basic modules required in many applications are not easy for digital circuits' implementation. However, the energy efficiency and performance can be improved significantly by using analog or mixed-signal computing. A two-dimensional reconfigurable architecture such as FPAA allows different computing modules to be connected for different equations and configured as required by the equations to obtain the desired results. Apart from the reduction in the energy consumption for computing ordinary differential equations, this architecture can also accelerate the computation of nonlinear partial differential equations as well as linear algebra [85, 86].

In addition to traditional applications such as linear algebra and differential equations required by scientific computing and physical simulation, neural networks can employ analog computing to achieve computing acceleration or lower energy consumption. Related research work in this area has gradually emerged in recent years. The computation pattern of neural networks can be represented by a large number of multiply–add and nonlinear activation function operations. In the field of analog computing, addition can be implemented by accumulating the currents on the vertices using Kirchhoff's Laws, while multiplication has many different designs, e.g., both DACs and multipliers can be implemented using multi-input floating gate structures [87], which can be used to accelerate neural networks. In addition, the activation function can be implemented by directly using the voltage or current transfer function of the circuit as mentioned above, or dynamically configured in memory compatible with analog computers. Meanwhile, the multiplications in neural networks can also be implemented by running current through a resistor, which is a research direction that combines analog computing with memory.

Although analog computing is getting more attention back, several issues, such as large signal interactions, high noise requirements, and difficulty in scaling, still need to be considered well in the design. Especially, due to the signals are realized by currents or voltages, analog computing cannot reach the high accuracy of digital computing due to noise and other problems, which limits the application of analog computing. Meanwhile, there is also a lack of efficient approach to store analog signals. In addition, analog computing may not provide the high performance as digital computing, because the latter can improve performance by stacking computing resources with higher parallelism, while it is not easy for analog computing.

2) In-memory Analog Computing

Memory is not a typical digital structure. For example, DRAM actually converts data to electric charge stored on a capacitor, defines the state with or without electric charge as 0 and 1, and requires a sense amplifier to read out the data. Although SRAM uses an end-to-end invertor structure to store data, it still involves signal amplification and stability problems during reading and writing. However, the resistive random access memory (ReRAM) works on a more "analog" principle, which determines the resistance value by applying a fixed voltage to both ends of the resistor and detecting the current flowing through the resistor. This requires not only a DAC for converting digital voltage to analog voltage, but also an ADC for judging the current value. In-memory analog computing has been a promising research direction in recent years, partly because memory has a large number of repetitive structures, which can solve the problem of insufficient parallelism in analog computing. More importantly, if computing can be performed directly in memory, it avoids the power consumption and performance loss caused by frequent data movement between the processor and memory, as well as the limitation in memory bandwidth for high-throughput computing.

The study [88] exploited the SRAM-based analog computing for the acceleration of binary neural networks. The motivation is that if multiple word lines of SRAM are activated simultaneously, the units on the same bit line all contribute to the bitline's current, and this operation is an analog implementation of addition. Furthermore, considering the voltage on the word lines, this operation can be taken as a binary vector multiplication, while matrix–vector multiplication is further enabled on the SRAM array. Consequently, the multiply–add operations of the neural network can be accelerated. In addition, neural networks generally require an activation operation after each multiply–add operation. The activation operation can often be done by simply fetching the sign bits. Therefore, only a single-bit ADC is required to sample the bitline's current and perform the activation operation after the multiply–add operation without the need of high-precision ADC, which avoids the area and power consumption caused by additional high-speed high-precision ADCs.

The analog computing capability is inherently supported in ReRAM. Each word line of the ReRAM's resistor array can be fed with a specific voltage, and eventually the currents flowing through the resistors on each bit line will be accumulated. Thus, such an operation on the ReRAM array is a matrix–vector multiplication, where each element of the matrix is the conductance of the ReRAM array unit, and the vector is the input voltage on the word line. Thus, the multiply–add operations in neural networks can also be accelerated using the ReRAM array [89]. In addition, ReRAM with some modifications can support arbitrary logic operations and can functionally be used as a computing core. As shown in Fig. 3.47, the desired logic function can

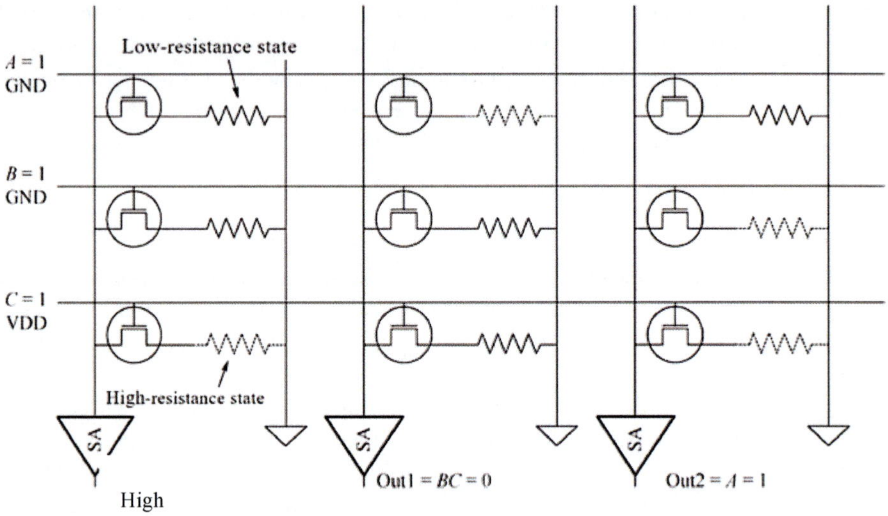

Fig. 3.47 ReRAM array implementation for function computation

be obtained by pre-coding the resistors. An array can implement different functions at the same time, so the degree of parallelism is very large. In summary, if part of the ReRAM array is allocated to memory and the other part is used as a computing core, the whole ReRAM array can form an in-memory GPP [90]. This in-memory processor can dynamically configure the computation on the location of the data to enable local computation on data, which can substantially improve the energy efficiency of the system and systematically solve the problem of insufficient access bandwidth.

Embedding the analog computing in memory remains to be further explored. As mentioned above, there have been many informative explorations. Analog computing is one of the possible solutions of software-defined architectures. If it can be merged with digital computing in the spatial computation model, it is expected to constitute a new efficient acceleration architecture completely different from the traditional computing architecture, greatly alleviating the memory and power walls in the current architecture and playing an important role in a variety of applications.

Although analog computing has demonstrated its advantages on energy efficiency and performance for many different applications, it still faces many challenges. Essentially, the fundamental problem is to define an effective boundary between digital and analog signals. Most of the research in analog computing that has emerged in recent years are not analog architectures due to their requirement of digital circuits for functional assistance. A mixed-signal system necessarily faces the partition between digital and analog signals. The analog unit is effective as an arithmetic accelerator, but if its granularity is so fine that digital-to-analog and analog-to-digital conversions are required each time, it may not yield good result. Conversely, if the granularity is too coarse, large-scale analog systems lack reliable transmission methods over long distances, and functions such as memory are difficult to implement without digital circuits, making system design very difficult.

2. Integration with SDCs

Figure 3.46 shows a typical architecture of analog computing chips, whose two-dimensional spatial array is very similar with that of SDCs. Analog computing PEs may have great energy efficiency and very low power consumption when processing certain tasks, so if some PEs of SDCs are customized as analog computing modules, the overall performance may be improved. Originally, the FPAA was designed to accommodate both digital and analog computing PEs in an array. However, the communication between its digital and analog computing PEs requires signal conversion. Analog signals are often carried in currents because they have less interaction and many functions are easier to implement in analog computing, while digital signals are carried in voltages. Not only that, the interconnect structure of the FPAA can take advantage of the floating gate, while the interconnect structure of the SDC requires to be dynamically configurable, thus producing different interconnect designs. The integration of analog computing into SDCs still has a lot to discuss and practical difficulties, but it is also a possible direction.

3.3.3 Exploration of Approximate Computing

With the rapid development of big data and artificial intelligence, the increasing requirements for massive data processing, super computing power and high energy efficiency have brought unprecedented challenges to computing circuits. It not only promotes the development of traditional technologies such as high-performance and energy-efficient GPPs and ASICs, but also inspires people to explore emerging technologies. Therefore, new architectures or methods such as multi-core technology, heterogeneous computing, and approximate computing have emerged to alleviate a series of problems such as "dark silicon" encountered in the development of traditional CMOS technology [91] and provide more efficient hardware platforms for big data processing and complex computing. Among them, approximate computing takes full advantage of the fault tolerance of applications or algorithms such as multimedia processing, data mining, and machine recognition, sacrificing some accuracy in exchange for significant improvements in performance and power consumption. The fault tolerance of an application can be interpreted as the fact that a certain deviation in the intermediate results of the computation process has a negligible impact on the final result of the application, or the impact will not be recognized by human's perception system. Many natural laws and algorithm characteristics contribute to the fault-tolerant performance of the application, mainly due to the existence of the following phenomena: ① the human perception system has a limited resolution and cannot distinguish small deviations or errors in images, audio and video; ② the input data of the digital signal system is the quantized data, whose quantization error is related to the system sampling rate as well as the natural noise; ③ since redundancy exists in the explosive growing data, the processing results are still acceptable as long as the error does not lead to the loss of key information; ④ in the probability-based computation, the computed data or results stick to the principle of probability computation, and a limited error will not have an important impact on the results; ⑤ machine learning and other algorithms are often based on iterative refinement principles, and errors in the computation process can be compensated by training or iterations. Therefore, witnessing the prevalence of fault-tolerant applications and the urgent need for high-performance computing, approximate computing has grown reapidly in the last decade, reflected by the emergence of a large number of approximate computing circuit design [92].

Although there are numerous designs of approximate computing circuits such as adders, multipliers, and dividers, there are only a few cases of applying approximate computing PEs to specific applications or processors, which greatly limits the further development of approximate computing. This is primarily because the diversity of design concepts and methods results in significant differences in the error characteristics of different designs, and the error characteristics are closely related to external factors such as input data characteristics, internal connections and system structures. Therefore, even if an approximate computing design is selected, the final accuracy of the processing results may vary when the structure of the system, the connection method between PEs, or the statistical characteristics of the input data change. It can be seen that an approximate computing design with fixed error characteristics may be effective for domain-specific dedicated chips, but it cannot meet the accuracy requirements of chips with multiple data sets, internal connection methods and architectures, and is difficult to apply to processors that require high flexibility. Therefore, combining software-based control with approximate computing to dynamically configure the accuracy of computing circuits through software or select approximate modules that meet the accuracy requirements of applications, can take advantage of the high performance and energy efficiency of approximate computing, while ensuring system accuracy and enhancing system reliability. Thus, the SDC provides a new possibility for the design of approximate computing circuits and brings a new hope for breaking through the application bottleneck of approximate computing.

In summary, in order to meet the needs of different applications, data characteristics, interconnection methods, and system structures, approximate computing systems need to provide diverse computational accuracies as required. Combined with SDC technology, approximate computing can be defined dynamically at two levels, namely unit and system. At the unit level, the accuracy and energy consumption of independent PEs can be changed by adjusting the supply voltage and internal circuit structure of PEs through software-based control; in the system level, multiple approximate computing PEs can be integrated into the system computing module and the appropriate approximate computing PEs can be dynamically selected through software. Compared with the definition at the unit level, the definition at the system level has a wider range of options, which improves the computational flexibility, but also increases the area consumption of the hardware.

1. Overview of Approximate Computing

To cope with the constraints on nonlinear computations such as division on system performance and energy efficiency, the design of approximate computing PEs has emerged since the 1960s. The approximate computing in this period was mainly designed for complex nonlinear computations using iterative optimization algorithms [93] and logarithmic approximation methods [94] to mathematically simply the PEs and reduce the operation cycle and complexity. However, approximate computing had been largely stagnant for the next few decades, although some simple and intuitive approximation methods emerged, such as cutting off less important bits from the multiplication result to construct a fixed-width multiplier [95]. It was not until the early 2000s that the concept of approximate computing first appeared in the circuit design of adders and multipliers [96]. Since then it attracts more attention with the development of big data. Currently, approximate computing has grown by leaps with the acknowledgement of fault-tolerant computing.

The scope of approximate computing is very broad, ranging from algorithms, circuits, architectures to programming languages [97]. The approximate computing circuits, such as approximate adder, multiplier, and divider, which act as the basis of computing, have developed particularly rapidly. Current approximation of computing circuits mainly takes the following design approaches, namely, voltage over-scaling (VOS), circuit-level simplification on conventional circuit, and reducing the computing complexity by using mathematical approximations such as Taylor series expansions.

Approximate computing circuits are currently mainly used in fault-tolerant applications such as image processing and machine learning. They are difficult to integrate as basic units in GPPs due to their constraints of accuracy and strict application conditions. However, they have shown significant advantages in terms of performance and energy efficiency in the design of dedicated chips [98].

2. Software-Defined Approximate Computing PEs

The following briefly discusses three design methods for approximate computing circuits and discusses the possibility and necessity of combining approximate computing PEs with SDC technology respectively. The possible impact of SDC technology on approximate computing circuits is also analyzed here.

As the simplest approximation method, VOS reduces the supply voltage of the computing circuit in a disproportionate way, which can reduce the power consumption intuitively without changing the circuit structure. However, the reduction of the supply voltage also brings in certain uncertainties. First, the circuit delay becomes larger as the voltage decreases, so the resulting timing errors may affect the high-order computation results, thus seriously affecting the computational accuracy; second, when the supply voltage is reduced close to the threshold voltage, the flip characteristics of the triode are affected accordingly, thus leading to uncertainty in the computation results errors. Therefore, a stable control mechanism is required in VOS to ensure the stability of its error characteristics at runtime. Considering hardware-based control management might cause great increasement on the power consumption, the software-based control is preferred to not only ensure the stability of VOS circuits, but also achieve better balance between the hardware consumption and precision loss without losing flexibility.

Figure 3.48 shows a VOS-based approximate adder [99], which chunks the adder in order to meet the timing requirements at low supply voltages. These modules are connected using a multiplexer, which selectively cuts off the critical path of the adder by controlling the multiplexer signal. Moreover, the multiplexer can choose whether to transmit the carry of the low-order module to the high-order one depending on the supply voltage, so as to control the length of the carry transmission and the critical path. When the carry of a module is "1" and it is not transmitted to the high-order module, the computation result is wrong. In order to reduce accumulation errors caused by this, the design introduces an error compensation mechanism to count carries that may generate errors. When the counter overflows, a computation clock is added to compensate for the error. Thus, the software-based control of the supply voltage and the multiplexer allows dynamic configuration of the error and energy consumption characteristics of this approximate adder. In addition, the circuit delay is also affected by factors such as ambient temperature and aging effect. These factors also should be considered simultaneously to provide more precise control during the management of supply voltage [100].

It is more common to design approximate computing circuits by modifying, removing or adding some basic circuit units based on conventional circuits. For example, an approximate adder with low power consumption and high speed can be obtained by removing some triodes from the mirror adder [101]; circuits can be simplified by modifying or simplifying the truth table or Karnaugh map [102]. The approximate computing circuits designed using this approach have deterministic error characteristics and do not produce unexpected errors. However, since this design is based on the original computational principles and architecture, its improvement on the hardware is not significant when the accuracy requirement is high; when higher hardware improvement is pursued, it is bound to generate great

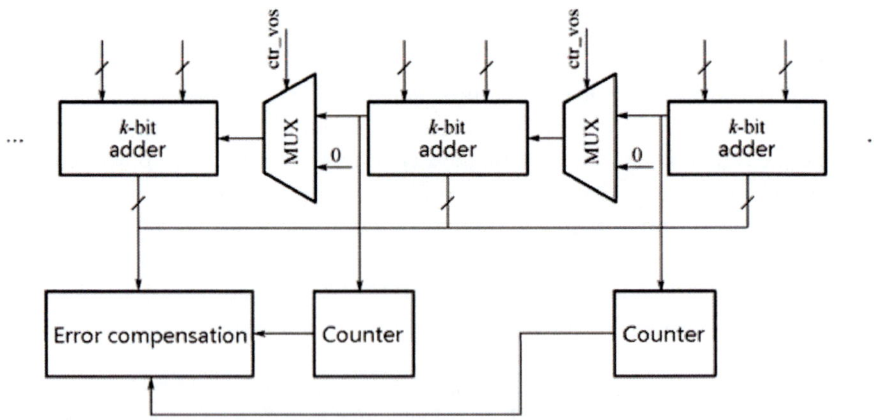

Fig. 3.48 VOS-based dynamic error compensation adder

computational errors. Therefore, it is difficult for an approximate computing design to meet the flexibility requirements of the application. The software-based control approach can facilitate the application and development of approximate computing PEs by making flexible configurations of the basic units and changing the internal connections of the circuit.

In order to achieve the generality of the approximate computing PE, a configurable approximate floating-point multiplier was designed in the literature [103], which uses an error regulation module to control the maximum computational error of the multiplier, i.e., the exact multiplier will be activated when the error exceeds the limit. Since the design integrates both approximate and accurate multipliers, the area of the circuit is larger than that of a conventional floating-point multiplier. Meanwhile, the voltage gating technique is also used to gain some improvement in average performance and energy efficiency. Another commonly used design method that enables both approximate and accurate computing is to use some of the circuits of the accurate computing PE to accomplish approximate computing. The literature [104] proposes the design structure of the adaptive divider and square root extractor, in which some less significant bits in the input operands are selectively discarded. And a smaller accurate computing PE is used to process the remaining part of the data to ensure higher accuracy using less hardware. Figure 3.49 shows the basic circuit structure of the adaptive approximate divider, which mainly encompasses the leading one position detector (LOPD), pruning circuit, lower-order accurate divider, subtractor, and the shift register. In addition, the divider uses an error compensation unit to correct the computational errors caused by overflows. In the approximate division computation, first, the LOPD and pruning circuit are used to prune the input operand, and the pruned result is transmitted to the smaller divider for processing; then, the output of the divider is shifted using a shift register, where the direction and shifting number are calculated by the subtractor; finally, an error compensation unit is used to compensate for some of the errors in the approximate computing. In combination with the SDC, the selected bits of the input operands can be controlled by software. Moreover, the voltage gating technology is adopted to change the sacle of the accurate divider. Therefore, the error range of the approximate division is controlled and the maximum delay and power consumption of the circuit are changed to meet the generality requirements while improving the average performance and energy efficiency.

Compared to the basic PEs such as addition, subtraction, and multiplication, the circuit implementation structures for nonlinear operations such as division, extraction, and exponentiation are more complex. These operations are often approximated at the algorithmic level by approximation algorithms include Taylor series expansions, Newton–Raphson algorithms, and Goldschmidt algorithms [105]. This method simplifies the PE in terms of the basic structure, so that the performance and energy efficiency can be significantly improved at the expense of marginal accuracy loss. Most of the approximation algorithms are based on the principle of iterative optimization, thus the computational accuracy is closely related to the number of iterations. Therefore, the advantages of this method can be fully used in combination with the SDC technology to dynamically control the number of iterations of the algorithm

Fig. 3.49 Adaptive approximate divider

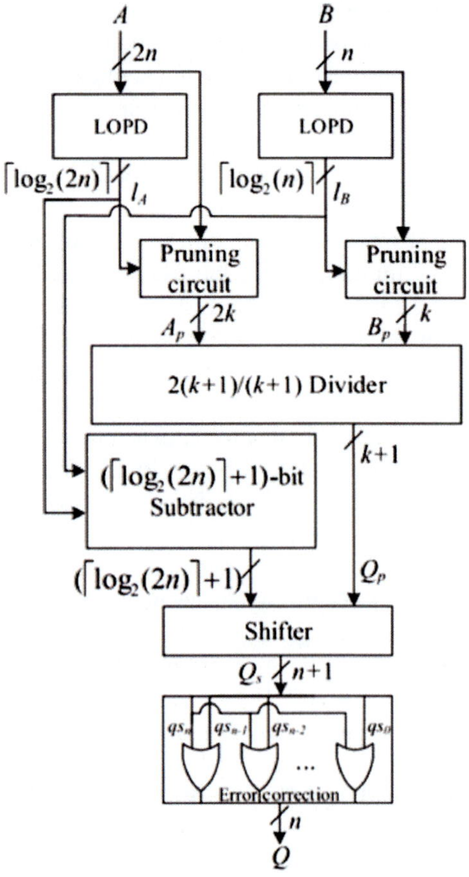

by software, so that the accuracy and speed of the PE can be controlled online, thus significantly improving the computational flexibility.

Taking the Goldschmidt divider as an example, the following equation can be used to perform the division $Q = \frac{N}{D}$

$$Q = \frac{N \cdot F_0}{D \cdot F_0} = \frac{N_0}{D_0} \cdots = \frac{N_{i-1} \cdot F_i}{D_{i-1} \cdot F_i} = \frac{N_i}{D_i} \tag{3.1}$$

where, $F_i = 2 - D_{i-1}$. Equation (3.1) can be used to transform the nonlinear division operation into the multiplication operation in a cycle. When D_i obtained in cycle i equals to 1, N_i is the result of the division. In the computation process, F_i in the next cycle can be adjusted according to the size of error $\epsilon_i = 1 - D_i$. If D_i is exactly 1, a large number of cycles may be required. For this reason, in the actual use of this algorithm, a limit error μ should first be pre-defined. When $\epsilon_i \leq \mu$, N_i is considered to be the result of the division. The circuit structure of the Goldschmidt divider is shown in Fig. 3.50, and the entire computation process is controlled by an FSM. Therefore, the SDC technology can be employed to control the FSM in real time

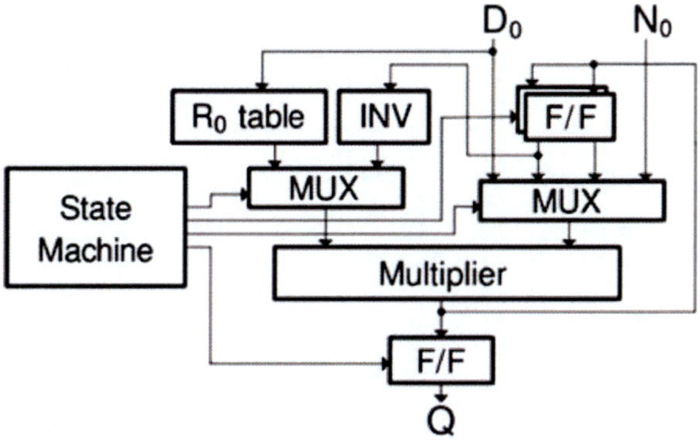

Fig. 3.50 Goldschmidt divider

by software. Moreover, the computational accuracy of the division, i.e., the number of loop iterations, as well as the delay and energy consumption of the divider, can be changed as required, so as to better balance the accuracy of the divider and the hardware consumption.

3. Software-Defined Approximate Computing Systems

Software-defined PEs can precisely configure the computing circuits with finer definition granularity independently. The circuit for configuration and control is more complex if taking finer granularity, leading to larger static power consumption and latency. Meanwhile, the independent definition of each PE in complex computing systems also leads to long configuration time and costly power consumption. Therefore, in complex or compute-intensive systems, software-defined approximate computing systems can be used to define the entire computing module or PEA in a more coarse-grained manner. In recent years, with the development of coarse-grained reconfigurable architectures and approximate computing, two reconfigurable approximate computing arrays have been proposed [106, 107].

As a pioneer of reconfigurable approximate computing PEAs, the literature [107] first proposed the polymorphic approximate coarse-grained reconfigurable architecture (PX-CGRA), whose computational functions are mainly implemented by many heterogeneous PX-CGRA modules (as shown in Fig. 3.51). Each PX-CGRA module with certain accuracy range integrates a polymorphic approximated ALU cluster (PAC) array. The PACs inside the array are connected by a two-dimensional network structure. The software configures the context memory within the accuracy, performance and energy constraints of the application. Then, it selects different PX-CGRAs for hardware mapping while using the voltage gating control module to turn off other unselected PX-CGRA modules, thus reducing static power consumption.

Figure 3.52 shows the internal structure of the PAC, consisting of the accurate, approximate, and precision-adjustable ALUs and switch boxes. Each ALU implements basic addition and subtraction operations, as well as logical and shifting operations. The addition and subtraction operations are performed by accurate, approximate, or precision-adjustable adders and multipliers. The approximate computing PEs use several existing designs with different errors and hardware characteristics.

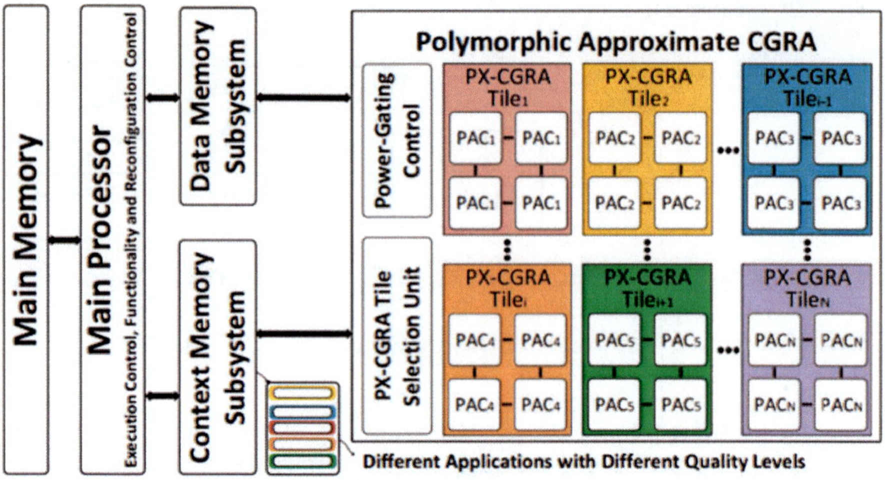

Fig. 3.51 Basic structure of PX-CGRA

The software configures the ALU by rewriting the context register. For the ALU with fixed accuracy, the context word consists of 14-bit binary numbers, as shown in Fig. 3.52b. In particular, the 5-bit Opcode is used to select the function of the ALU; the two input sources of the ALU are determined by the 3-bit MUXA and MUXB; the 3-bit WR specifies the output direction of the ALU. Meanwhile, the OM accuracy configuration bit is added in precision-adjustable approximate computing PE, while the configuration length depends on the design of the approximate computing PE.

In 2018, another coarse-grained reconfigurable architecture based on approximate computing was proposed in the literature [106], as shown in Fig. 3.53a, where the functional modules of this design are interconnected in a one-dimensional feed-forward manner using crossbar switches. In particular, the data between modules can be transferred only from the bottom up. The accuracy of each module is deterministic, while the modules in the same column have the same accuracy. Moreover, a 3×3 ALU array is integrated inside the module, and each ALU has the same accuracy, as shown in Fig. 3.53b. In addition to the basic addition, multiplication, logic and shift operations, the ALU in this design also integrates a divider.

The approximate module only integrates approximate adders, multipliers and dividers, and other operations are the same as those in the accurate module. The ALUs in the computing module are interconnected by crossbar switches to achieve a one-dimensional feed-forward interconnection with fixed data transfer direction. In addition, the clock-gating is also taken in this design to manage each module independently. Compared with PX-CGRA, this design has a simple interconnection structure between modules and a single dataflow direction, thus enabling fast configuration on chip.

The above two CGRA architectures based on approximate computing have reconfigurable designs for the whole computing system in terms of PEs, making it possible

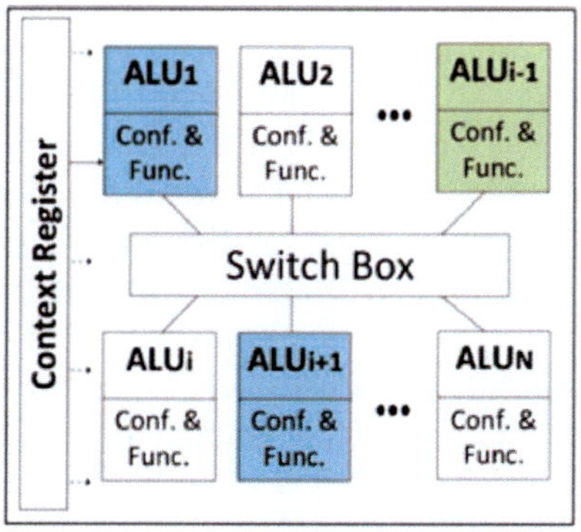

(a) PAC internal circuit unit

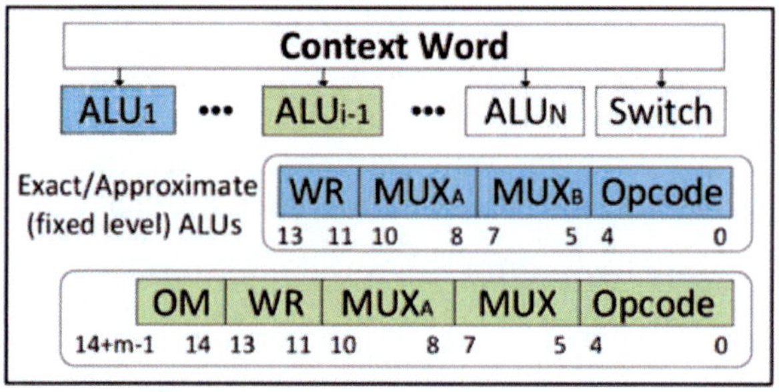

(b) PAC context word

Fig. 3.52 Basic structure of PAC

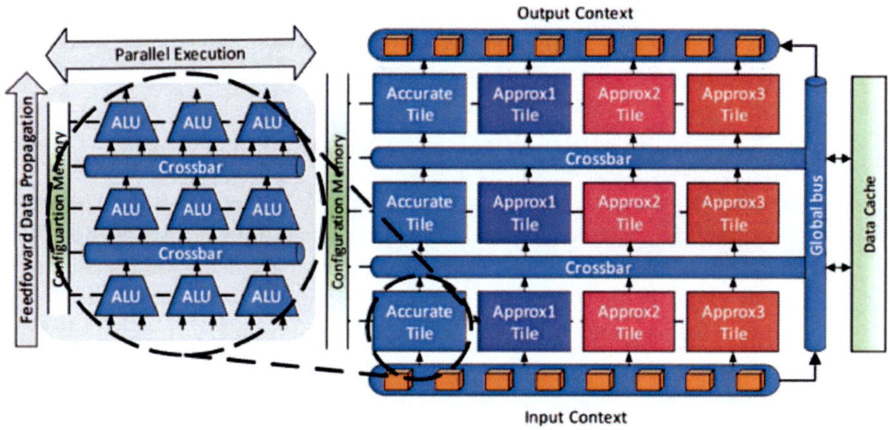

(b) (a)

Fig. 3.53 Design of the basic structure and modules of CGRAs based on approximate computing PEs to satisfy both accurate applications with high accuracy requirements and fault-tolerant applications with certain fault tolerance. They improve the module's flexibility as well as the energy efficiency of the system. However, since the design integrates approximate computing modules with different accuracies, the area of the circuit may increase significantly and the utilization of each module will be reduced, while the difficulty and time of configuration will also be affected accordingly. Therefore, in order to fully leverage the features of SDCs and approximate computing, the system needs to be evaluated comprehensively to achieve desirable trade-off among all aspects, such as the architecture, PEs, interconnection and mapping.

3.3.4 Exploration of Probabilistic Computing

Probabilistic computing usually refers to the procedure of processing and evaluating the probabilistic information in intermediate data, processes or results based on the basic principles of probability. Different from traditional binary computing, probabilistic computing not only deals with fuzzy information with higher robustness, but also serves as a new design approach to reduce the hardware cost such as area and power consumption [108]. Machine learning-based probabilistic computing is mainly to obtain robust inference results using machine learning model algorithms based on uncertain, ambiguous, or conflicting input information in nature. Stochastic computing encodes data into a sequence of 0 and 1 with certain probability characteristics according to the principles of probability. This redundancy-intensive coding scheme can greatly suppress the impact of circuit faults such as bit flips on the computation results, and makes the circuits simpler. Combining the SDC technology with the above two kinds of probabilistic computing, can improve the flexibility of the probabilistic computing system while optimizing the performance and energy efficiency.

1. Software-Defined Machine Learning-based Probabilistic Computing

In machine learning-based probabilistic computing, it is necessary to use probability principles to build a generative model based on probability distribution, conditional probability, prior knowledge and empirical data, etc. There are various algorithms for generative models, such as Monte Carlo algorithm and variational inference algorithm. Different algorithms and application domains may yield different probabilistic inference models, thus requiring different neural network structures of the model and algorithm for hardware implementation. Therefore, traditional accelerators or dedicated chips with fixed network structures cannot meet the demand of probabilistic computing for high flexibility. Using the SDC technology, with a single neuron as the basic PE, the machine learning-based probabilistic computing system is designed with efficient and flexible interconnections controlled by software, thus obtaining different neural network structures, implementing different probabilistic inference model algorithms, and meeting the requirements for probabilistic computing of different algorithms and applications.

2. Software-Defined Stochastic Computing

In stochastic computing, the data is coded by a long sequence of 0 and 1, i.e., the data is mapped to the probability of 1 appearing in the sequence, which not only increases the robustness of computation results, but also greatly simplifies the computing circuit. For example, a traditional binary multiplier requires hundreds of transistors to implement; in a random circuit, a sequence representing p1 × p2 can be obtained by performing an AND logical operation to two sequences representing p1 and p2, i.e., a random multiplier can be implemented using one AND gate. However, existing stochastic computing mainly uses pseudo-random sequences to encode data, and the computation accuracy depends on the sequence length and the generation of random numbers, while the correlation between different sequences also affects the computation results, which makes it difficult for stochastic computing to play their own advantages in applications requiring high accuracy. Therefore, combining stochastic computing with the SDC technology can effectively make up for its own accuracy disadvantage and give full play to its hardware advantages. That is, the software is used to dynamically adjust the length of the random sequence [109], the method of generating random numbers and the seed of the random number generator, so as to dynamically control the accuracy, delay and energy efficiency of stochastic computing, increase the flexibility of stochastic computing, expand its application scope, and maximize the advantages of stochastic computing such as high reliability and low power consumption.

3.3.5 Summary

This section focuses on the application of the circuit-level SDC technology, aiming to explore the design space on the circuit and the advantages brought by SDC technology. First, the limitations of circuits with fixed supply voltage and operating clock are analyzed, and the feasibility and advantages of software-based dynamic voltage and clock are discussed, while some existing designs are briefly introduced and analyzed; then, the corresponding computing circuits and computation models are analyzed at the software-defined level for a variety of emerging computational techniques, including analog computing, approximate computing, and probabilistic computing. The design space of the SDC technology in PEs is explored, as well as the overall optimization in terms of accuracy and resource consumption that may be brought to the computing module or system.

References

1. Fisher JA, Faraboschi P, Young C. Embedded computing: a VLIW approach to architecture, compilers and tools. San Francisco: Morgan Kaufmann Publishers; 2005.

2. Shen JP, Lipasti MH. Modern processor design: fundamentals of superscalar processors. Long Grove: Waveland Press; 2013.
3. Sharma A, Smith D, Koehler J, et al. Affine loop optimization based on modulo unrolling in chapel. In: International conference on partitioned global address space programming models; 2014. p. 1–12.
4. Rau BR. Iterative modulo scheduling. Int J Parallel Prog. 1996;24(1):3–64.
5. Lam M. Software pipelining: an effective scheduling technique for VLIW machines. In: ACM SIGPLAN 1988 conference on programming language design and Implementation; 1988. p. 318–28.
6. Ebeling C. Compiling for coarse-grained adaptable architectures. Technical Report UW-CSE-02-06-01. Washington: University of Washington; 2002.
7. Park H, Fan K, Kudlur M, et al. Modulo graph embedding: mapping applications onto coarse-grained reconfigurable architectures. In: International conference on compilers, architecture and synthesis for embedded systems; 2006. p. 136–46.
8. Annaratone M, Arnould E, Gross T, et al. The warp computer: architecture, implementation, and performance. IEEE Trans Comput. 1987;C-36(12):1523–38.
9. Cong J, Huang H, Ma C, et al. A fully pipelined and dynamically composable architecture of CGRA. In: The 22nd Annual international symposium on field-programmable custom computing machines; 2014. p. 9–16.
10. Nowatzki T, Gangadhar V, Ardalani N, et al. Stream-dataflow acceleration. In: The 44th Annual international symposium on computer architecture; 2017. p. 416–29.
11. Mishra M, Callahan TJ, Chelcea T, et al. Tartan: evaluating spatial computation for whole program execution. ACM SIGARCH Comput Arch News. 2006;34(5):163–74.
12. Kagotani H, Schmit H. Asynchronous PipeRench: architecture and performance evaluations. In: The 11th Annual IEEE symposium on field-programmable custom computing machines; 2003. p. 121–9.
13. Goldstein SC, Schmit H, Moe M, et al. PipeRench: a coprocessor for streaming multimedia acceleration. In: International symposium on computer architecture; 1999. p. 28–39.
14. Singh H, Lee M, Lu G, et al. MorphoSys: an integrated reconfigurable system for data-parallel and computation-intensive applications. IEEE Trans Comput. 2000;49(5):465–81.
15. Takashi M, Kunle O. REMARC: reconfigurable multimedia array coprocessor. IEICE Trans Inf Syst. 1999;E82D(2):261.
16. Mei B, Vernalde S, Verkest D, et al. ADRES: an architecture with tightly coupled VLIW processor and coarse-grained reconfigurable matrix. In: International conference on field programmable logic and applications; 2003. p. 61–70.
17. Mirsky E, Dehon A. MATRIX: a reconfigurable computing architecture with configurable instruction distribution and deployable resources. In: IEEE International symposium on field-programmable custom computing machines; 1996. p. 157–66.
18. Nicol C. A coarse grain reconfigurable array (CGRA) for statically scheduled data flow computing. Wave Computing White Paper; 2017.
19. Govindaraju V, Ho C, Nowatzki T, et al. DySER: unifying functionality and parallelism specialization for energy-efficient computing. IEEE Micro. 2012;32(5):38–51.
20. Prabhakar R, Zhang Y, Koeplinger D, et al. Plasticine: a reconfigurable architecture for parallel patterns. In: The 44th Annual international symposium on computer architecture; 2017. p. 389–402.
21. Wu L, Lottarini A, Paine TK, et al. Q100: the architecture and design of a database processing unit. ACM SIGARCH Comput Arch News. 2014;42(1):255–68.
22. Burger D, Keckler SW, McKinley KS, et al. Scaling to the end of silicon with EDGE architectures. Computer. 2004;37(7):44–55.
23. Voitsechov D, Etsion Y. Single-graph multiple flows. ACM SIGARCH Comput Arch News. 2014;42(3):205–16.
24. Parashar A, Pellauer M, Adler M, et al. Triggered instructions. ACM SIGARCH Comput Arch News. 2013;41(3):142–53.

25. Swanson S, Michelson K, Schwerin A, et al. Wave scalar. In: IEEE/ACM international symposium on microarchitecture; 2003. p. 291–302.
26. Voitsechov D, Port O, Etsion Y. Inter-thread communication in multithreaded, reconfigurable coarse-grain arrays. In: The 51st Annual IEEE/ACM international symposium on microarchitecture; 2018. p. 42–54.
27. Liu L, Li Z, Chen Y, et al. HReA: An energy-efficient embedded dynamically reconfigurable fabric for 13-dwarfs processing. IEEE Trans Circuits Syst II Express Briefs. 2018;65(3):381–5.
28. Gobieski G, Nagi A, Serafin N, et al. MANIC: a vector-dataflow architecture for ultra-low-power embedded systems. In: IEE/ACM International symposium on microarchitecture; 2019. p. 670–84.
29. Lee W, Barua R, Frank M, et al. Space-time scheduling of instruction-level parallelism on a raw machine. ACM SIGOPS Oper Syst Rev. 1998;32(5):46–57.
30. Mishra M, Goldstein SC. Virtualization on the tartan reconfigurable architecture. In: International conference on field programmable logic and applications; 2007. p. 323–30.
31. Park H, Fan K, Mahlke SA, et al. Edge-centric modulo scheduling for coarse-grained reconfigurable architectures. In: International conference on parallel architectures and compilation techniques; 2008. p. 166–76.
32. Ahn M, Yoon JW, Paek Y, et al. A spatial mapping algorithm for heterogeneous coarse-grained reconfigurable architectures. In: Design automation & test in Europe conference; 2006. p. 6.
33. Yoon J, Ahn M, Paek Y, et al. Temporal mapping for loop pipelining on a MIMD-style coarse-grained reconfigurable architecture. ISOCC 2006; 319–322.
34. Nowatzki T, Sartin-Tarm M, de Carli L, et al. A general constraint-centric scheduling framework for spatial architectures. ACM SIGPLAN Not. 2013;48(6):495–506.
35. Nowatzki T, Ardalani N, Sankaralingam K, et al. Hybrid optimization/heuristic instruction scheduling for programmable accelerator codesign. In: International conference on parallel architectures and compilation techniques; 2018. p. 1–15.
36. Tu F, Wu W, Wang Y, et al. Evolver: a deep learning processor with on-device quantization-voltage-frequency tuning. IEEE J Solid-State Circuits. 2020;56(2):658–73.
37. Mo H, Liu L, Hu W, et al. TFE: energy-efficient transferred filter-based engine to compress and accelerate convolutional neural networks. In: The 53rd Annual IEEE/ACM international symposium on microarchitecture, 2020: 751–65.
38. Culler DE, Schauser KE, von Eicken T. Two fundamental limits on dataflow multiprocessing. In: Architectures and compilation techniques for fine and medium grain parallelism; 1993. p. 153–64.
39. Weng J, Liu S, Wang Z, et al. A hybrid systolic-dataflow architecture for inductive matrix algorithms. In: 2020 IEEE International symposium on high performance computer architecture; 2020. p. 703–16.
40. Jacob B. The memory system: you can't avoid it, you can't ignore it, you can't fake it. Synth Lect Comput Archit. 2009;4(1):1–77.
41. Chang D, Lin C, Yong L. Rohom: requirement-aware online hybrid on-chip memory management for multicore systems. IEEE Trans Comput Aided Des Integr Circuits Syst. 2016;36(3):357–69.
42. Chang D, Lin C, Lin Y, et al. OCMAS: online page clustering for multibank scratchpad memory. IEEE Trans Comput Aided Des Integr Circuits Syst. 2018;38(2):220–33.
43. Song Z, Fu B, Wu F, et al. DRQ: dynamic region-based quantization for deep neural network acceleration. In: The 47th Annual international symposium on computer architecture; 2020. p. 1010–21.
44. Tsai P, Beckmann N, Sanchez D. Jenga: software-defined cache hierarchies. In: Annual international symposium on computer architecture; 2017. p. 652–65.
45. De Sutter B, Raghavan P, Lambrechts A. Coarse-grained reconfigurable array architectures[M]. New York: Springer; 2019. p. 427–72.
46. Culler D, Singh JP, Gupta A. Parallel computer architecture: a hardware/software approach. San Francisco: Morgan Kaufmann Publishers; 1999.

47. Sorin DJ, Hill MD, Wood DA. A primer on memory consistency and cache coherence. Synth Lect Comput Archit. 2011;6(3):1–212.
48. Stuecheli J, Blaner B, Johns CR, et al. CAPI: A coherent accelerator processor interface[J]. IBM J Res Dev. 2015;59(1):1–7.
49. Jerger NE, Krishna T, Peh L. On-chip networks. Synth Lect Comput Archit. 2017;12(3):1–210.
50. Zhang Y, Rucker A, Vilim M, et al. Scalable interconnects for reconfigurable spatial architectures. In: The 46th Annual international symposium on computer architecture; 2019. p. 615–28.
51. Aisopos K, Deorio A, Peh L, et al. Ariadne: agnostic reconfiguration in a disconnected network environment. In: 2011 International conference on parallel architectures and compilation techniques; 2011. p. 298–309.
52. Dally WJ, Towles BP. Principles and practices of interconnection networks. San Francisco: Morgan Kaufmann Publishers; 2004.
53. Pager J, Jeyapaul R, Shrivastava A. A software scheme for multithreading on CGRAs. ACM Trans Embed Comput Syst. 2015;14(1):1–26.
54. Park H, Park Y, Mahlke S. Polymorphic pipeline array: a flexible multicore accelerator with virtualized execution for mobile multimedia applications. In: IEEE/ACM international symposium on microarchitecture; 2009. p. 370–80.
55. Atak O, Atalar A. BilRC: an execution triggered coarse grained reconfigurable architecture. IEEE Trans Very Large Scale Integr (VLSI) Syst. 2012;21(7):1285–98.
56. Schüler E, Weinhardt M. Dynamic system reconfiguration in heterogeneous platforms. New York: Springer Science & Business Media; 2009.
57. Ye ZA, Moshovos A, Hauck S, et al. Chimaera: a high-performance architecture with a tightly-coupled reconfigurable functional unit. ACM SIGARCH Comput Archit News. 2000;28(2):225–35.
58. Hauck S, Fry TW, Hosler MM, et al. The Chimaera reconfigurable functional unit. IEEE Trans Very Large Scale Integr (VLSI) Syst. 2004;12(2):206–17.
59. Vaishnav A, Pham KD, Koch D, et al. Resource elastic virtualization for FPGAs using OpenCL. In: The 28th International conference on field programmable logic and applications; 2018. p. 1111–7.
60. Shaojun W, Leibo L, Shouyi Y, et al. Reconfigurable computing. Beijing: Science Press; 2014.
61. Dadu V, Weng J, Liu S, et al. Towards general purpose acceleration by exploiting common data- dependence forms. In: IEEE/ACM International symposium on microarchitecture; 2019. p. 924–39.
62. Chen T, Srinath S, Batten C, et al. An architectural framework for accelerating dynamic parallel algorithms on reconfigurable hardware. In: The 51st Annual IEEE/ACM international symposium on microarchitecture; 2018. p. 55–67.
63. Koeplinger D, Prabhakar R, Zhang Y, et al. Automatic generation of efficient accelerators for reconfigurable hardware. In: The 43rd Annual international symposium on computer architecture; 2016. p. 115–27.
64. Gao M, Kozyrakis C. HRL: efficient and flexible reconfigurable logic for near-data processing. In: 2016 IEEE International symposium on high performance computer architecture; 2016. p. 126–37.
65. Farmahini-Farahani A, Ahn JH, Morrow K, et al. NDA: Near-DRAM acceleration architecture leveraging commodity DRAM devices and standard memory modules. In: The 21st International symposium on high performance computer architecture; 2015. p. 283–95.
66. McDonnell MD. Training wide residual networks for deployment using a single bit for each weight. arXiv preprint arXiv:1802.08530; 2018
67. Thoma F, Kuhnle M, Bonnot P, et al. Morpheus: heterogeneous reconfigurable computing. In: 2007 International conference on field programmable logic and applications; 2007. p. 409–14.
68. Bahr R, Barrett C, Bhagdikar N, et al. Creating an agile hardware design flow. In: The 57th ACM/IEEE design automation conference (DAC); 2020. p. 1–6.
69. Mullapudi RT, Adams A, Sharlet D, et al. Automatically scheduling halide image processing pipelines. ACM Trans Graphics. 2016;35(4):1–11.

70. Li J, Chi Y, Cong J. HeteroHalide: from image processing DSL to efficient FPGA acceleration. In: The 2020 ACM/SIGDA international symposium on field-programmable gate arrays; 2020. p. 51–7.
71. Durst D, Feldman M, Huff D, et al. Type-directed scheduling of streaming accelerators. In: ACM SIGPLAN Conference on programming language design and implementation; 2020. p. 408–22.
72. Barrett C, Stump A, Tinelli C. The satisfiability modulo theories library (SMT-LIB) [EB/OL]. http://smtlib.cs.uiowa.edu/language.shtml
73. Daly R, Truong L, Hanrahan P. Invoking and linking generators from multiple hardware languages using CoreIR. In: Workshop on open-source EDA technology; 2018. p. 1–5.
74. Bachrach J, Vo H, Richards B, et al. Chisel: constructing hardware in a Scala embedded language. In: Design automation conference; 2012. p. 1212–21.
75. Binkert N, Beckmann B, Black G, et al. The gem5 simulator. ACM SIGARCH Comput Archit News. 2011;39(2):1–7.
76. Weng J, Liu S, Dadu V, et al. Dsagen: synthesizing programmable spatial accelerators. In: The 47th annual international symposium on computer architecture; 2020. p. 268–81.
77. Nowka K, Carpenter G, MacDonald E, et al. A 0.9V to 1.95V dynamic voltage-scalable and frequency-scalable 32b PowerPC processor. In: 2002 IEEE International solid-state circuits conference. Digest of Technical Papers; 2002. p. 340–1.
78. Sinangil ME, Verma N, Chandrakasan AP. A reconfigurable 8T ultra-dynamic voltage scalable (U-DVS) SRAM in 65 nm CMOS. IEEE J Solid-State Circuits. 2009;44(11):3163–73.
79. Lee S, John LK, Gerstlauer A. High-level synthesis of approximate hardware under joint precision and voltage scaling. In: Design, automation & test in Europe conference & exhibition (DATE); 2017. p. 187–92.
80. Tschanz J, Kim NS, Dighe S, et al. Adaptive frequency and biasing techniques for tolerance to dynamic temperature-voltage variations and aging. In: 2007 IEEE International solid-state circuits conference. Digest of Technical Papers; 2007. p. 292–604.
81. Verma N, Chandrakasan AP. A 65 nm 8T sub-Vt SRAM employing sense-amplifier redundancy. In: 2007 IEEE International solid-state circuits conference. Digest of Technical Papers; 2007. p. 328–606.
82. Chapiro DM. Globally-asynchronous locally-synchronous systems. Stanford: University of Stanford; 1984.
83. Hall TS, Twigg CM, Gray JD, et al. Large-scale field-programmable analog arrays for analog signal processin. IEEE Trans Circuits Syst I Regul Pap. 2005;52(11):2298–307.
84. Guo N, Huang Y, Mai T, et al. Energy-efficient hybrid analog/digital approximate computation in continuous time. IEEE J Solid-State Circuits. 2016;51(7):1514–24.
85. Huang Y, Guo N, Seok M, et al. Analog computing in a modern context: a linear algebra accelerator case study. IEEE Micro. 2017;37(3):30–8.
86. Huang Y, Guo N, Seok M, et al. Hybrid analog-digital solution of nonlinear partial differential equations. In: The 50th Annual IEEE/ACM international symposium on microarchitecture; 2017. p. 665–78.
87. Zhao Z, Srivastava A, Peng L, et al. Long short-term memory network design for analog computing. ACM J Emerg Technol Comput Syst. 2019;15(1):1–27.
88. Zhang J, Wang Z, Verma N. In-memory computation of a machine-learning classifier in a standard 6T SRAM array. IEEE J Solid-State Circuits. 2017;52(4):915–24.
89. Chi P, Li S, Xu C, et al. Prime: a novel processing-in-memory architecture for neural network computation in reram-based main memory. ACM SIGARCH Comput Archit News. 2016;44(3):27–39.
90. Zha Y, Nowak E, Li J. Liquid silicon: a nonvolatile fully programmable processing-in-memory processor with monolithically integrated reram for big data/machine learning applications. In: Symposium on VLSI circuits; 2019. p. C206-7.
91. Shafique M, Garg S, Henkel J, et al. The EDA challenges in the dark silicon era: temperature, reliability, and variability perspectives. In: The 51st annual design automation conference; 2014. p. 1–6.

92. Jiang H, Santiago FJH, Mo H, et al. Approximate arithmetic circuits: a survey, characterization, and recent applications. Proc IEEE. 2020;108(12):2108–35.
93. Goldschmidt RE. Applications of division by convergence. Cambridge: Massachusetts Institute of Technology; 1964.
94. Mitchell JN. Computer multiplication and division using binary logarithms. IRE Trans Electron Comput. 1962;4:512–7.
95. Lim YC. Single-precision multiplier with reduced circuit complexity for signal processing applications. IEEE Trans Comput. 1992;10:1333–6.
96. Lu S. Speeding up processing with approximation circuits. Computer. 2004;37(3):67–73.
97. Liu W, Lombardi F, Shulte M. A retrospective and prospective view of approximate computing point of view. Proc IEEE. 2020;108(3):394–9.
98. Yoo B, Lim D, Pang H, et al. 6.4 A 56 Gb/s 7.7 mW/Gb/s PAM-4 wireline transceiver in 10 nm FinFET using MM-CDR-based ADC timing skew control and low-power DSP with approximate multiplier. In: 2020 IEEE International solid-state circuits conference; 2020. p. 122–4.
99. Mohapatra D, Chippa VK, Raghunathan A, et al. Design of voltage-scalable meta-functions for approximate computing. In: 2011 Design, automation & test in Europe; 2011. p. 1–6.
100. Amrouch H, Ehsani SB, Gerstlauer A, et al. On the efficiency of voltage overscaling under temperature and aging effects. IEEE Trans Comput. 2019;68(11):1647–62.
101. Gupta V, Mohapatra D, Raghunathan A, et al. Low-power digital signal processing using approximate adders. IEEE Trans Comput Aided Des Integr Circuits Syst. 2012;32(1):124–37.
102. Kulkarni P, Gupta P, Ercegovac M. Trading accuracy for power with an underdesigned multiplier architecture. In: The 24th international conference on VLSI design; 2011. p. 346–51.
103. Imani M, Peroni D, Rosing T. CFPU: configurable floating point multiplier for energy-efficient computing. In: The 54th ACM/EDAC/IEEE design automation conference; 2017. p. 1–6.
104. Jiang H, Lombardi F, Han J. Low-power unsigned divider and square root circuit designs using adaptive approximation. IEEE Trans Comput. 2019;68(11):1635–46.
105. Kong I, Kim S, Swartzlander EE. Design of Goldschmidt dividers with quantum-dot cellular automata. IEEE Trans Comput. 2013;63(10):2620–5.
106. Brandalero M, Beck ACS, Carro L, et al. Approximate on-the-fly coarse-grained reconfigurable acceleration for general-purpose applications. In: The 55th ACM/ESDA/IEEE design automation conference; 2018. p. 1–6.
107. Akbari O, Kamal M, Afzali-Kusha A, et al. PX-CGRA: polymorphic approximate coarse-grained reconfigurable architecture. In: Design, automation & test in Europe conference & exhibition; 2018. p. 413–8.
108. Alaghi A, Qian W, Hayes JP. The promise and challenge of stochastic computing. IEEE Trans Comput Aided Des Integr Circuits Syst. 2018;37(8):1515–31.
109. Kim K, Kim J, Yu J, et al. Dynamic energy-accuracy trade-off using stochastic computing in deep neural networks. In: The 53nd ACM/EDAC/IEEE design automation conference (DAC); 2016. p. 1–6.

Chapter 4
Compilation System

Compiler writers must evaluate tradeoffs about what problems to tackle and what heuristics to use to approach the problem of generating efficient code [1].
—Alfred V. Aho et al. Compilers: Principles, Techniques, and Tools, 2006.

Compute-intensive applications such as AI, bioinformatics, data centers and IoT have become the hot topic of our time, and these emerging applications are becoming increasingly demanding in terms of the computing power of SDCs. To meet the demanding computing power requirements of these applications, the scale of programmable computing resources in SDCs has increased rapidly. Therefore, how to use these computing resources conveniently and efficiently has gradually become a key issue affecting the application of SDCs. With the advent of large-scale programmable computing resources and emerging applications, the cost of manual mapping has seriously affected the development efficiency of users. Therefore, it is urgent to develop an automated compilation system for SDCs.

The SDC compilation system is a software system that translates the behavior or function described by users in high-level languages into a piece of binary machine code with equivalent functions that can be recognized by hardware. An excellent compilation system can effectively exploit the hardware potential of SDCs without affecting programmer productivity very much, and provide users with a more convenient and efficient way to use chip hardware resources. With the feature of dynamic reconfiguration, the SDC can employ dynamic compilation techniques to further improve resource utilization. Compiler developers have to make trade-offs when designing an SDC compilation system. This chapter starts with the traditional static compilation technique and dynamic compilation supported by SDCs to discuss how to make a trade-off between time overhead, compilation quality, and ease of use in the design of compilation systems, thus utilizing the hardware resources of the SDC efficiently.

Section 4.1 starts with a general overview of the SDC compilation system and briefly describes the key phases in the static and dynamic compilation process. Section 4.2 details the traditional static compilation. Specifically, it starts with the introduction of intermediate representations (IRs), abstracts and models the mapping problem, introduces commonly used solution and optimization methods, including

S. Wei et al., *Software Defined Chips*,
https://doi.org/10.1007/978-981-19-6994-2_4

modulo scheduling, software pipelining, and integer linear programming, and finally discusses the mapping problem of irregular tasks. Since the SDC supports dynamic reconfiguration, users can apply dynamic compilation to further improve the energy efficiency of the chip. Therefore, Sect. 4.3 discusses dynamic compilation of SDCs. Specifically, it first introduces the basis for dynamic compilation of SDCs hardware resource virtualization, and then describes how to generate and transform configurations dynamically.

4.1 Overview of the Compilation System

An SDC compilation system is a software tool chain capable of translating high-level programming language describing the functions of an application into a machine language that can be recognized by the underlying hardware. The SDC has substantial programmable computing resources (PEs) that can perform computations in parallel and fully exploit multiple parallelisms in applications. However, a highly-parallel hardware architecture cannot have both the high performance and ease of use of the programming model, and more details can be referred to in Section 1 of Volume II. To fully utilize the computing advantages of the rich hardware resources in SDCs will certainly lead to a reduction in the ease of use of the chip. Manual compilation can achieve the highest computational performance and energy efficiency. However, due to the diverse architectures of SDCs and rapid development of applications, manually compiling the applications for each architecture will significantly extend the development time and greatly increase the NRE cost of SDC software development. Therefore, in the development process of SDCs, the design of the compilation system is as important as that of the hardware architecture. However, the current SDC compilation systems still require a lot of manual assistance to ensure compilation quality and lack versatility, and automated implementation is still today's research hotspot.

Compilation techniques can be divided into static compilation and dynamic compilation according to the temporal characteristics of the compiler [2]. Static compilation is the mainstream compilation technique nowadays, which is adopted by almost all GPPs and FPGA compilers (e.g. GCC and Vivado). The compilation is completed offline before the application runs. Due to the lack of dynamic information at runtime, even in chips that support multi-task and multi-thread, static compilation can only ensure high energy efficiency of a single application. The chip cannot dynamically adjust itself after it starts working. Dynamic compilation means that the optimization and adjustment of the way the application is executed at runtime can be made based on dynamic information at runtime for further performance improvements. Java, Microsoft's .NET Framework, and NVIDIA's NVRTC all adopt dynamic compilation. With the enormous hardware resources and dynamic reconfiguration, the SDC can employ dynamic compilation techniques to improve hardware utilization. Figure 4.1 shows the workflow of an SDC compilation system. The following briefly describes the main phases in the static and dynamic compilation process in the compilation system.

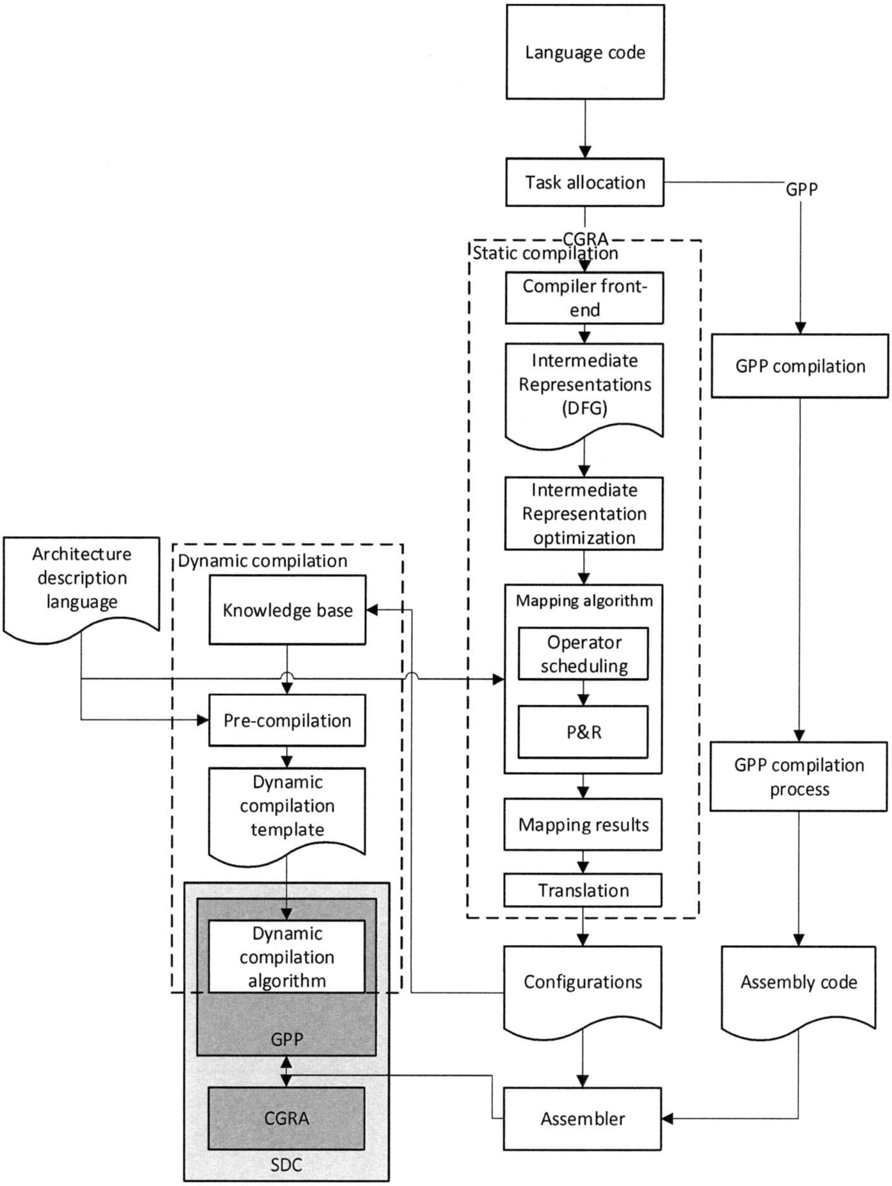

Fig. 4.1 Workflow of an SDC compilation system

4.1.1 Static Compilation Process

Static compilation is a process of translating application code snippets described in a high-level language into a binary machine language that can be recognized by the SDC before execution. As described in Chap. 3, the SDC generally uses a coarse grained reconfigurable architecture (CGRA) as an accelerator to be coupled with the GPP, and its static compilation process includes the compilation processes of the GPP and CGRA. Since the compilation technique of the GPP has been relatively mature, this section focuses on the static compilation process of the CGRA.

1. Task Allocation

The SDC typically uses a coarse-grained reconfigurable PEA as the accelerator engine, and the entire computing system consists of the GPP and CGRA. The hardware architecture of the CGRA contains a large number of parallel PEs, which is particularly suitable for data-parallel and compute-intensive code areas, but control statements in the application can cause control conflicts in the pipeline, making it less efficient. Therefore, it is a common practice to allocate the basic blocks to the CGRA for execution and the control code other than the basic block to the GPP for execution. For most applications, loop statements tend to have the largest percentage of execution time [3]. Therefore, loop statements are often executed on the CGRA. The allocated code is translated by the CGRA compiler and GPP compiler respectively into the corresponding machine code (configuration contexts of the CGRA and assembly code of GPP). Then, it is translated by the assembler into a machine language that can be recognized by the SDC system to finally complete the static compilation process. However, too frequent data synchronization and communication between the GPP and CGRA may seriously reduce the gains of such allocation. Therefore, the above task allocation method can result in gains provided that the overhead of communication between the GPP and CGRA is less than the performance improvements that the CGRA can provide [4]. For control-intensive applications or those with limited parallelism, another way of allocating tasks across basic blocks can be used. Specifically, multiple basic blocks are combined with control statements into a single hyperblock [5] before being allocated to the CGRA for execution, in order to increase the parallelism in the program and reduce the overhead of communication between the GPP and CGRA. The computational tasks within the basic blocks of an application are referred to as regular tasks, and the tasks across the basic blocks are referred to as irregular tasks. The mapping of regular and irregular tasks on the CGRA will be described in detail later.

2. Compiler Front-end

After the application code is allocated to the CGRA, it will first be processed by the front end of the compiler. The compiler front-end usually performs lexical analysis, syntax analysis and semantic analysis, with the aim of resolving all operations and actions described in the code snippets of the source program and the dependences between them. After being processed by the front end of the compiler, the code snippets described in a high-level language (e.g. C/C++) will be translated into

IRs. The IR is an equivalent representation of the source program and serves as a bridge between the source code and the target code. It can intuitively describe the control dependences and data dependences in the source program and is independent of the hardware architecture. The compiler can optimize the target code by modifying the IR. An IR may take one of several forms: assembly language, data flow graph (DFG), control/data flow graph (CDFG), or abstract syntax tree (AST). An ideal compiler would optimize the target machine code by optimizing the IR alone without changing the behavior description of the source program, whereas it is a popular practice to optimize the final hardware architecture by adding compiler directive statements to the source code, such as the HLS tool of Vivado. Despite the performance improvements to the program, this increases the cost of software development.

3. Mapping Algorithm

The mapping algorithm of the CGRA maps all operations and dependences in the optimized IR to the corresponding target hardware structures. The behavioral functions described by the IR are implemented on the hardware and the execution is optimized as required. This means that the mapping algorithm has a decisive impact on the performance and power consumption of the CGRA, which makes it the most important part of the compilation system. The mapping quality can be measured by the initialization interval (II). The II is defined as the interval between the start of two consecutive iterations of the kernel. A smaller II means shorter average execution time per iteration and higher performance. The mapping algorithm usually uses the software pipelining technique [6] to make operations in different loop iterations be executed at the same time and reduce the II by overlapping the running time of multiple iterations. As described in Chap. 3, there are two execution models of the CGRA: static scheduling and dynamic scheduling. The static-scheduling CGRA requires the compiler to allocate operators to the fixed control steps, and the operations on the same control steps are executed synchronously. The dynamic-scheduling CGRA, similar to an OoO superscalar processor, uses a special hardware structure to schedule the operators. The former hands scheduling over to the compiler, thus simplifying the hardware and improving energy efficiency. The latter exploits parallelism at runtime through mechanisms such as dataflow, which simplifies the design of compilation algorithms but increases the hardware design overhead. Currently, the static-scheduling CGRA is the most common design method, which has higher requirements on the compilation algorithm. Therefore, this chapter focuses on the mapping algorithm of the static-scheduling CGRA.

The mapping problem of the CGRA has proved to be NP-complete [7], which cannot be solved by an algorithm with polynomial time complexity. Therefore, the mapping algorithm usually adopts the following solution strategies: Greedy algorithms, stochastic algorithms, heuristic algorithms, and integer linear programming. The mainstream mapping algorithms can be divided broadly into two types: decomposed and integrated [8]. The decomposed mapping algorithm executes operator scheduling, place & route (P&R), and register allocation step by step, and divides a big problem into several smaller ones to reduce the time overhead. Essentially, this mapping strategy reduces the solution space through problem division, thus increasing the solving speed. The disadvantage is that some high-quality solutions may be lost while reducing the solution space. Therefore, the decomposed mapping algorithm mainly sacrifices the solution quality for the speed. In contrast, the integrated mapping algorithm builds the whole mapping problem into a unified model and finds the globally optimal or suboptimal solution. Although this mapping strategy can produce better-quality mapping results, it may introduce a high time overhead and cannot obtain a feasible solution within an acceptable time overhead. In summary, the design of the mapping algorithm requires a trade-off between mapping quality and time overhead.

4.1.2 Dynamic Compilation Process

The dynamic compilation process is to compile the application online under the constraints of dynamic information (e.g., computing resource utilization, dynamic power requirements, and hardware failures) to generate configuration contexts. The dynamic compilation process consists of offline and online phases. To shorten the running time of the dynamic compilation algorithm and reduce the impact on the execution time, the compiler simplifies the problem model in the offline phase or generates guidance information (e.g., templates) for the online phase in advance. The algorithm with low time complexity is required in the online phase to ensure that it does not affect the performance of the overall SDC system.

1. Pre-compilation

In order to bring performance gains from dynamic compilation, the compilation system should ensure that the time overhead of the dynamic compilation algorithm should not be too high. Therefore, the complexity of the dynamic configuration generation algorithm should be as low as possible. However, as mentioned in Sect. 4.1.1, the mapping algorithm of the CGRA is an NP-complete problem. Thus, the dynamic configuration generation algorithm cannot directly use the mapping method of the static compilation algorithm. The dynamic compilation process usually requires running the pre-compilation process offline, and by leveraging the results (templates) of the pre-compilation process and the basic configuration contexts generated by static compilation for virtual hardware resources. As shown in Fig. 4.1, the pre-compilation generates templates based on the hardware architecture description and the knowledge base. The knowledge base is an archive that includes the configuration contexts characteristics, and the configuration contexts generated by static compilation for all applications supported by the SDC system are added to this knowledge base. By virtue of the static configuration in the knowledge base, the pre-compilation process can produce better results, thus enabling better performance in dynamic compilation. There are also some mainstream CGRA dynamic compilation techniques [9, 10] that do not use static pre-compilation or task allocation, but directly use less complex algorithms such as greedy algorithms to transform GPP instruction streams into configuration context streams to achieve transparent programming.

2. Dynamic Compilation Algorithm

In the CGRA supporting multithreading, the controller dynamically monitors the state of each PE and returns the real-time resource constraint information (e.g., long idle PEs) to the GPP. The GPP then invokes the dynamic compilation algorithm to generate configurations that meet the resource constraints for the application to be executed to improve the overall hardware utilization. The dynamic compilation algorithm of the CGRA is divided into the dynamic generation algorithm based on instruction streams and dynamic transformation algorithm based on configuration streams. The former requires dynamic analysis of the dependences between instructions, resulting in lower quality of the configuration contexts. The latter uses the statically generated mapping results and templates to dynamically convert the basic configuration contexts generated by static compilation into the configuration contexts for virtual hardware resources.

In summary, the compilation system, as an automated tool chain for efficiently using SDCs to execute applications with great ease of use, short compilation time, and high compilation quality (high performance). Since SDCs usually adopt the GPP + CGRA heterogeneous architecture, it is important to properly divide the application code into different tasks in order to exploit their performance advantages in control-intensive and compute-intensive tasks respectively. The rich parallel computing resources in SDC have the potential to improve instruction parallelism, but the mapping problem becomes a complex NP-complete problem. Meanwhile, it is necessary to design an IR for the compilation system to resolve the incompatibility with imperative high-level languages. The dynamic reconfiguration characteristics enable SDCs to support hardware virtualization, and the use of dynamic compilation techniques can further improve the energy efficiency of SDCs and allow SDCs to adapt to the changing environment.

4.2 Static Compilation Methods

The design of the static compilation method requires a trade-off between time overhead, compilation quality, and ease of use. Among them, time overhead is the time consumed in the static compilation process. Compilation quality is usually measured by energy efficiency (ratio of performance to power consumption). Ease of use is the productivity of the user during use of the chip. This section will discuss in detail how to design the static compilation method for SDCs to improve these three key design elements in the static compilation method.

4.2.1 IR

Today's high-level languages (e.g., C/C++, Python, and Java) are usually designed to describe serial semantics for the classic von Neumann architecture, which is clearly in contradiction with the parallel computing model of SDCs. Since the first compiler using high-level languages, the Fortran compiler, was introduced in 1957, people have been working for decades to get a better GPP compiler. Since the first EDA workstation platform, Apollo, was built in 1983, various synthesis tools using hardware description languages such as VHDL, Verilog, and SystemC have emerged. However, the integrated compilation tools combining the above two methods (e.g., HLS tools) cannot generate circuits with satisfactory performance. This is mainly because high-level languages are usually imperative languages, which fit well with the von Neumann architecture used by GPPs. In fact, the popular high-level languages are designed for the von Neumann architecture. The main goal of hardware description languages is to better express the connection relationships between the underlying logic modules, and their expression forms are often more complex and difficult to modify and optimize. The IR is an intermediate layer to better bridge the high-level language and the underlying modules, and it serves as a link between the front end and back end of the compiler. A good IR can accurately represent the functions and behaviors of the application source code and is independent of any specific language or hardware architecture.

Almost all of the current mainstream SDC compilation techniques use IRs. Table 4.1 lists the IRs used by major SDC compilers or HLS tools (collectively referred to here as compilers) in recent years. IRs can be classified into software IRs (or high-level IRs) and hardware IRs (or low-level IRs). Software IRs (such as those used in the DySER [11], RAMP [12], Legup [13], and CASH [14] compilation frameworks) are generally the intermediate layers that are transformed from high-level languages (e.g., C/C++, Java) and are based on static single-assignments (SSAs) or DFGs. Hardware IRs (e.g., FIRRTL [15] and LNAST [16]) refer to the intermediate layers that are transformed from hardware construction languages (HCLs) and are closer to the underlying hardware. In particular, μIR, is special in that it is a hardware IR generated by a software IR. The software IR focuses on the behaviors of the application and expresses the same functions as a high-level language does, regardless of functional implementations using hardware. The hardware IR focuses only on the structure of the underlying hardware modules and how they are connected, aiming at optimizing the hardware structure of the circuit. For SDCs with certain hardware architectures, software IRs are usually used instead of hardware IRs because the optimization of low-level IRs transformed through HCLs is too flexible and may go beyond the representation of the target hardware architecture. For SDCs with uncertain hardware architectures (e.g., agile hardware design and HLS), both software IRs and hardware IRs can be used. Adopting IRs as the intermediate layer of the compilation system can bring the following benefits:

Table 4.1 Mainstream compilers and IRs

Compiler	Compiler type	Front-end language	IR structure	IR type
DySER [11]	SDC compiler	C/C++	SSA/DFG	Software IR
CASH [14]	HLS	C	SSA/DFG	Software IR
Chisel [17]	HCL	Chisel	FIRRTL [15]	Hardware IR
μIR [18]	HLS	C/C++	μIR	Hardware IR
EPIMap [7]	SDC compiler	C/C++	SSA/DFG	Software IR
REGIMap [19]	SDC compiler	C/C++	SSA/DFG	Software IR
RAMP [12]	SDC compiler	C/C++	SSA/DFG	Software IR
Legup [13]	HLS	C/C++	SSA/DFG	Software IR
SPATIAL [20]	HCL	SPATIAL	PATTERN	Hardware IR
LiveHD [16]	HCL	HCL	LNAST	Hardware IR

1. Improved Ease of Use

Programming languages are moving toward higher-level and more abstract expressions, thus making programmers more efficient in writing programs with complicated functions. Hardware architectures are tending to be more complex (e.g., multi-core, multi-CPU, and GPUs). As the abstraction gap between applications and hardware widens, it is increasingly difficult for compilers to automatically utilize hardware resources for optimal performance. The reason is that high-level languages lack the semantic information needed to efficiently translate coarse-grained execution blocks to low-level hardware. High-level languages describe processes that are executed sequentially, while hardware architectures work in parallel. For better performance of the target hardware architecture, programmers have to use low-level and architecture-specific programming models (e.g., CUDA for GPUs and MPI for processor clusters) [21]. SDCs are very different from GPPs in terms of hardware architecture. Therefore, the compiler designs (especially the back end) for them must also be very different. This makes it impossible to directly apply the programming languages, programming models, and code optimizations that are originally suitable for use on GPPs to SDCs.

To achieve higher performance, the programmer must first be familiar with the characteristics of the target hardware architecture, and then rewrite the program code or add directive statements to the code (Fig. 4.2). The optimized code is related to the target hardware architecture (e.g., cacheline size, number of processor cores, and number of processor pipeline stages), and this will make the source program difficult to maintain and transport, reduce the ease of use of the compilation system, and greatly increase the time cost of software development. By optimizing the program at the IR level, it is possible to improve the efficiency of programs running on SDCs while maintaining user productivity.

Fig. 4.2 Compiler front-end requiring manual guidance

```
void （int * mem）{
    mem[512] = 0;
    For （int i=0;i<512;i++）
        mem[512] += mem[i];
}
```

(a) Behavioral description of the application

```
# define ChunkSize （sizeof（MPort）/sizeof（int））
# define LoopCount （512/ChunkSize）
void （MPort* mem）{
    MPort buff[LoopCount];
    memcpy（buff, mem, LoopCount）;
    int sum=0;
    For（int i=1; i<LoopCount; i++）{
    #pragma PIPELINE
        for（int j=0; j<ChunkSize; j++）{
        #pragma UNROLL
            sum += （int）（buff[i]>>j*sizeof（int）*8 ）;}}
    Mem[512] = sum;
}
```

The Constructing Hardware in a Scala Embedded Language (Chisel) and Flexible Intermediate Representation for RTL (FIRRTL) were created as compilation systems to improve the productivity of hardware development. The UC Berkeley developed Chisel as an HCL while designing the open-source hardware RISC-V to enable more agile development and reuse of hardware designs. Based on Scala, Chisel can be compiled into Verilog HDL and then synthesized into the domain specific language (DSL) for circuit netlists. Chisel enabled the design team to significantly increase their hardware design productivity, and a team of 12 graduate students could complete the tapeout of over 10 chips in 3 years. However, they found that Chisel was difficult to be well used outside the team. Common users did not understand the principles of the Chisel compiler, and the program writing efficiency and final circuit performance were not satisfactory. Therefore, the design team at the UC Berkeley added an IR FIRRTL between the source language (Chisel) and the target language (Verilog), and proposed optimization methods at the FIRRTL level (e.g., constant propagation, common-sub-expression elimination, and dead code elimination). In summary, adding IRs to the existing compilation system can decouple the source program description from the hardware-oriented optimization and improve the compilation efficiency without modifying the source program description, thus making the compilation system significantly easier to use.

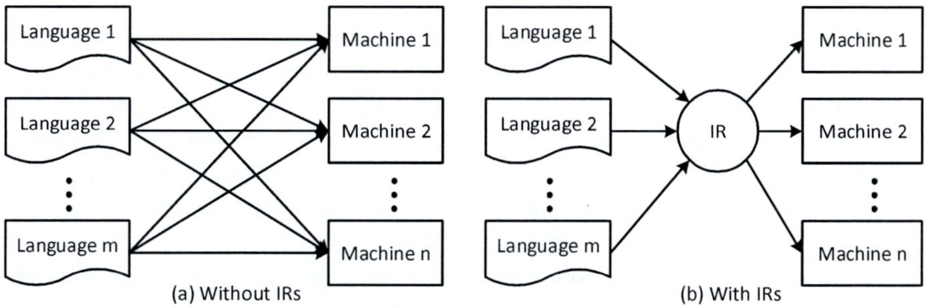

Fig. 4.3 Reduced design overhead of the compilation system by IRs

2. Reduced Design Overhead of the Compilation System

Using IRs as the intermediate layer of the compiler can also effectively reduce the compiler design effort. First, the IR can divide the compilation system into the front end and back end, which not only makes the compilation system logically clearer, but also makes it easier for the compilation system designer to design and develop the front-end and back-end algorithms at the same time. Second, the IR can enhance the scalability of the compilation system and make it easy to be reused. Suppose it is required to design compilers with n types of chips for m types of languages respectively, as shown in Fig. 4.3a, if the traditional compiler design method without IRs as the intermediate layer is used, $m \times n$ independent compilers need to be designed. As shown in Fig. 4.3b, if an appropriate IR is defined and the "front-end + IR + back-end" compiler design method is adopted, only m front-end compilers and n back-end compilers need to be designed, which significantly reduces the compiler design effort.

Stanford University Intermediate Format (SUIF) [22] is an open-source compiler infrastructure developed by Professor Qianling Lin's team at Stanford University. It has subsequently followed by the SUIF2 and MachSUIF. SUIF2 is the front end of the entire compilation system, while MachSUIF is the back end of the compiler, independent of the target machine. SUIF2 provides developers with processing programs for code optimization (also known as passes), or developers can customize their own passes. The application is first transformed into a high-level IR (high-SUIF) by the pass and then into a low-level IR (low-SUIF) by a low pass. Then, the low-SUIF is fed into MachSUIF to generate a virtual machine language (suifvm) that is independent of the architecture of the target machine, and the instruction translation is done for the specific machine to obtain the target machine language. Many SDC compilers use the SUIF compiler infrastructure, such as REMUSC [23], XPPVC [24], and Garpcc [25].

LLVM [26] is an open-source compiler infrastructure that originated from an open-source project initiated by the University of Illinois at Urbana-Champaign in 2000. In 2005, Chris Lattner, one of the main authors of LLVM, developed C/C++/Objective-C compiler front end for Apple, which could take the place of GCC and was released after being integrated with LLVM 2.6. For LLVM's significant contributions to the industry, Chris Lattner received Software System Award from the ACM in 2012. The LLVM IR and its optimizer are independent of the programming language and target machine architecture, so that the front-end designer does not have to worry about the implementation of the back end, and similarly, the back-end designer does not have to worry about the language used in the front end. This enables the compiler infrastructure extensible with modular design. A compilation system for a new architecture can reuse the open-source front end and only requires developing new compiler back-end algorithms, greatly shortening the design time of the compilation system. Many of today's mainstream SDC compilers (e.g., RAMP [12], HPRC [8], and CGRA-ME [27]) are developed based on the LLVM infrastructure.

3. Improved Mapping Quality

An excellent IR should also improve the compilation quality, i.e., the energy efficiency of the source program running on an SDC. As shown in Fig. 4.4, the IRs generated by the front end from different languages are optimized for SDCs with different hardware architectures, and the IRs that can improve the mapping efficiency of the source program on the target hardware architecture are generated. Therefore, in the design of the IR structure, whether its data structure is helpful for optimization should be taken into account. The IR structure can be broadly divided into the following categories.

Both three-address code (TAC) and static single assignment form (SSA) are common compiler IRs. The TAC is similar to the assembly language in that it creates some temporary intermediate variables and then converts the source program into multiple three-address instructions. The TAC usually use quadruple or triple data structure. The difference between the SSA and TAC is shown in Fig. 4.5. All assignment instructions in the SSA are to assign values to variables with different names. Although they both can represent the correct dependences, the anti-dependency existing in the original code can be eliminated because the SSA renames the assigned variables. As shown in Fig. 4.5a, instructions 4 to 6 of the TAC are not functionally dependent on the first three instructions, and if the SSA in Fig. 4.5b is used as an

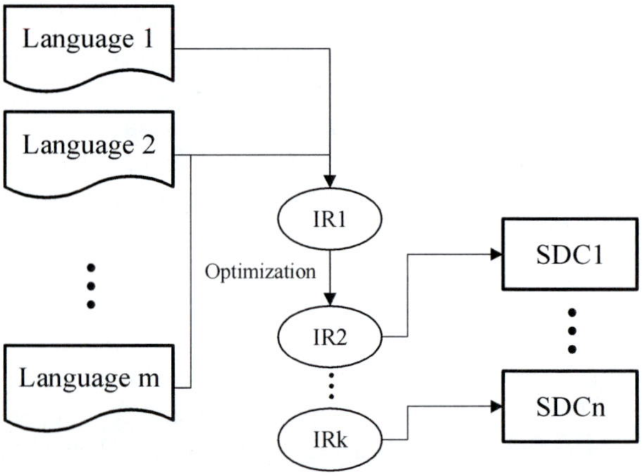

Fig. 4.4 Different optimization methods provided by IRs for different architectures

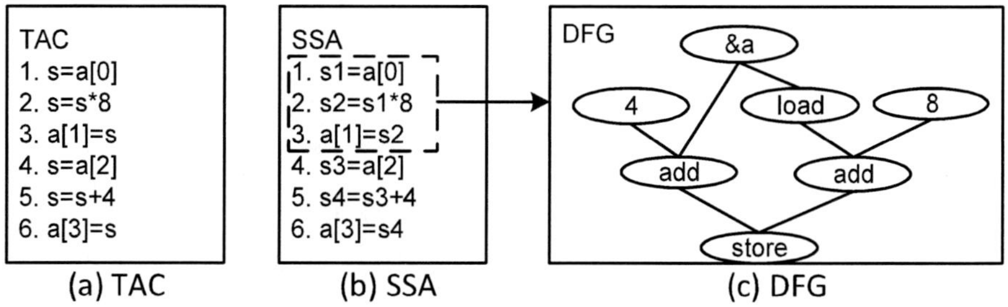

Fig. 4.5 Three IRs

IR, the anti-dependency can be eliminated. This can increase the instruction level parallelism and help improve the mapping efficiency of SDCs. The SSA can transform control dependences into data dependences by using function ϕ, i.e., by using predicate execution, when handling branch instructions in a program, which also increases the instruction level parallelism. The SSA can also be simply transformed into the more intuitive equivalent DFG or CDFG. Both IRs will be described in detail in the subsections later in this chapter. Currently, many mainstream compilers use SSAs as IRs, such as the GCC and LLVM.

μIR is another IR for easy optimization of the hardware architecture generated by the HLS tool. μopt is an optimization method for μIR, which supports pipelining, parallelization, and high-order computation fusion. μIR can be generated by the LLVM IR and connects time-variant task blocks through a latency-insensitive interface (e.g., FIFO). In this way, each task block is relatively independent, and each of these task blocks can be optimized without considering the impact on the other modules, thus greatly simplifying the optimization of IRs.

4. Reduced Compilation Time Overhead

Since the static compilation of the SDC is an NP-complete problem, the task mapping time may reach an unacceptable level when a large task is allocated to the PEA of the SDC. The most common solution to the above problem is to further divide the tasks into subtasks before mapping. Since the time complexity of the NP-complete problem grows exponentially with the problem size, this "divide and conquer" method can effectively reduce the compilation time overhead. However, in this way, the tasks of the source program are allocated to the GPP and CGRA respectively, and the tasks allocated to the CGRA are further divided into multiple subtasks, which are allocated to each subarray of the CGRA. The direct and fine division of tasks is error-prone because dependences and task sizes are not intuitively represented in the original code. Since the IR can fully represent all the operations in the source program and all the dependences between them, moving the fine task division to the IR level will make it easier and more efficient for users to complete the subtask division. A more balanced task division not only significantly improves the mapping efficiency, but also reduces the time overhead of mapping subtasks to target subarrays. For this reason, the SDC employs two mechanisms to reduce the task size.

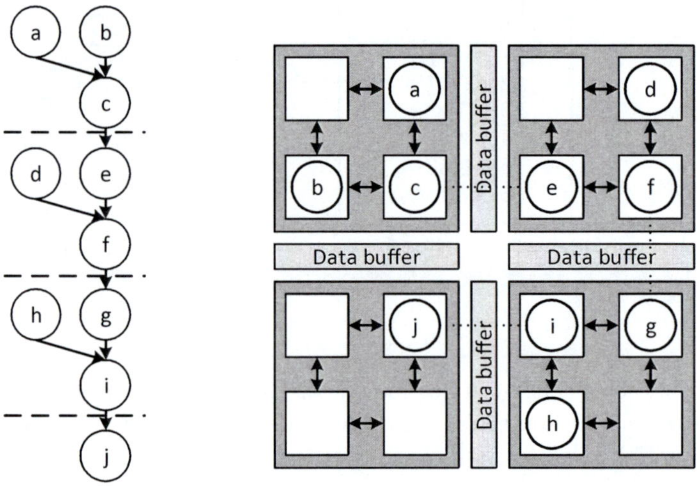

Fig. 4.6 Spatial partitioning of the DFG

The first is to perform spatial partitioning of IRs by utilizing the hardware resources of the SDC, as shown in Fig. 4.6. Many compilers, such as Autoscaler [28] and CHAMPION [29], have adopted the spatial partitioning mechanism. The PEAs are divided into multiple "islands" manually or using virtualization techniques, and there is usually a public data buffer for communication between islands. Thus, the tasks allocated to the CGRA are further allocated to islands, and an acceptable compilation time can be obtained by mapping each subtask to an island separately. In general, the read speed between PEA islands is slow. Therefore, the number of edges of the islands should be minimized during spatial partitioning in order to obtain mapping results with higher performance.

The other one is to perform temporal partitioning of IRs by using the dynamic reconfiguration characteristics of SDCs, and an IR is divided into multiple subtasks in a TDM manner. Figure 4.7 shows how to divide the DFG. Specifically, the compilation system generates a set of configuration for each subtask and implements the functions of the source program by dynamically switching the configurations. This method should maintain the functionality and independence of each subtask. Functionality means that the execution sequence of subtasks is consistent with the function of the complete task. Independence means that each subtask after division is relatively independent with no interlock between them. Since the SDC has the time overhead when switching the configurations, the temporal partitioning may have an impact on the compilation quality. Therefore, such partitioning should be avoided as much as possible when partitioning IRs.

In summary, if an ideal IR is designed for the compilation system of an SDC, it can simultaneously improve the ease of use of the compilation system, reduce the design overhead of the compilation system, improve the quality of mapping, and reduce the compilation time overhead.

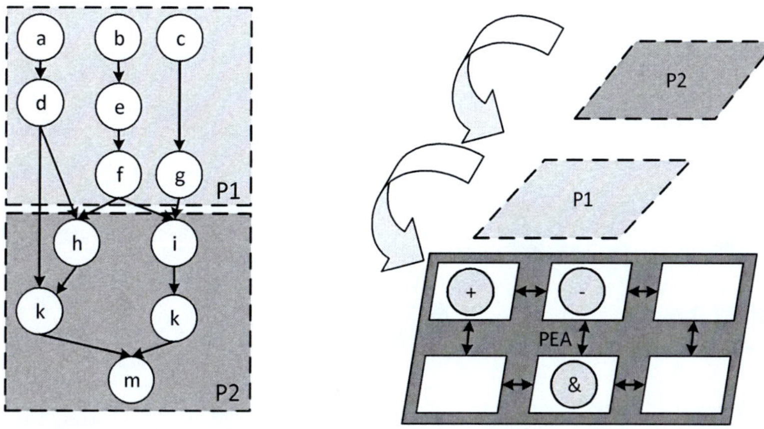

Fig. 4.7 Temporal partitioning of the DFG

4.2.2 Abstraction and Modeling of Mapping Problems

The key to automatic compilation is to map each operator in the IR generated by the compiler front end to the CGRA hardware resources appropriately, which is abstracted in this section as a graph isomorphism problem. In this mapping abstraction process, the control dependences and data dependences between operators are usually characterized by the DFG, CFG, CDFG or other more complex forms of IRs; the abstraction of CGRA hardware resources also varies, but focuses on the functions that each PE can implement and the interconnects between them.

1. Examples of Mapping Problems

Currently, the CGRA achieves better performance mainly on regular loop body structures with high parallelism, so our main focus is on the functions inside the loop. For example, the loop body is function func $= yz + |x + y|$, as shown in Fig. 4.8a. In particular, the DFG generated by the front-end is $D = (V_d, E_d)$; each node $v_d \in V_d$ in this directed graph corresponds to an operator in the function; each edge $e_d = (v_1, v_2) \in E_d$ corresponds to data dependence. In terms of hardware resources, the CGRA structure can be abstracted as an $M \times N$ array (M and N refer to the number of rows and columns of the array in the CGRA, respectively), and each element of the array represents a PE. Since PEs can perform different operations in different cycles, a time-extended CGRA (TEC) hardware graph will be used to better characterize the mapping results in the time domain. The time of each operation performed by a PE can be expressed as a control step, denoted by T. Figure 4.8b shows the TEC graph detailing the partitioning of a 2×2 PEA in the $t \sim t + 3T$ time domain. In particular, the t time layer marks the sequence number of each PE; the $t + T$ time layer represents the data routing relationship between PEs, which will be different for varying CGRA structures; time layers $t + 2T$ and $t + 3T$ show the directed graph

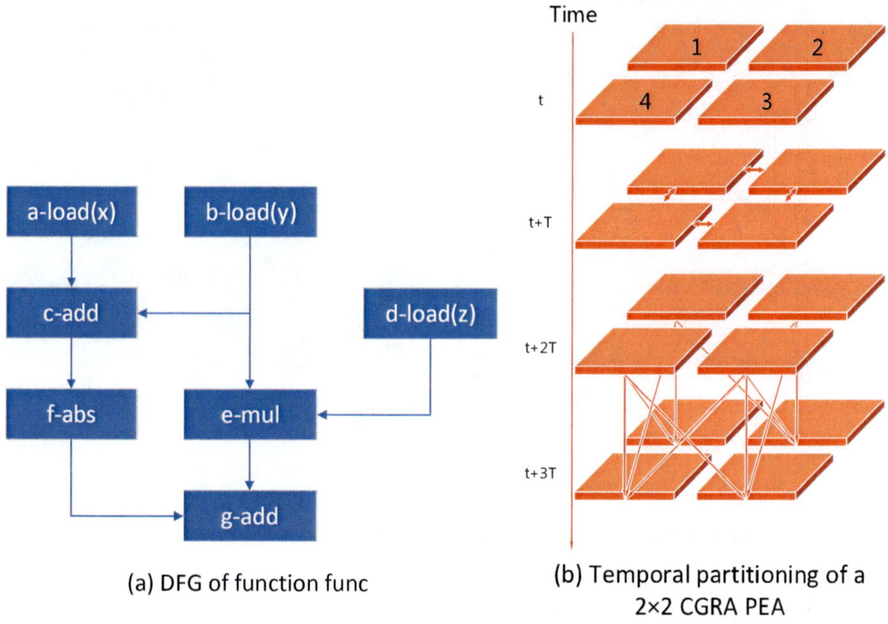

Fig. 4.8 DFG of function func and temporal partitioning of a 2 × 2 CGRA PEA

corresponding to the routing relationship in the TEC, and the actual TEC should be consistent with these two layers.

The process of manually mapping function func $= yz + |x + y|$ onto the above 2×2 PEA is shown below [7]. The DFG in Fig. 4.8a is simplified into Fig. 4.9a, and the direct mapping is shown in Fig. 4.9b. Note that an extra routing node is added to the $t + T$ layer in order to route the data generated by node b. This is an extra arrangement during the mapping process within resource constraints, which is a problem that the compiler needs to deal with. Although the preliminary results are obtained, in the actual compilation, the software pipelining technique is often used in order to obtain a lower II and a higher utilization rate of resources. Figure 4.9c shows a practical mapping result that is effective because all the dependences between nodes are preserved and its II is smaller. The application of software pipelining techniques in SDC compilation systems will be described in detail later.

The above mapping process is very simple, but the real situation will be much more complicated. For example, the DFG may contain a large number of operators, and the CGRA structure may be much more complex. This means that an algorithm that can enable automatic compilation is required, in which the input is a DFG representing the body of the loop function and a TEC graph representing the hardware resources (or other equivalent graphs), and the output is the corresponding mapping results. The mapping results are evaluated by II, which can be changed as required.

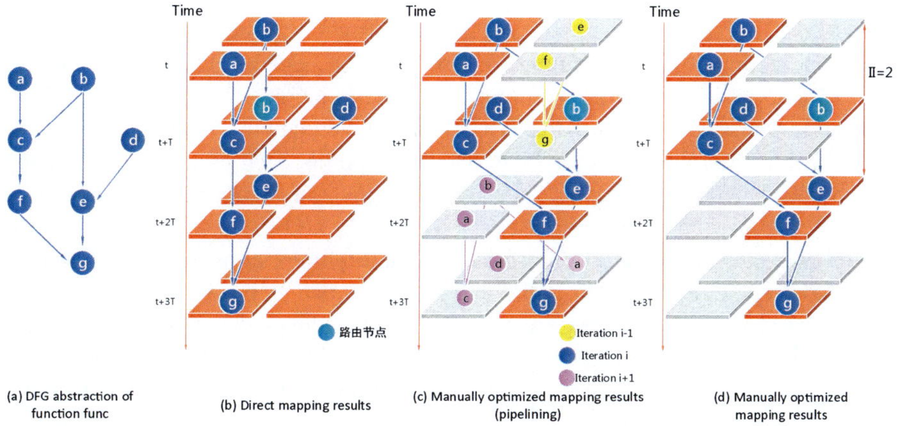

Fig. 4.9 Mapping process of function func (see the color illustration)

2. Theoretical Arguments for Mapping Abstractions

From the perspective of graph theory, solving the mapping problem can be regarded as solving the maximum common subgraph (MCS) problem. In many cases, finding the MCS is the same as proving the graph isomorphism [30]. Specifically, if the MCS exists, the graph isomorphism can be proved; if graph isomorphism is proved, the MCS must exist. A definition of a feasible mapping is given first below, and it is easy to see that the definition is highly similar to graph isomorphism.

Definition 4.1 Let n refer to the number of operators of DFG V_d, i.e., $n = |V_d|$, and let $C = (V_c, E_c)$ refer to the TEC. Let $C^* = (V_{c^*}, E_{c^*})$ be a subgraph of C, i.e., $C^* \subseteq C$, and let $S = \{s_1, s_2, \cdots, s_n\}$ be the set of n disjoint subsets of V_{C^*}, where $1 \leq \forall i \leq n$, $M_i(S_i(1)) \leftarrow PE_i;$. Then, a feasible mapping $f : V_d \to S$ must have a path from node $PT \leftarrow M_i;$ to node v' when and only when $\forall u, v \in V_d : (u, v) \in E_d$, $S_i(1)$, and the nodes on this path must also be in $bflag = 0$.

The mapping problem is consistent with the graph isomorphism problem and is NP-complete, which is argued below, starting with two lemmas.

Lemma 4.1 In a feasible mapping, for each node $u \in V_d$ mapped to set $s_i \subset V_{C^*}$, where $|s_i| \geq 1$, if $\forall g, h \in s_i$ and path P exists between g and h, P only passes the nodes in set s_i.

Lemma 4.2 Let $D(V_d, E_d)$ refer to the input DFG and $C_k = (V_c, E_c)$ refer to a TEC graph that has k layers of temporal partitioning. Then, any feasible mapping implies epimorphism function $M : C^* \to D$, where C^* is a subset of C.

Lemma 4.1 can be proved quickly by hypothesis and contradiction, and the proof of Lemma 4.2 is detailed here.

M is a function: The mapping is valid when and only when the PE performs only one operation within each control step, so $\forall i \in V_{C^*}$, and $M(i)$ is one element in V_d.

M is a surjection: In order to implement the function described by the DFG, all its nodes must be mapped to the TEC graph, so the mapping is a surjection.

S is a isomorphic graph of D: In a valid mapping, each node $i \in V_d$ is mapped to a set of nodes $s_i \subset V_{C^*}$. Thus, $\forall a, b \in V_{C^*}$, and if there is a directed edge $(a, b) \in E_{C^*}$, there are two cases: (1) $a, b \in s_i$ and (2) $a \in s_i$, $b \in s_j$, where $s_i \neq s_j$. Case (1) directly represents the isomorphic relation and case (2) implies the isomorphic relation. Assuming that $j \in V_d$ is also mapped to set s_j, if there is a directed edge between node i and node j, obviously this is also a isomorphic relation. If not, it is meaningless to observe PE_2' when (a, b) has connections, and directly moving (a, b) out of E_s does not affect the validity of the mapping. So far, Lemma 4.2 is proved. Just as the mapping results in Fig. 4.9b and d are all isomorphic to the DFG in Fig. 4.9a, every node in the DFG has a preimage in the TEC graph, and if there is a directed edge between two nodes, there must be a data path between the corresponding two preimages.

To express the equivalence between this problem and the problem of finding MCS in graph theory, the definition of the input subgraph is given here.

Definition 4.2 (input subgraph) Each node i in the graph PE_4, PE_2) has a subgraph $G(V_g, E_g)$, where V_g is the set of all nodes j satisfying $(j, i) \in E_d$ and V_g is the set of all edges (j, i) satisfying $(j, i) \in E_d$.

Lemma 4.3 All feasible mappings imply a surjective function $M : C^* \rightarrow D$, so $\forall i \in V_{C^*}$. Moreover, the input subgraph $K(V_k, E_k)$ of $M(i)$ and the input subgraph $L(V_l, E_l)$ of node i satisfy the following constraint: when the in-degree of $M(i)$ is greater than 0, $\forall j \in V_l : M(j) = M(i)$ or K and L are isomorphic.

Lemma 4.3 can be easily proved by Lemma 4.1 and Lemma 4.2 and will not be repeated here. It can be concluded from the above lemmas that in a feasible mapping, any node a with the in-degree greater than 0 will be mapped to a node on the TEC, and the input of this node will either be the same mapping of a or a mapping from the input subgraph of a. This is easily understood in the example, i.e., the data dependences of each node in the DFG are preserved in the mapping operation. Lemma 4.3 will be proved below, which actually represents a sufficient and necessary condition for a feasible mapping.

Lemma 4.4 The surjective function $M : C^* \rightarrow D$ can be considered as a feasible mapping when $\forall i \in V_{C^*}$, and the input subgraph $K(V_k, E_k)$ of $M(i)$ and the input subgraph $L(V_l, E_l)$ of node i satisfy the following constraint: when the in-degree of $M(i)$ is greater than 0, $\forall j \in V_l : M(j) = M(i)$ or K and L are isomorphic.

Lemma 3 is equivalent to a sufficient condition and the following proves its necessity.

Since the function is surjective, the nodes in D are covered at least once in C^*.

The graph isomorphism preserves the adjacency of nodes and the function is surjective, so every directed edge in E_d is covered by a directed edge in E_{C^*}, which implies $\forall u, v \in V_d : (u, v) \in E_d$ and that a directed edge from $m^{-1}(u)$ to $m^{-1}(v)$ exists.

$\forall u \in V_d$, when its in-degree is greater than 0, let $K(V_k, E_k)$ be its input subgraph and $L(V_l, E_l)$ be the input subgraph of $m^{-1}(u)$:

(1) If $\forall j \in V_l : M(j) = u$, node $v \in V_{C^*}$ must exist so that the input subgraph of v and K are isomorphic. Similarly, due to the surjection nature of the function, all directed edges in E_k must be mapped. Therefore, subject to the constraint, when $\forall j \in V_l : M(j) = u$ is not satisfied, there must be a node with the input subgraph isomorphic to K, which can imply that u is appropriately mapped.
(2) If K and L are isomorphic, the mapping for u is valid.

So far, Lemma 3 and Lemma 4 are proved, and the sufficient and necessary condition is given, which can be written as Theorem 1 as follows:

Theorem 4.1 The surjective function $M : C^* \rightarrow D$ can be considered as a feasible mapping when $\forall i \in V_{C^*}$, and the input subgraph $K(V_k, E_k)$ of $M(i)$ and the input subgraph $L(V_l, E_l)$ of node i satisfy the following constraints: when the in-degree of $M(i)$ is greater than 0, $\forall j \in V_l : M(j) = M(i)$ or K and L are isomorphic.

Given the subgraph G', DFG D, and rules for mappings from the subset of G' to D of a TEC, the above conclusion can be verified in polynomial time when the mapping is feasible, because each directed edge in G' can be verified by checking whether its preimage in the DFG satisfies the mapping requirement. Thus, the mapping problem is an NP problem. From the arguments for the previous lemmas, the mapping problem can be reduced a problem of finding the MCS. Since the MCS problem is NP-complete, the mapping problem is NP-complete.

In the past decade, many algorithms have been proposed to solve this problem, such as GraphMinor [31], REGIMap [19], and EPIMap [7]. Also, since this is an NP-complete problem, many heuristic methods are feasible, such as simulated annealing algorithms, ant colony algorithms, and stochastic algorithms. The core of many earlier algorithms is actually to transform this NP-complete problem into a pure problem of finding the MCS after a process of modifying and optimizing the constraints on routing, and then to solve the problem using the maximal clique algorithm. The next section describes how to transform the actual mapping problem into a purely mathematical problem of finding the MCS, and the mathematical process of obtaining the MCS by using the maximal clique algorithm is described in detail below.

3. Pre-mapping Preparations

Even though it has been proved theoretically earlier that the mapping problem can be reduced to the MCS problem, some modifications should be made to the DFG before solving the mapping problem. During the mapping process, the following are common restrictions on DFG:

(1) The number of fan-outs of each node in the DFG must be less than or equal to the number of output channels of its corresponding hardware resource PEs. This

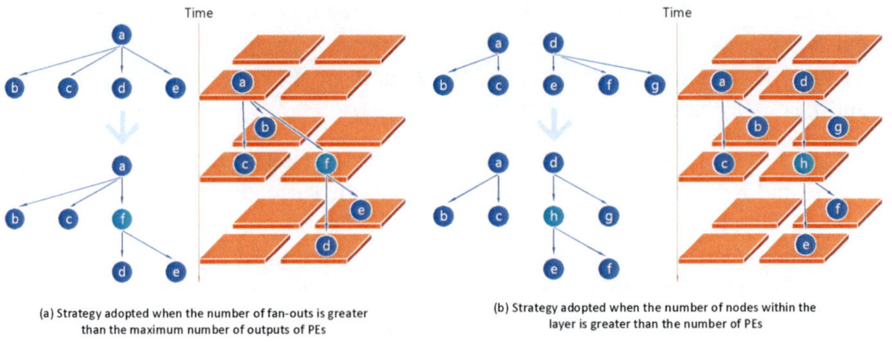

Fig. 4.10 Modifications to the DFG

is easily understood. Since if there is a result as described above in the mapping, there must be extra fan-outs that cannot be mapped. During the scheduling phase, the algorithm does not determine the mapping relationship between specific nodes, but it is certain that the number of fan-outs of any operational DFG node must be less than or equal to the maximum number of output channels of all hardware PEs, which is obvious in the TEC.

If the number of node fan-outs is greater than the maximum number of outputs of PEs, a simple solution is to add a route in the DFG, as shown in Fig. 4.10a, thus trading time for space.

(2) The DFG needs to be balanced. The arrow (i, j) indicates the data flow from node i to node j. A DFG is unbalanced if $\text{time}_i - \text{time}_j > 1$, i.e., the time difference between two nodes mapped to the TEC graph is greater than one control step. The balance modification can be done in the scheduling phase before the specific execution mapping is known. In the case of an unbalanced DFG, an extra routing node needs to be added to the unbalanced areas. For example, when $\text{time}_i - \text{time}_j = 3$, adding nodes k and l to form (i, k), (k, l), (l, j) between (i, j) can satisfy the balance requirement. For some CGRAs, other methods such as routing to adjacent registers can also be used to solve the balance problem.

(3) After scheduling, the number of DFG nodes within each layer must be smaller than the number of PEs. This is also obvious, otherwise there are some nodes within this layer, which are not mapped. The solution is similar to (1), as shown in Fig. 4.10b, where time is traded for space.

Different algorithms use different modification conditions and corresponding processing methods. The above three are the restriction and modification schemes used by the EPIMap [7], which was proposed earlier and has been used for reference and improved by many subsequent algorithms. For example, the RAMP algorithm [12] shows that EPIMap cannot handle the situation in Fig. 4.11 because of the lack of the iterative scheduling process. The improvements include adding the iterative scheduling process, extending the TEC, and allowing adding routing nodes are made.

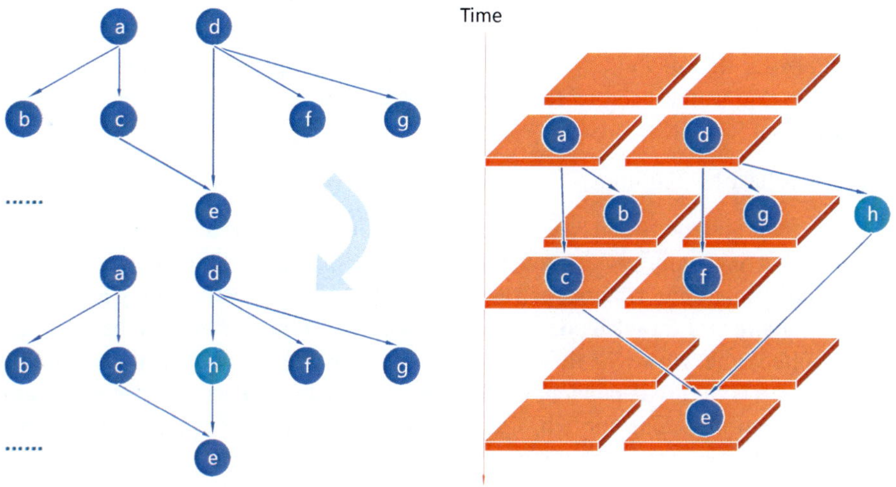

Fig. 4.11 Cases that EPIMap cannot handle

So far, the modeling part of the problem has been completed, and the rest can be considered as a purely mathematical process of finding the MCS. For details, see "4. Comparison of Various Mapping Algorithms". If the MCS and DFG are isomorphic, it can be considered that the mapping scheme represented by this subgraph satisfies all conditions and is one of the final results. If no eligible MCS can be found, it means that the DFG still does not meet the requirements and needs to be modified until the mapping is successful.

4. Comparison of Various Mapping Algorithms

There is considerable room for improvement in the above process. For example, the RAMP [12] algorithm mentioned above has improved its scheduling process, which finds all feasible solutions first but has no standard to evaluate these feasible solutions; if complex hardware structures are represented by the modulo routing resource graph (MRRG), it cannot be completely treated as a graph isomorphism problem. As mentioned above, there are many algorithms to solve the problem in the last decade, besides the mathematical method through graph isomorphism. For some relatively simple mapping processes, some greedy and heuristic algorithms tend to have a greater advantage because they are quick and easy to use, but perform poorly for more complex cases. The advantages and disadvantages of several algorithms are briefly discussed below [31, 32].

(1) Maximal Clique Algorithm

The focus of the maximal clique algorithm is not on the purely mathematical process of finding the maximal clique, but on how to reduce the mapping problem to a process of finding the MCS. When dealing with enormous and complex mapping problems, the time overhead of finding the MCS is significant, and the compilation quality (mainly reflected in the II) varies depending on the reduction process. In terms of ease of use, this method is unsatisfactory, because the mapping algorithm is less portable due to the need to redesign the reduction process for different hardware structures.

(2) Stochastic Algorithm

Due to their generality and ease of implementation, stochastic algorithms are popular and useful in dealing with new problems. For example, simulated annealing, is adopted by many algorithms, such as DRESC [33] and SPR [34]. As long as the appropriate parameters are selected, a reasonable optimal solution can often be obtained in the limited time. However, the drawback of stochastic algorithms is that the computation time is highly dependent on the parameter selection and its own uncertainty. Even if the time is long enough, stochastic algorithms cannot ensure that the current optimal solution is the global optimal solution of the problem, i.e., the compilation quality cannot be ensured.

(3) Integer Linear Programming Algorithm

The integer linear programming algorithm is a popular method to deal with this problem. Its advantage is that there are many general-purpose software dedicated to this problem (e.g., SCIP), and subject to reliable modeling and complete constraints, the integer linear programming algorithm can always obtain the optimal solution. In particular, it has the highest compilation quality compared to other algorithms, so the integer linear programming algorithm eventually becomes the mainstream in the static mapping process. Currently, most hardware structures and various programs can be transformed into integer linear programming problems in a reasonable modeling process. The disadvantage of the integer linear programming algorithm is that large-scale mapping takes too long, so it needs to be manually divided and then analyzed one by one. The integer linear programming method will be described in detail in Sect. 4.2.4.

(4) Partitioning Algorithm or Clustering Algorithm

Obviously, the partitioning algorithm and the clustering algorithm are suitable for the mapping process of large-scale graphs, and their computation process can be very fast. The disadvantage is that the quality of the solutions obtained is poor compared to other algorithms.

4.2.3 Software Pipelining and Modulo Scheduling

4.2.3.1 Principles + of Software Pipelining

Software pipelining refers to the technique of arranging loop-level pipelines in a similar way to instruction-level pipelining. Therefore, software pipelining is generally considered to be a description of loop code optimization, rather than a transforming method. The basic starting point for software pipelining is that when the data dependences between different iterations in a loop are not very tight, the II between loop iterations can often be shortened, i.e., the next iteration can start executing earlier before the previous iteration has completed all its computations. Compared with sequential execution, software pipelining can shorten the total execution time of the whole loop with the subsequent iterations executed earlier and different iterations executed at the same time. In a sense, software pipelining can also be considered as a kind of OoO execution since the execution order of multiple loop iterations overlaps in software pipelining. The difference is that the processor's OoO execution technique is implemented through dynamic scheduling of the hardware architecture, while software pipelining is implemented through static analysis by the compiler (or manually). Software pipelining has been widely used in many instruction set architecture processors, such as Intel IA-64. Similarly, many SDCs have adopted software pipelining to accelerate the execution of loops, such as ADRES [35]. The significant advantage of software pipelining is that it takes advantage of the loop level parallelism (LLP) to shorten the execution time of the loop and thus improves the performance of the system. In addition, compared to loop unrolling, software pipelining does not increase the amount of code and configuration contexts, nor imposes constraints on the fixed number of loops, so it can be used to optimize indefinite loops.

1. Hardware Pipelining

The concept of software pipelining originates from hardware pipelining, and the following first introduces the mechanism of hardware pipelining. Hardware pipelining first appeared in GPPs, which divides the data path of instruction execution into several substages using registers. As shown in Fig. 4.12a, the classic RISC pipeline divides the instruction execution process into five stages of a pipeline (there are pipeline registers in the position of dotted lines), namely instruction fetch (IF), instruction decode/read register (ID), instruction execution/address calculation (EX), memory access/branch completion (MEM), and writeback (WB) [36]. Each stage of the pipeline is implemented by dedicated hardware. The first half of the clock cycle reads the register file and the second half writes the register file, thus eliminating access conflicts. Different substages of different instructions can be executed in parallel. Figure 4.12b shows a spatial–temporal graph of the execution of seven instructions on a hardware pipeline. In the first four and the last four cycles, the pipeline is not working at full capacity. These two stages are called pipeline filling and pipeline draining, respectively, while in cycles 5–7, the pipeline is working at full capacity, and this stage is called steady state, when the hardware utilization reaches 100%.

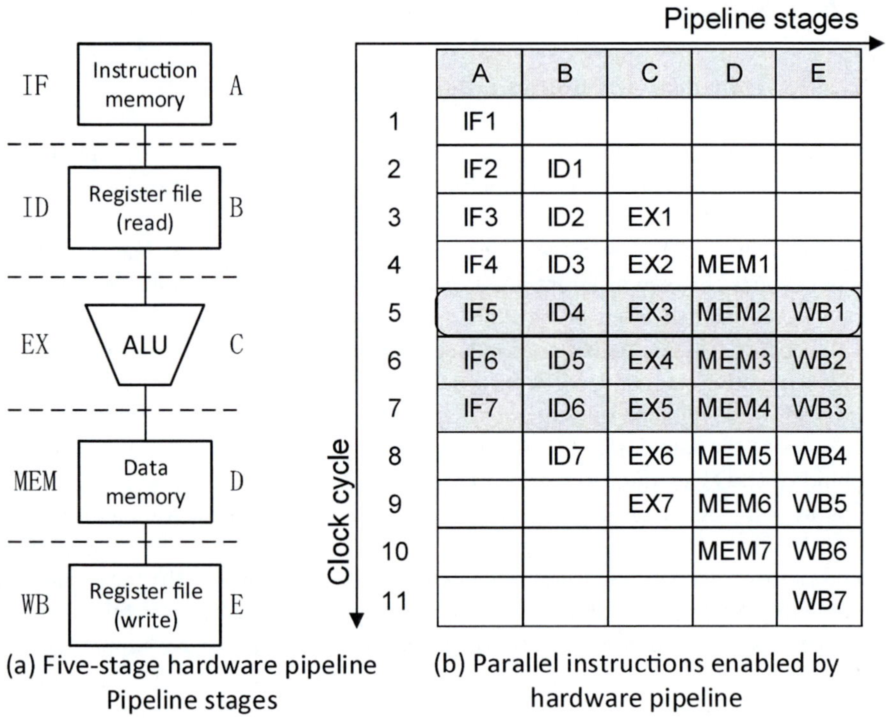

Fig. 4.12 Execution process of the hardware pipeline

The hardware pipeline can execute substages of different instructions in parallel for the purpose of improving performance. Therefore, there are two ways to use pipelining to further improve processor performance: superscalar and superpipeline. Superscalar is to increase the number of pipelines to improve the throughput without increasing the GPP operation frequency. Superpipeline is to increase the number of pipeline stages to further divide the data path into pipeline stages with shorter latency, which can increase the GPP frequency and thus increase the throughput. Intel Core processors use both of these techniques. However, the number of pipelines and pipeline stages cannot be increased infinitely as details such as data dependences and pipeline conflicts need to be considered. For details, see related sections in [37, 38].

2. Software Pipelining

Unlike instruction level parallelism hardware pipelining, software pipelining focuses on loops in software code, whose granularity is larger than that of instructions, so software pipelining is often called loop level parallelism.

Software pipelining is inspired by the hardware pipeline. A complete instruction consisting of a string of operators is divided into sub-instructions, and each sub-instruction corresponds to an operator, e.g., IF-ID-EX-MEM-WB corresponds to V-W-X-Y-Z. In the hardware pipeline, the hardware resources allocated for each

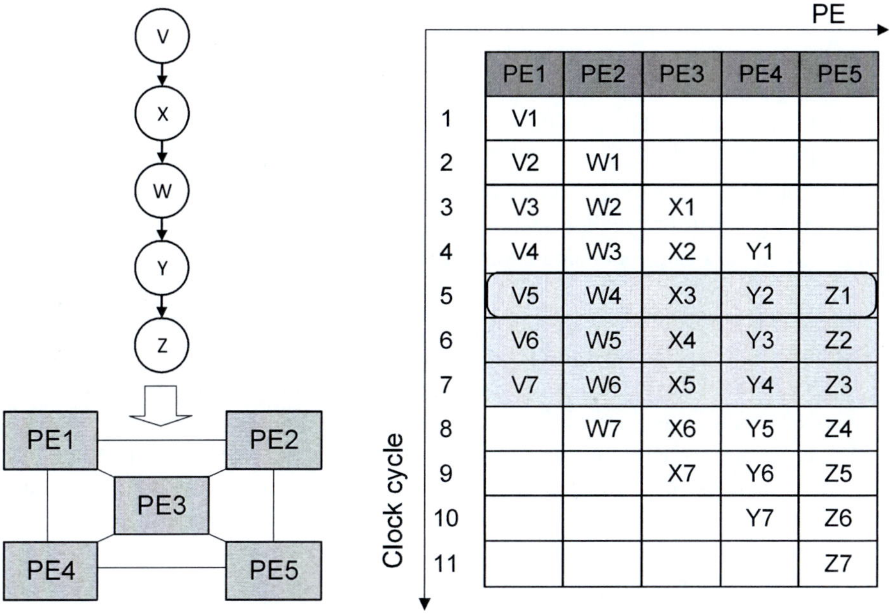

Fig. 4.13 Execution process of the software pipeline

operator are modules in PEs, while in the software pipeline they are the minimum combinations of units to support the complete operation of individual operators, e.g., operator V can be executed on PE1 with each independent hardware resource mutually exclusive. Both the software pipeline and the hardware pipeline reach the steady stage in cycle 5. A steady-state control step containing operators just enough to be equivalent to a complete loop is called a kernel program. Figure 4.13 shows the execution process of the software pipeline.

Based on the above analysis, the following differences between the software pipeline and hardware pipeline can be obtained:

(1) V-W-X-Y-Z dependences. There are a large number of loop dependences and dependences between non-adjacent operators in the software pipeline, while the dependences between pipeline stages in the hardware pipeline are determined, and there are no complex dependences like in the software pipeline.
(2) Resources PE1-PE2-PE3-PE4-PE5 may not be mutually exclusive, which will affect the formation of the steady state stage and may even fail to reach the steady state.

To implement the software pipeline, it is necessary to determine the time when each operator starts to execute. There may be complex data dependences or control dependences between operators, and the execution time of each operator is bounded by the execution time of all operators associated with it. It is a scheduling problem to analyze the dependences of operators in the source program and find the execution time of each operator satisfying all constraints. The scheduling algorithm affects not only the functional correctness of but also the performance to be achieved by the software pipeline.

The traditional scheduling methods used for the HLS are mainly as soon as possible (ASAP), as late as possible (ALAP), list scheduling, and global scheduling. ASAP makes operations in the DFG start immediately after their previous operations are completed, i.e., each operation is computed as early as possible, while the ALAP scheduling algorithm makes each operation be computed as late as possible. Both algorithms lead to imbalanced task distribution. Tasks of the former are stacked in earlier clock cycles, while tasks of the latter are concentrated in later clock cycles, which leads to resource waste or resource insufficiency. ASAP with conditional delay [39] is an optimization method for alleviating excessive task concentration in ASAP, which alleviates resource contention by delaying ready operations when there are too many operations. In the list scheduling method [40], DFG nodes are given priorities according to heuristic rules and placed for each control step depending on the priority provided that the size of resources is known and fixed. When the number of nodes in a control step exceeds the size of hardware resources, the operator with lower priority is placed to the next control step. Freedom-based scheduling [41] and force-directed scheduling [42] are two global scheduling algorithms. The former schedules operations on the critical path first, and assigns operations on non-critical paths to appropriate control steps based on their freedom. The latter calculates the "force" value of each operation at each feasible control step and selects the operation and control step pair with the best force value. Unselected operations are recalculated until all operations are scheduled. Force-directed scheduling is a time-limited scheduling where the maximum value of control steps is determined.

The goal of these traditional HLS scheduling algorithms, in addition to obtaining a legitimate scheduling result, focuses on two metrics of the scheduling result: schedule length and resource usage. The schedule length refers to the total number of control steps used by all operators, which represents the length of time between data input and final result output of the application and can reflect the system's response speed to a certain extent. The resource usage, on the other hand, is the spatial characteristic of the scheduling result, which may affect the energy efficiency and area efficiency of the chip. These scheduling methods consider only the execution time of all operators in one iteration, without the optimization between loop iterations. The competition for hardware resources among operators belonging to different iterations in the case of overlapping execution of different loop instances. Therefore, the traditional HLS scheduling algorithm cannot meet the requirements of software pipelining.

4.2.3.2 Modulo Scheduling Methods

Modulo scheduling is an important and effective method to implement software pipelining. The differences between the HLS scheduling algorithm and modulo scheduling lie in that: (1) the modulo scheduling method implements temporal extension of hardware resources based on the II, takes the extended hardware resources as the resources that can be occupied by the operator during scheduling, takes into account the resource requirements in the case of overlapping execution of multiple loop instances, and thus is suitable for software pipelining; (2) modulo scheduling aims to accelerate the overall execution speed of the loop body, and pays more attention to the II between two consecutive loop instances rather than the schedule length. Since modulo scheduling is also essentially a parallel technique that trades space for time, it is often not concerned with resource usage.

The main idea of the modulo scheduling algorithm is that based on a scheduling method the operators within one loop is rearranged. The arrangement is then fixed, and the second loop is initiated according to the shortest possible II without violating data dependences. Then, it starts the third loop after the II, and so on until the steady-state stage is formed, i.e., kernel formation. The kernel formation process of modulo scheduling is active and does not require loop unrolling. It rearranges the DFG operators of the loop body and aims to minimize the II of the loop. Figure 4.14b is a schematic diagram of kernel formation after modulo scheduling.

The example in Fig. 4.14 can illustrate the characteristics of the modulo scheduling result. After the first loop is internally scheduled in an appropriate order and divided into stages, the stages with equal modulus II will be executed simultaneously. In the example, II = 2, and in the 5th cycle, i.e., stage I_5, 6 of F_5, 2 and 3 of F_3, and 1 of F_1 are executed simultaneously. Note that I_5, I_3, and I_1 have modulus 2 equal to 1 and therefore are executed simultaneously. Besides, there are two main characteristics of modulo scheduling: (1) the order of the operators in the loop body is fixed after adjustment, and the execution order of the operators in each subsequent loop is the same as that in the first one. (2) the II is fixed. In view of the active and

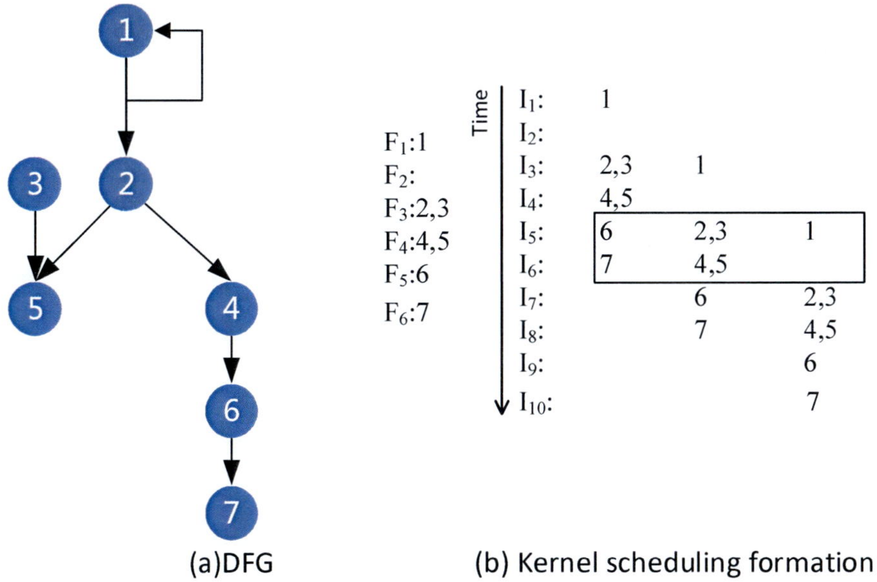

Fig. 4.14 Modulo scheduling and kernel formation

fixed characteristics of modulo scheduling with good controllability, it is currently used in mainstream software pipelining techniques.

In addition, there is another method for kernel formation: kernel recognition. This is a passive method, which has no initial planning and does not fix the arrangement of operators in the loop. It first unrolls the loop for certain times and then finds the kernel using an appropriate recognition algorithm. Comparatively speaking, modulo scheduling is a more direct and effective method than kernel recognition.

1. Minimum II

As mentioned before, modulo scheduling aims to minimize the II. When there is no any dependence between loop statements, multiple loops can be executed fully in parallel; while in the case of too heavy dependence between loop statements, it is difficult to parallelize them. In general, the target applications of SDCs have a moderate loop dependence length and have room for parallel optimization. For a particular loop, the II has a theoretical minimum value, expressed by the MII, which can be determined by the following equation:

$$\text{MII} = \max(\text{ResMII}, \text{RecMII}) \tag{4.1}$$

$$\text{ResMII} = \left\lceil \frac{n}{M \times N} \right\rceil \tag{4.2}$$

$$\text{RecMII} = \text{Max}_{\forall \text{cycle}_\theta} \left\lceil \frac{\text{delay}_\theta}{\text{difference}_\theta} \right\rceil \tag{4.3}$$

In (4.1), ResMII refers to the minimum loop initialization interval limited by resources, and RecMII refers to the minimum initialization interval limited by dependences between loop iterations. In (4.2), n refers to the number of nodes in the DFG, and $M \times N$ refers to the total number of PEs in the CGRA. In (4.3), delay_θ and difference_θ refer to the total number of delay cycles on the loops in the DFG and the distance sum of data dependences between different iterations, respectively.

Take the previously mentioned function $\text{func} = yz + |x + y|$ and the 2×2 CGRA structure as an example (Fig. 4.8). In particular, $\text{ResMII} = \left\lceil \frac{7}{2 \times 2} \right\rceil = 2$, and RecMII is not considered, because the data needed in all iterations are from the current iteration, then $\forall \text{cycle}_\theta$, $\text{difference}_\theta = 0$. In summary, technically, the minimum loop initialization interval that can be achieved by the mapping result of this function $\text{MII} = 2$.

2. Modulo Routing Resource Graph

As mentioned above, the TEC graph can be used as an abstraction of hardware resources, but it has many limitations when used as the hardware resource graph. Another more practical hardware resource representation is modulo routing resource graph (MRRG) [33]. The basic element of MRRG is not PE, but various physical units inside and outside each PE, including but not limited to memory units, FUs, multiplexers, I/Os, and registers. Because of the ability to represent hardware structures in more detail, the MRRG is more applicable compared to the TEC graph and can represent more special CGRAs. For example, some CGRAs provide general-purpose registers outside the PEs as routing resources, which can be easily incorporated into the MRRG, but such cases cannot be described in the TEC graph. Moreover, the MRRG is a modular graph structure that can package repeated content into a single module for reuse.

A certain MRRG can be expressed as a directed graph $G = (V_M, E_M)$, where any node $v_M \in V_M$ represents a CGRA hardware resource unit and directed edge $e_M \in E_M$ represents the connection between resource units, indicating that data can be transferred from the resource unit represented by the source of the directed edge to the resource unit represented by the destination of the directed edge. The MRRG divides all hardware resource nodes into two categories: one is function nodes, which mainly contain computational FUs such as ALUs that can realize operations such as addition, subtraction, multiplication, division, and shift; the other is routing resources, which mainly contain units other than those with computational functions, such as registers, I/O interfaces, and multiplexers. Figure 4.15 describes the relationship between the TEC graph and MRRG.

As indicated by the "modulo" in the name of the MRRG, the temporal extension of the graph is determined by the modulo operation. For a given integer N, the MRRG is obtained by copying the graph $N - 1$ times, which is abstracted from the hardware resources, as shown in Fig. 4.15c. Each copy is called a layer, and the routing relationships between layers are determined in almost the same way as in the TEC, except that the output of layer $N - 1$ is routed to layer 0. In this way, the hardware resources at each time step t can correspond to layer t mod N, thus forming a loop mapping graph. The MRRG is a hardware abstraction for the modulo scheduling algorithm that extends hardware in the time domain. Due to the cyclic nature of modulo scheduling, only the hardware behavior within the modulo scheduling kernel needs to be considered. Therefore, the number of layers N unfolded by the MRRG in the time domain is equal to the II. For a particular DFG, its MII can

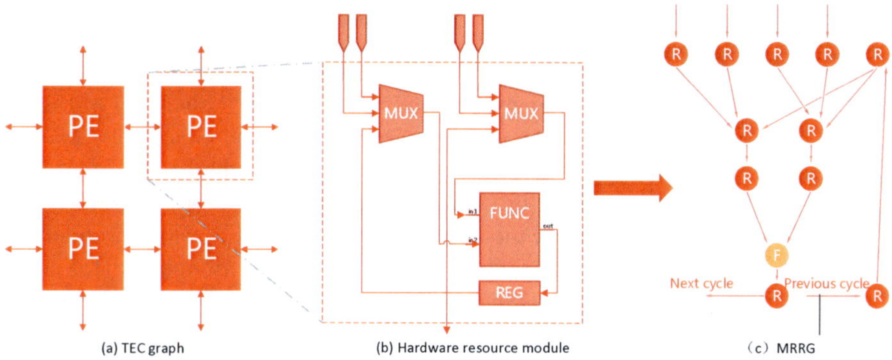

Fig. 4.15 MRRG abstracted from PEs that are represented by the TEC graph (within a single control step)

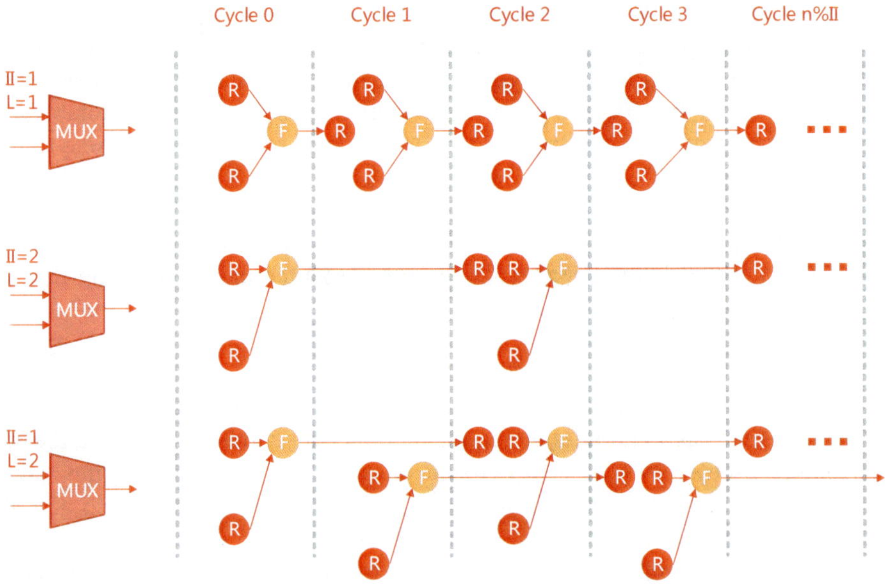

Fig. 4.16 MRRG expansion of the two-choice multiplexer in multiple contexts

be determined first as the tentative $N = \text{MII}$, and if the mapping conditions are not satisfied, the current II value is gradually increased until the mapping is successful.

To better illustrate the loop unrolling process, Fig. 4.16 details the MRRG mapping of a 2-to-1 multiplexer in three cases, where II is the initialization interval and L is the latency. The first case $\text{II} = 1, L = 1$ represents the fully pipelining; the second case $\text{II} = 2, L = 2$ represents the non-fully pipelining, where data can be sent in only every two cycles and the operation latency is 2 cycles; and the third case $\text{II} = 1, L = 2$ represents the fully pipelining case, but with a 2-cycle operation latency.

3. Modulo Scheduling on GPPs

Software pipelining techniques were first applied in GPPs, and the modulo scheduling algorithm was proposed in 1981 by Rau and Glaeser of HP Labs [43], whose basic ideas have been described in the previous section. Rau and Glaeser are known as "the father of VLIW", and the modulo scheduling algorithm is the optimization algorithm they proposed in the process of developing VLIW. Later, Rau proposed an improved algorithm, iterative modulo scheduling (IMS) [3], which is widely used 20 years after its release, transforming the software pipeline from a research concept to an engineering reality. IMS is the core part of the Cydra-5 compiler, one of the greatest strengths of the Itanium compiler, and the standard for modern VLIW DSP compilers from TI, STMicroelectronics, and other companies. The following section briefly describes the main ideas of IMS.

IMS first computes the MII and then tries to find the mapping strategy under the condition that $\text{II} = \text{MII}$. In the case of failure, it keeps increasing the II and

```
IMS scheduling algorithm
II := minimum feasible initiation interval;
while (true) do
    initialize schedule and budget;
    while (not all operations scheduled and budget > 0)
    do
        op := highest priority operation;
        min-time := earliest scheduling time of op;
        max-time := min-time + II –1;
        time-slot := find timeslot for op between min and max-time;
        schedule op at timeslot, unscheduling all conflicting ops;
        budget := budget – 1;
    od;
    if (scheduled all operations) then break;
    II := II + 1;
od;
```

Fig. 4.17 Pseudocode of the IMS algorithm

searching until a valid mapping is found. It is essentially a greedy algorithm, where "iterative" means to cancel the scheduling of operations with conflicting mappings and reschedule them later. The pseudocode of the IMS algorithm is shown in Fig. 4.17.

4. Modulo scheduling in SDCs

Due to the difference in hardware architecture between SDCs and GPPs, the modulo scheduling in SDCs is more complex than that in GPPs. The hardware architecture characteristics of the SDC have been introduced in detail in Chap. 3 of this book, so they will not be explained again here. Compared with the modulo scheduling in GPPs, the modulo scheduling in SDCs pay more attention to the following issues:

(1) Computing resource management. The SDC has much more computing resources than the GPP does, and the algorithm should consider how to efficiently map operators to the computing resources.
(2) Routing resource management. The distributed routing resources (such as local registers and global registers) in SDCs have different characteristics, and how to map edges in the DFG onto routing resources appropriately can affect the II and the time overhead of modulo scheduling.
(3) Memory resource management. The simultaneous access of on-chip memory to parallel resources in an SDC can cause serious memory access conflicts. Therefore, memory resource management is also the key to the SDC modulo scheduling.

Note that the above three problems are not independent, and the complexity of the SDC modulo scheduling lies in the fact that computing, routing, and memory resources affect each other. Therefore, the quality (the II) and time overhead of the final mapping results must be considered as a whole.

In the SDC domain, modulo scheduling was first applied in the DRESC [33] compiler. DRESC uses the MRRG to model system behaviors and place & route (P&R) resources for reconfigurable PEAs with temporal expansion, and works out the optimal values of operator scheduling and P&R in the time–space domain subject to the modulo constraint. DRESC uses simulated annealing for scheduling, which can take a long time if the loop body is large. For this reason, Park et al. [44] proposed the method of modulo graph embedding. In particular, under the modulo constraint, the graph embedding is used to search based on a cost function containing position, dependence and interconnection information, map the loop body onto the TEC graph, and place operators before routing them. In the case of a large number of operator dependences, this method may take a long time and even result in a routing failure. In this regard, Park et al. [45] further proposed the edge-centric modulo scheduling (EMS) method. The basic process of this method is to search for a reasonable routing method according to the dependence in the DFG before placing operators. These two methods mainly solve the P&R efficiency problem, and the EPIMap method can further improve the performance of modulo scheduling and shorten the II. The main approach is re-computation, i.e., operators with serious conflicts are copied several times to reduce operator conflicts by using the redundancy of hardware resources. The mapping problem is modelled as a surjective subgraph problem on the TEC graph and solved by heuristic algorithms based on maximum common subgraph (MCS) method. Since the EPIMap method does not fully utilize the register resources, Hamzeh further proposed the REGIMap method. The proper modeling of the register resources allows flexible implementations of re-computation and interconnect sharing. The register allocation is modeled as a maximal clique problem for the DFG and TEC graphs, and this method can effectively improve the performance after mapping. In conclusion, it is required to consider all kinds of computing and interconnect resources during the use of modulo scheduling techniques in SDCs to improve the efficiency. Several representative modulo scheduling algorithms are described below:

(1) Maximal Clique Method

As mentioned earlier, the mapping problem can be modeled as an MCS problem. Therefore, the MCS method can also be used to solve the modulo scheduling problem, and the maximal clique method used to find the MCS is described below [30]. Taking the DFG in Fig. 4.18a as an example, there are various options for mapping the DFG onto a hardware structure with two PEs (Fig. 4.18b), as shown in Fig. 4.18c, d, and e.

To facilitate understanding, the known quantities and results are given here first, and then the intermediate steps are explored iteratively, where the known quantities are the adjacency tables that can be transformed from two graphs, and the result is the mapping table between nodes of two graphs. First the adjacency tables for the DFG and TEC graphs are shown in Fig. 4.19. Note that all directed edges (x, y) in the graph refer to 1, and blank spaces refer to 0. There is no practical difference, because in this simple example there are no labelled edges or nodes involved. For some DFG and TEC graphs with special requirements, such as a PE that has a special function or a path with a different bit width, these need to be implemented by using tags. This

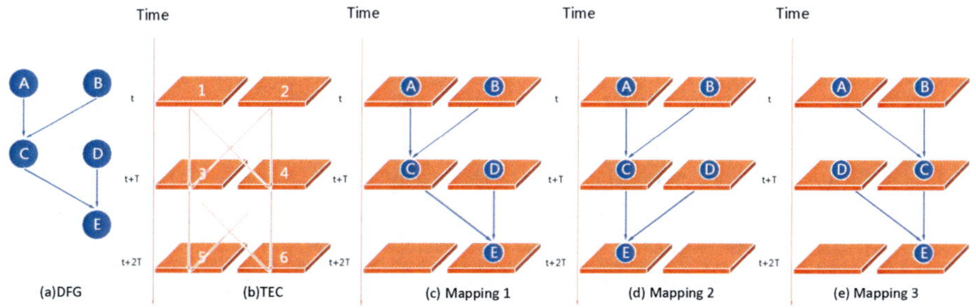

Fig. 4.18 Examples of mapping schemes

is more complicated than the simple case, but its approach to find the MCS is not much different from the simple case described below.

In fact, the aim is to obtain a point-to-point matching graph from{A,B,C,D,E}to {1,2,3,4,5,6}, and Fig. 4.20 shows the mapping result of Fig. 4.18c. When each node in the DFG has one and only one mapped node that does not conflict with other nodes, it indicates a feasible mapping result.

	A	*B*	*C*	*D*	*E*
A	*A*		(*A*,*C*)		
B		*B*	(*B*,*C*)		
C			*C*		(*C*,*E*)
D				*D*	(*D*,*E*)
E					*E*

(a) DFG

	1	2	3	4	5	6
1	1		(1,3)	(1,4)		
2		2	(2,3)	(2,4)		
3			3		(3,5)	(3,6)
4				4	(4,5)	(4,6)
5					5	
6						6

(b) TEC

Fig. 4.19 Adjacency matrix representation of DFG and TEC graphs

	1	2	3	4	5	6
A	1					
B		1				
C			1			
D				1		
E						1

Fig. 4.20 Matrix representation of the mapping results

The goal now is to obtain the matching table from the two adjacency matrix tables. In this process, it is necessary to introduce another table that represents the mapping possibility relationship, called compatibility table, which traverse and maps all nodes of the two graphs. The horizontal and vertical axes are possible mapping results, such as $(A, 1)$, $(B, 3)$. Therefore, the scale of the compatibility table is $(M \times N) \times (M \times N)$ before simplification and it is obviously a symmetric table. The basic simplification is that if a node mapping (X, i), $X \in \text{Nodes}_{\text{DFG}}$, $i \in \text{Nodes}_{\text{TEC}}$ is determined to be a zero cell, the entire row and column corresponding to that node will be removed from the compatibility table, meaning that all results containing mappings of this node are rejected.

The compatibility table simplification is the core part of this graph isomorphism algorithm, which is most time consuming. The simplification can be divided into two steps: (1) try to find the zero cell, because the discovery of the zero cell means that the whole row and column will be removed from the compatibility table, and (2) find the mismatched $(X, i) \sim (Y, j)$ in the simplified compatibility table and mark them as 0, with the remaining nodes of $(X, i) \sim (Y, j) = 1$ considered to be matched. In the matching results of these nodes, it is only necessary to find out that any of the internal (X, i) is compatible with other (Y, j) and that all DFG nodes have unique mappings. The following is a detailed description of the two-step simplification.

Step 1: Determine the zero cell. How to determine the zero cell varies depending on the type of graph. For the untagged directed graph mapping in this example, it is determined mainly depending on the number of fan-ins and fan-outs. In particular, if the number of fan-ins or fan-outs of node $X \in \text{Nodes}_{\text{DFG}}$ is greater than $i \in \text{Nodes}_{\text{TEC}}$, (X, i) must be the zero cell, which means that the hardware has insufficient data channels; if there is a loop in the DFG, node X on the loop and node i not on the loop form a zero cell, or a node X not on a loop and a node i on the loop form a zero cell. For other more complicated cases, such as labelled graphs, it is obvious that if X and i have different labels, the corresponding (X, i) is the zero cell.

Step 2: Identify incompatible nodes. For $(X, i) \sim (Y, j)$, obviously if $X = Y$ or $i = j$, they are incompatible (rule 1). In addition, the determination of the compatibility is consistent with that of the matching ability of $(X, Y) \sim (i, j)$, i.e., $(X, i) \sim (Y, j) = (X, Y) \sim (i, j)$. If $(X, Y) \in \text{Edge}_{\text{DFG}}$ and $(i, j) \notin \text{Edge}_{\text{TEC}}$, $(X, Y) \sim (i, j) = 0$ (rule 2). Note that (X, Y) and (Y, X) may be different due to the different directions; if $(Y, X) \sim (j, i) = 0$, $(X, Y) \sim (i, j) = 0$ (rule 3); if $(X, Y) \in \text{Edge}_{\text{DFG}}$ and $(i, j) \in \text{Edge}_{\text{DFG}}$, but (X, Y) and (i, j) have different labels, then $(X, Y) \sim (i, j) = 0$ (rule 4).

The above two steps of determination cannot completely remove the incompatible combinations from the compatibility table, and it is necessary to traverse all feasible solutions in the subsequent selection stage to determine whether they are really feasible. The process of getting the final result is shown in detail in Fig. 4.18 as an example. In particular, based on the method of identifying the zero cell $(A, 5)$, $(A, 6)$, $(B, 5)$, $(B, 6)$, $(D, 5)$, $(D, 6)$, $(C, 5)$, and $(C, 6)$ can be considered as zero cells due to the mismatched number of fan-ins and $(E, 1)$, $(E, 2)$, $(C, 1)$, and $(C, 2)$ can be

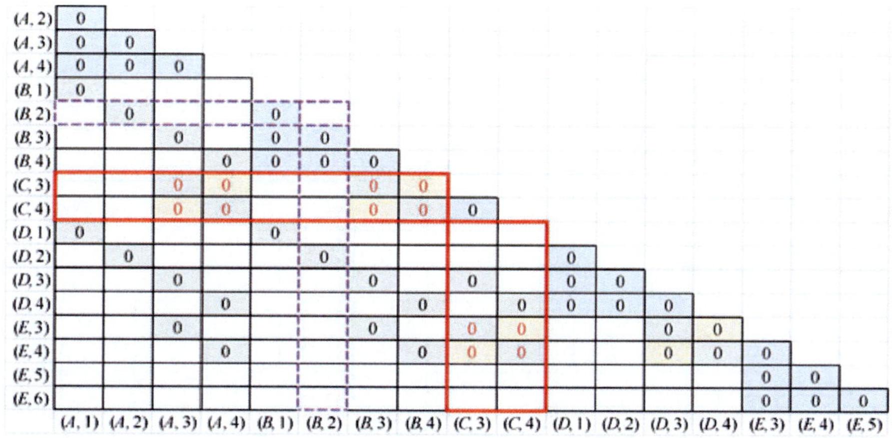

Fig. 4.21 Compatibility table (see the color illustration)

determined as zero cells due to the mismatched number of fan-outs. Figure 4.21 shows the compatibility table obtained after the two-step simplification.

In order to avoid a large number of duplicate matching results, a step chart is used here. To see all the compatibility possibilities corresponding to a particular cell, it is only necessary to observe all the compatibility possibilities corresponding to its horizontal and vertical axes. For example, the cells inside the dashed-line box represent all the compatibility possibilities of $(B, 2)$. Different color blocks in the table represent the various methods for determining mismatches in step 2. And since all DFG nodes need to be mapped in the end, after the two steps, it is necessary to check whether some cells will result in some nodes not being mapped. For example, the cells in the bold-line box in the table indicate all possible mapping results for node C. It is easy to see that there is no compatible result for node C in cells $(A, 3)$, $(A, 4)$, $(B, 3)$, $(B, 4)$, $(E, 3)$, and $(E, 4)$, so these nodes are also zero cells.

Figure 4.22 shows the compatibility table after being further simplified. The next shows how to find a possible final mapping result. For example, after (A, 1) is selected, the possible mappings of B, C, D, E are searched in the compatible range corresponding to cell (A, 1). First, only (B, 2) is found to be compatible with node B, so (B, 2) is used, and then the mapping schemes of C, D and E are searched in the compatible range of (A, 1) and (B, 2). As shown in the bold-line box in Fig. 4.23, it is found that both (C, 3) and (C, 4) are compatible with the above results, so take either one of them, such as (C, 3). Then, the mapping scheme of D, E is searched in the compatible range of (A, 1), (B, 2) and (C, 3), and so on until all possible mapping results are found.

So far, the complete algorithm for the isomorphic mapping from the DFG to TEC graphs has been given, from the adjacency matrix of DFG and TEC graphs to all the final feasible matching solutions. The algorithm has some modifications compared to the algorithm for purely obtaining isomorphic graphs, but the main idea is the same.

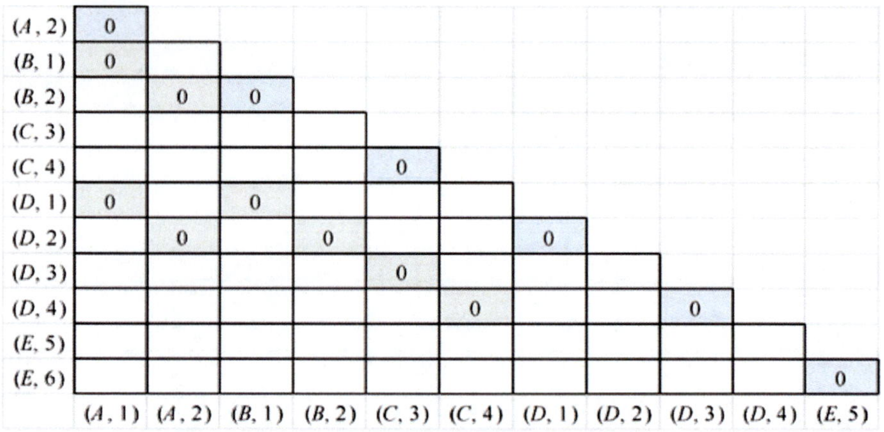

Fig. 4.22 Simplified step chart (see the color illustration)

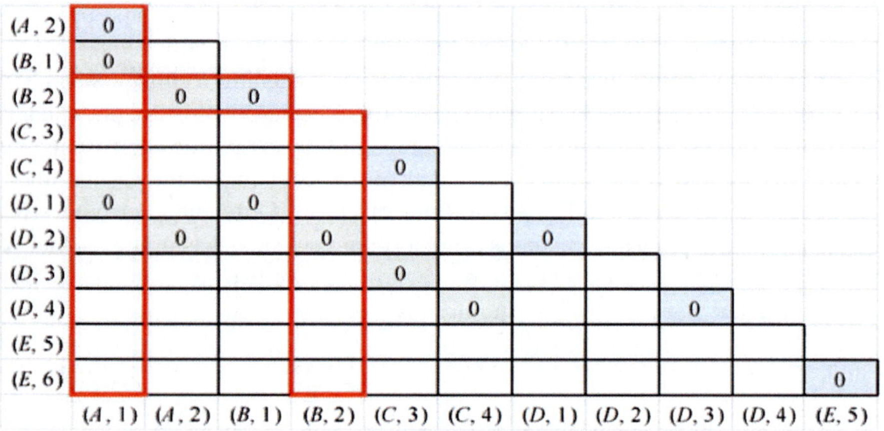

Fig. 4.23 Mapping scheme in the step chart (see the color illustration)

(2) TAEM Method

The transfer-aware effective loop mapping (TAEM) method [46] can effectively utilize the heterogeneous resources on the CGRA and accelerate the compilation effectively. Experimental results show that the TAEM method can have the same or even better performance of loop mapping results while greatly reducing the compilation time.

The performance and power consumption of the CGRA largely depend on whether the compilation and mapping methods can effectively utilize the various data transfer resources on the CGRA, including routing PEs, local register files (LRFs), global register files (GRFs), and on-chip data memory. The main problem to be solved in the DFG loop mapping process is how to efficiently transfer data through various resources on the CGRA, thus reducing the total execution time and power consumption of the loop. However, current studies either focus on only a small number of data transfer resources or require a long compilation time [7, 12, 19, 31, 45, 47, 48]. For example, EMS and EPIMap use only routing PEs for data transfer on PEAs, which usually results in wasted PE resources and may also fail to find a feasible mapping solution. GraphMinor and REGIMap use LRFs and routing PEs to transfer data, but the use of LRFs restricts the data transfer to be on the same PEs. MEMMap [48] uses memory or routing PEs for data transfer, but memory access may lead to poor computational performance and many additional access operations. Although the RAMP technique has taken multiple types of data transfer resources into account, including routing PEs, LRFs, GRFs, and memory, it treats the more efficient GRFs as the same as LRFs, and thus does not fully utilize GRFs. In addition, the resource selection strategy (repeatedly searching for different data transfer resources) and clique search algorithm used are both inefficient, thus leading to longer compilation time.

The TAEM method presents the following three main ideas:

(a) It provides an efficient loop mapping method, and all heterogeneous data transfer resources on the CGRA are considered, including routing PEs, LRFs, GRFs and memory. mapping-constrained LRFs and more flexible GRFs can be effectively distinguished, and these resources are fully utilized to find the optimal mapping result on the CGRA.
(b) It proposes a more efficient register allocation strategy based on the flexible transfer capability of registers. The data dependences after loop modulo scheduling are fully considered, and the occupied cycles of each register are accurately sorted out according to each DFG dependence edge. This leads to better utilization of GRFs and LRFs.
(c) It proposes a complete and fast loop mapping algorithm, which is based on the improved clique search method IBBMCX [49] and employs the optimal pruning strategy based on the greedy algorithm and the bit-based adjacency matrix computing strategy to increase the search speed.

The main goal of the loop mapping is to minimize the II as close to the MII as possible, so that the CGRA can execute the loop as many times as possible in the same time. Figure 4.24a gives an example of a 2 × 2 CGRA with four LRFs (one register per PE), where all PEs share two GRFs. Figure 4.24b gives an example of the DFG with a simple loop, where the edges of the DFG represent data dependences. The MII in this example is 2. Figure 4.24c shows how this DFG is extended by four different data transfer resources.

EMS/EPIMap, as shown in Fig. 4.24d, transfers data via PEs only. For example, node b, after getting the result at moment t, saves the result in PE4, which is represented by routing node b'. Similarly, node c needs to use PE3 to transfer routing node c'. The II is 3, which is greater than the MII. Therefore, this mapping strategy may insert too many routing nodes in one cycle wasting valuable PE resources, resulting in poor mapping results or mapping failure. GraphMinor/REGIMap, as shown in Fig. 4.24e, uses only PEs or LRFs to transfer immediate data on the PEA, and nodes

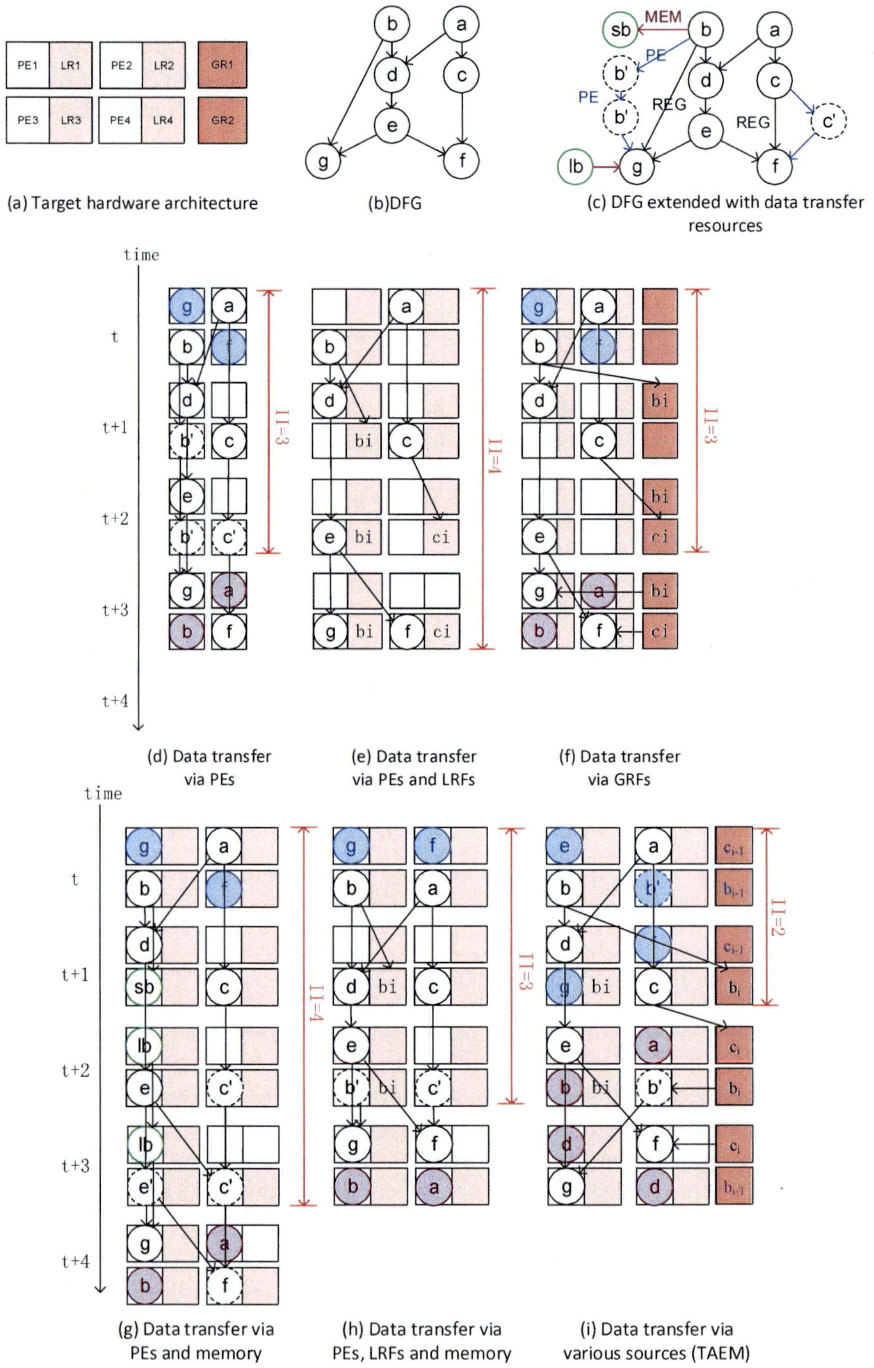

Fig. 4.24 Loop mapping results with different on-chip data transfer strategies on the CGRA (see the color illustration)

c and f are mapped onto a PE even if other PEs and registers are idle. The II of this mapping strategy is 4, which is also greater than the MII. Note that if LRFs are used to transfer data, two related operation nodes on an edge of the DFG must be mapped onto the same PE so that they can access the same LRF. Due to this limitation, the optimal mapping is usually not available. The use of GRFs can overcome some limitations of LRFs. As shown in Fig. 4.24f, node b needs to use GR1 to transfer the result to node g while node c needs to use GR2 to transfer the result to node f, when data is transferred via only GRFs. The II of this strategy is 3, which is still greater than the MII, but better than the GraphMinor/REGIMap method. MEMMap [48], as shown in Fig. 4.24g, uses routing PEs and on-chip memory for data transfer. When the edge length of the DFG is greater than or equal to 3, memory resources will be used to transfer data between different PEs. Since the memory access operation may consume more than two cycles, nodes lb and sb are inserted into the DFG to indicate the cycles occupied by read and store operations.

The II of this strategy is 4, which is larger than the MII. Since too many memory access operations are inserted, the performance is affected and it is difficult to obtain the optimal mapping result. As shown in Fig. 4.24h, the RAMP strategy uses various data transfer resources such as LRFs/GRFs (RAMP does not distinguish them), PEs, and memory. For example, routing node b' and GR4 are used in the data transfer between node b and node g. Its final II is 3, which is still greater than the MII. As shown in Fig. 4.24i, the TAEM method fully considers various data transfer resources, including routing PEs, LRFs, GRFs, and memory. In particular, a distinction is made between functionally constrained LRFs and flexible GRFs. For example, the result of node b is transferred via GR2 at moment t, and then the result obtained from GR2 at moment $t + 2$ is transferred to node g using PE3, which finally yields the II of 2, exactly equal to the MII. This means that the TAEM method can effectively utilize these data transfer resources and achieve higher performance compared with all the mapping techniques mentioned above.

The overall framework of the TAEM method is shown in Fig. 4.25, which consists of four steps: (1) search for all data transfer resources, (2) DFG extension and rescheduling, (3) DFG mapping, and (4) CGRA resource analysis. The priorities of the four main data transfer resources are sorted out based on flexibility and efficiency: GRFs > LRFs > routing PEs > memory. Since there is a performance gain by using modulo scheduling only when the II is smaller than the critical path of the traditional scheduling result, the TAEM method will only explore different resources in this case until an effective loop mapping scheme can be generated.

The algorithm in Fig. 4.26 gives the pseudocode of the IBBMCX-based TAEM algorithm. A data transfer resource is considered to be a feasible mapping solution if it successfully transfers all unmapped dependences for a given II. If no feasible mapping solution exists, it means that the DFG cannot be mapped onto the current TEC graph due to resource constraints. TAEM will increase the II and then try again. Although the TAEM algorithm may seem a bit complicated, it uses several effective strategies to accelerate the compilation process. First, the modified optimal pruning strategy based on the greedy algorithm can improve the search efficiency. Second, graph search method is optimized to prioritize nodes with a larger degree to accelerate

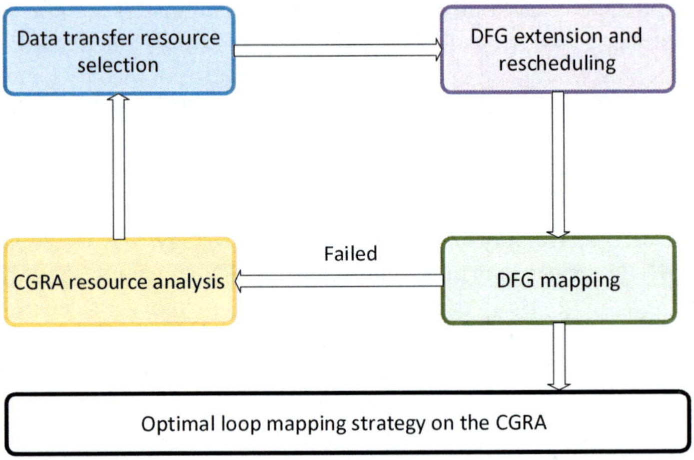

Fig. 4.25 Schematic diagram of the TAEM algorithm

the BBMCX search. This method allows the search order of the graph to quickly make the objective function reach the upper bound, which can reduce the size of the search tree. Thirdly, the bitwise operation is used to accelerate the adjacency matrix computation of BBMCX. For example, if the INT64 variables are used in the TAEM algorithm, the bitwise operation can reduce the time and space overhead by a factor of nearly 64.

(3) Conflict-Free Loop Mapping Based on Multi-bank Memory [50]

The pipeline stall caused by memory access conflicts can affect the performance of the loop pipeline in the CGRA mapping. Therefore, a novel SDC architecture with multi-bank memory and a loop mapping strategy without memory access conflicts is proposed in [50], dual-force directed scheduling (DFCS) algorithm. The main feature of the SDC with multi-bank memory is to store data in several memory banks to improve the parallelism of data access. How to make full use of the multi-bank memory structure, allocate data appropriately, and achieve high performance can be modeled as the following mapping problem:

Given a DFG $G_d = (V_d, E_d)$ and a CGRA C with N_b memory banks, find the kernel access pattern P such that:

(a) An effective loop mapping $M_{\text{op}} : (V_d, E_d) \rightarrow (V_r, E_r)$ exist, where (V_r, E_r) is the corresponding TEC graph with a height of the II.
(b) N_b memory banks store the valid data of all array elements.
(c) The II be minimized.

The II is the initialization interval with access conflicts considered. To illustrate the meaning of the kernel access pattern P, the following notations are given:

(a) Data domain: given a finite n-dimensional vector A, the address $x \in X$ of any data element can be represented by $x^A = (x_0^A, x_1^A, \cdots, x_{n-1}^A)$, where $x_i^A \in$

```
TAEM
Inputs: DFG D, CGRA C
Outputs: Mapping result CC or Failed
MII ← Calcul;ate MII(D,C);
II ← MII;
Threshold ← Maximum depth of nodes in D;
while II <= Threshold do
    D' ← Schedele(D,C);
    CG ← CalcCompGraph(D', C);
    CC ← IBBMCX_Clique(CG);
    if |V_CC| = |V_D'| then return CC;
    end if
    CC' ← TransferViaGRF (CC', D',C);
    CC' ← TransferVisLRF (CC', D',C);
    if |V_CC| = |V_D'| then return CC;
    end if
    while ∃ not mapped edge E' with length > 1 do
        CC' ← TransferViaPE (CC', D', C);
        if |V_CC| = |V_D'| then rerurn CC;
        end if
    end while
    CC' ← TransferViaMemory (CC', D', C);
    if |V_CC| = |V_D'| then return CC;
    end if
    II ← II + 1;
end while
return Failed
```

Fig. 4.26 Pseudocode of the TAEM algorithm

$[0, w_i^A - 1]$, $1 \leq i \leq n$, and w_i^A refers to the width of the data domain in the ith dimension.

(b) Single access pattern: access operations in each control step of the kernel form a data access pattern formed by, marked as P^A, containing m_r adjacent data elements where the data is constant.

(c) Kernel access pattern P_{kernel}: is formed by the access patterns of all control steps in the kernel.

Figure 4.27 shows the process of memory access pattern extraction. The SDC containing two data memory banks and source code are as shown in Fig. 4.27a and b, respectively. The single access pattern of each control step can be obtained according to its scheduled DFG. For example, the single access pattern of x in the 4th control step of the rescheduled DFG for access pattern changes (Fig. 4.27f) is $P^x = \{(1, 0)^T, (1, 1)^T\}$. The single access patterns of all control steps constitute the kernel access pattern P^x_{kernel} and $P^x_{\text{kernel}} = \{\emptyset, \emptyset, \{(0, 0)^T\}, \{(1, 0)^T, (1, 1)^T\}\}$ (Fig. 4.27g).

DFCS algorithm can find a suitable memory access pattern and an effective loop mapping pattern simultaneously. The algorithm adjusts the access patterns to achieve effective data placement and mappings, and expands the search space of the kernel access pattern by expanding the critical path. The "dual-force" refers to the force on the memory bank and that on the PE. Since the kernel is formed by modulo scheduling, the mobility of all operations and the extended length of the critical path are limited to the range of [0, II]. The number of memory access operations in the kernel constitutes the force on the memory bank. The more access operations there are, the more likely a bank access conflict will occur, i.e., the greater the force on the memory bank. In the mapping problem, the freedom of computing operations also needs to be considered, so the number of operations in the kernel constitutes the force on the PE.

The pseudocode of the DFCS algorithm is shown in Fig. 4.28. Since the scheduling of the non-memory access operations will affect the distribution of the dual-force computation of the memory access operations, these non-memory access operations are first scheduled according to the distribution graphs of PE resources and force (lines 1–5), and then the memory access operations are scheduled by calculating the two forces. F^B determines which control step the read operation is placed on first and F^{PE} determines which read operation to be scheduled. First, the force F^B is computed. Assuming that the force caused by a single read operation is 1, the distribution graph $\text{DG}(t_1)$ of the memory bank can be obtained by calculation, where $\text{DG}(t_1)$ indicates the resource occupation of control step t_1. Then, the impact $v(t_0, t_1)$ is calculated, which represents the impact of assigning node n to control step t_0 on $\text{DG}(t_1)$. For example, if operation L_2 in Fig. 4.27 is assigned to the 4th control step, since the probabilities for L_2 appearing in control steps 4 to 7 are all 1/4, $L_2(1, t_1)$ is 3/4, -1/4, -1/4, -1/4 when t_1 is 4 to 7, respectively. From this, the force can be computed by using the equation: F^B $F_v(t_0) = \sum_{t_1} \text{DG}(t_1)v(t_0, t_1)$.

After calculating the forces F^B of all memory access operations, the scheduling algorithm will select the operation with the smallest force and assign it to the control step. However, since the read operation affects both forces, F^{PE} still needs to be computed in the case that multiple read operations have the smallest force F^B. Note that routing operations may be generated during the adjustment of operation mobility, and since these operations can be mapped onto PEs or registers in PEs, the forces are not uniformly distributed. If the force caused by a computation or memory access operation is defined to be 1, the force caused by a routing operation can be expressed as $1/(1 + N_r)$, where N_r refers to the number of registers in each PE. Therefore, the probability of operation L_2 being scheduled to the 4th and 7th control step is

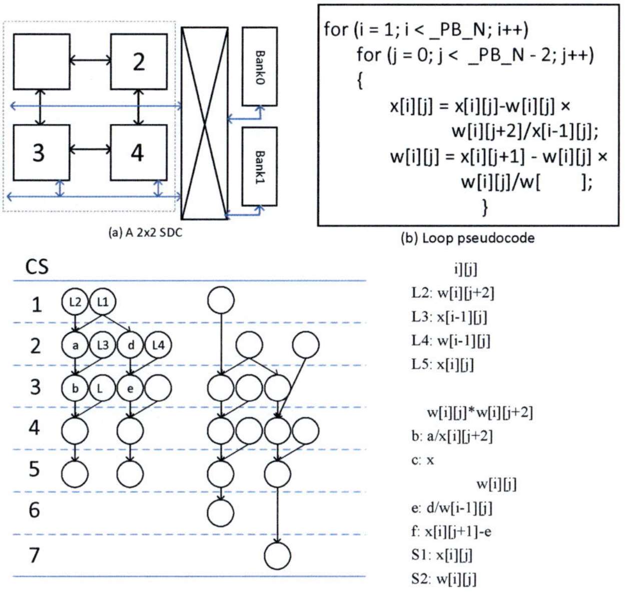

(c) Original DFG

(d) Rescheduled DFG whose memory access pattern has been changed

	Access Pattern	
	w[i][j+2]	w[i][j]
2	x[i-1][j]	w[i-1][j]
3	x[i][j]	x[i][j+1]
4		

(e) Memory access patterns of the original DFG

CS	Access Pattern	
1	w[i][j+2]	
2	w[i][j]	w[i-1][j]
3	x[i-1][j]	
4	x[i][j]	x[i][j+1]

(f) Memory access patterns of the new DFG

CS	Access Pattern	
	x	w
1	Ø	w[i][j+2]
2	Ø	w[i][j] w[i-1][j]
3	x[i-1][j]	Ø
4	x[i][j] x[i][j+1]	Ø

(g) Kernel access patterns of the ne

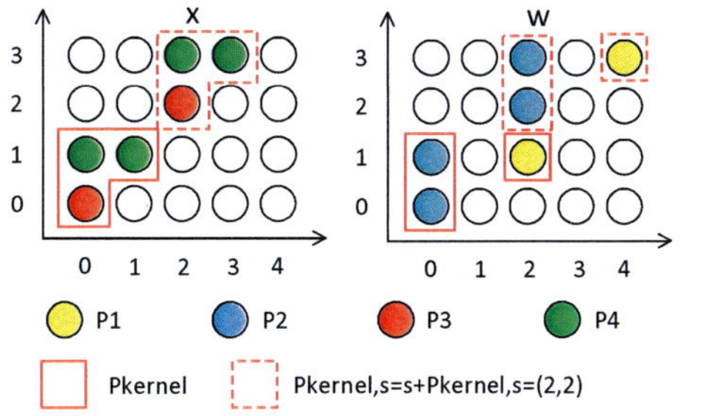

(h) Memor cess patter n data field form indicating the ac attern of the rth co step)

Fig. 4.27 Example of memory access pattern extraction (see the color illustration)

Dual-FDS

Input V_M, V_{Sch}, G_D II, CGRA C, Output $\{P^{A_1}_{kernel}, \cdots, P^{A_q}_{kernel}\}$

V_M: All the memory access operations to be scheduled.

V_{Sch}: All the operations (except V_M) to be scheduled.

1. $F^{PE} \leftarrow \forall v \in V_{Sch}$; // Calculate force for PE
2. While V_{Sch} is not $\varnothing$ do
3. $F^{PE} \leftarrow \text{Min}(F^{PE})$;
4. Allocate control step for node v_{Sch} with F^{PE};
5. $V_{Sch} \leftarrow V_{Sch} \setminus v_{Sch}$;
6. $(F^B, F^{PE}) \leftarrow \forall v \in V_M$; // Calculate dual-force for banks and PE
7. while V_M is not $\varnothing$ do
8. $(F^B, F^{PE}) \leftarrow \text{Min}(F^B, F^{PE})$;
9. UpdatePattern$\{P^{A_1}_{kernel}, \cdots, P^{A_q}_{kernel}\}$;
10. while $r{=}1; r \leqslant \text{II}; r{+}{+}$ do
11. $C_r \leftarrow \text{Count}\,(\{P^{A_j}_r | 1 \leqslant j \leqslant q\} != \varnothing)$;
12. if MaxDif$(\{C_r | 1 \leqslant r \leqslant \text{II}\}) \geqslant 2$ then
13. RecoverPattern$\{P^{A_1}_{kernel}, \cdots, P^{A_q}_{kernel}\}$;
14. Continue;
15. Allocate control step for node v_M with (F^B, F^{PE});
16. $V_M \leftarrow V_M \setminus v_M$;

Fig. 4.28 Pseudocode of the DFCS algorithm

the same and the corresponding force can be computed after scheduling. The force distribution of L_2 can be obtained by multiplying the probability and force. Next, the impact $v(t_0, t_1)$ is calculated and the force F^{PE} is obtained. Then, among the control steps and nodes with the minimum F^B, the node with the minimum F^{PE} is selected and assigned to that control step (line 8). After that, a pattern update is performed, which will temporarily add the selected node to the current access pattern (line 9). If the difference in the number of memory access operations in each step is greater than 1, this is not a good solution. This unbalanced memory access pattern will eventually result in N_f not being the minimum. Therefore, C_r is used to count the number of memory access operations in each control step (line 11). This algorithm checks the obtained memory access pattern by counting C_r $(1 \leq r \leq \text{II})$ (line 12). If the largest difference is greater than 1, the unnecessary paths will be pruned out of the search space. After that, the pattern recovery is performed to remove the temporarily added node (line 13) and continue to try another configuration (line 14). If all memory access operations are scheduled completely, the final generated memory access pattern is returned.

4.2.4 *Integer Linear Programming*

As mentioned above, the most popular approach to the NP-complete problem of mapping operators onto hardware is to abstract it as an integer linear programming problem, and then use software dedicated to solving integer linear programming problem for further processing. The reason is that the optimal solution can always be obtained by adjusting the parameters as required. Generally speaking, the II is used as the main metric, and hardware resource utilization, throughput, and the priority of the computation of some data are also taken into account in practice.

4.2.4.1 Modeling Process of Integer Linear Programming

As a mathematical model for system optimization, integer linear programming mainly consists of three parts [51]: (1) key variables used to describe the output results, (2) equations or inequalities used to describe constraints on the above variables, and (3) an objective function to rank feasible solutions according to requirement. Although the composition of all ILPs can be summarized into the above three parts, various ILP approaches may differ in many ways. In terms of modeling, the "compute node-compute node", "routing node-routing node", and "link-link" mapping relationships are generally used as variables. However, there are other ways to select variables, such as selecting the mapping relationship of each downstream path of the routing nodes as a variable. The variation in constraint equations is more significant. For example, some hardware structures allow multiple operators to be mapped onto the same PE and implement dynamic scheduling, and part of PEs of some hardware structures do not support certain operators. The objective function has been described before, even with the same hardware and software code, the rank of feasible solutions is different subject to different requirements.

The ILP can handle most of the CGRA mapping problems and can modify the definition of variables and objective functions in the ILP according to the target hardware characteristics and mapping requirements. Because of its wide range of applications, the ILP can be used to build a CGRA automatic compilation system. CGRA-ME [27] is a relatively complete CGRA simulation system at present, which adopts ILP as one of the core algorithms of graph mapping. The ILP part of CGRA-ME will be described in detail later.

The core part of the ILP is the establishment of constraints, because the constraints reflect the mapping rules and hardware limitations, thus reflecting the essence of the problem. Subject to the same constraint, different variable definitions simply characterize different abstraction perspectives. Although this affects the form of the constraint equations, the inherent meaning of the constraint is the same. If each constraint scheme is regarded as a set, giving the union of all these sets is an effective way to understand the constraint schemes. The following describes the various constraints of the ILP from five aspects.

Five intuitive abstractions in the form of graph mappings can be used to describe the problems encountered during scheduling and mapping, which fit with the hardware primitives, each representing a class of work that needs to be done by the compiler. Therefore, appropriate constraints can also be found from these five aspects [51]:

(1) Placement of computation, representing the mapping of operators to PEs, corresponds to the underlying hardware resource organization.
(2) Routing of data, representing the implementation of computational semantics that can reflect internal network allocation, data channel contention, corresponds to the network connecting hardware resources.
(3) Managing event timing, representing the timing relationships between operators, especially in the case of loop formation, corresponds to the timing and synchronization of hardware resources.
(4) Managing utilization, representing the utilization of hardware resources, usually is used as one of the optimization objectives. The allocation of hardware resources in the case of parallel computing and resource reuse needs to be considered involving the parallelism within and between code blocks.
(5) Optimization objective is the required optimization direction and strategy to meet the performance requirements.

For better illustration, a simple mapping example is first given as shown in Fig. 4.29, which represents the manual mapping of the DFG of a simple function $z = (x+y)^2$ (Fig. 4.29a) to a simplified version of DySER [11]. The CGRA structure is shown in Fig. 4.29b and individual nodes are labelled in the figure. The triangles represent the inputs and outputs; the circles represent the computing resources; the squares represent routing resources, and arrows indicate the direction in which data can flow. This manual mapping is not optimal compared to the results obtained by the ILP method which will be discussed later.

The first step in ILP problem solving is to build the model and name each key variable. Here, in the DFG, V (vertex) refers to the set of vertices, and $v \in V$; E (edge) refers to the set of directed edges, and $e \in E$; G refers to the set $G(V \cup E, V \cup E)$ of vertices and edges, and this variable also characterizes the relationship between vertices and directed edges. $G(v, e) = 1$ indicates that the directed edge e is the output of the vertex v, and $G(e, v) = 1$ indicates that the directed edge e is the input of the vertex v. In the hardware abstraction graph, N (node) refers to the set of nodes, and $n \in N$; L (link) refers to the set of data paths connecting the hardware nodes, and $l \in L$; similarly, H characterizes the set $H(N \cup L, N \cup L)$ of nodes and links; $H(l, n) = 1$ indicates that the link l is the input of node n and $H(n, l) = 1$ indicates that link l is the output of node n. Some vertices and nodes represent units that are not of computational functions. For example, routing resources such as inputs and outputs are distinguished by a matching table $C(V, N)$, in which $C(v, n) = 1$ indicates that vertices v and nodes n can be mapped; for the routing nodes constituting

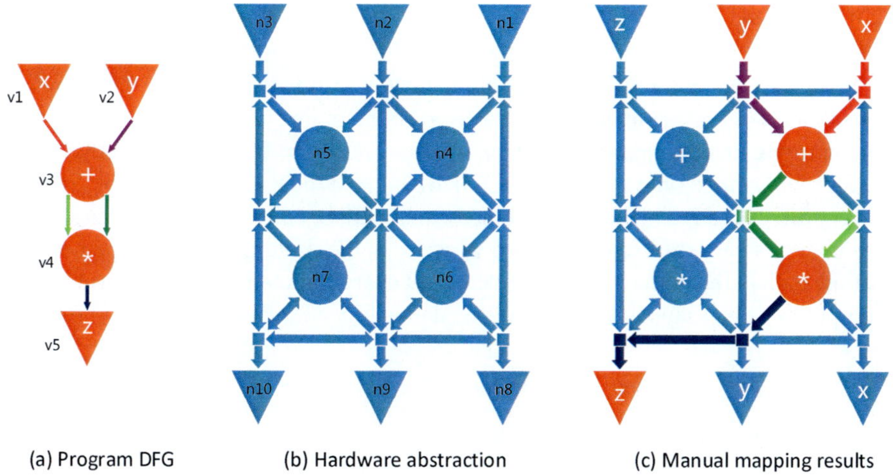

Fig. 4.29 Example of manual mapping results (see the color illustration)

a network, which may be registers, multiplexers, etc., the set $r \in R$ of these units is represented uniformly by R, corresponding to the small square in Fig. 4.29.

The whole constraint system is described below in terms of the five aspects that the scheduler needs to accomplish [51].

1. Placement of Computation

The first work that the scheduler needs to accomplish is the direct one-to-one mapping of vertices to nodes, i.e., $V \to N$. All results of this mapping are saved in the matrix $M_{vn}(V, N)$, i.e., $M_{vn}(v, n) = 1$ means that vertex v is mapped onto node n, while $M_{vn}(v, n) = 0$ means that it is not. An obvious constraint is that all vertices in the DFG can be mapped only once and must be mapped to the matched nodes; mapping is not allowed for mismatched nodes, i.e.,

$$\forall v, \quad \sum_{n|C(v,n)=1} M_{vn}(v, n) = 1 \tag{4.4}$$

$$\forall v, \quad n|C(v, n) = 0, \quad M_{vn}(v, n) = 0 \tag{4.5}$$

As shown in Fig. 4.29a and b, the constraint reflected in this example is $M_{vn}(v_1, n_1) = 1, M_{vn}(v_2, n_1) = 0, M_{vn}(v_3, n_4) = 1$, and $M_{vn}(v_3, n_5) = 0$. Note that there is no restriction here that each hardware node n can be mapped only once, i.e., reuse of routing and computational resources is allowed, which is clearly possible. However, many compilers will include this as one of the constraints during automatic compilation, as shown in Eq. (4.6).

$$\forall n, \quad \sum_{n|C(v,n)=1} M_{vn}(v, n) = 1. \tag{4.6}$$

2. Routing of Data

The scheduler also needs to map the data flow represented by the DFG onto the data path network connecting hardware resources, i.e., $E \rightarrow L$. The result of this mapping is saved in matrix $M_{el}(E, L)$, i.e., $M_{el}(e, l) = 1$ indicates that the directed edge e is mapped onto the link l, and $M_{el}(e, l) = 0$ indicates that it is not. Similarly, every directed edge in the DFG should be mapped. Specifically, if vertex v is mapped to node n, every directed edge with vertex v as the input must be mapped onto one of the data paths with data from node n, and every directed edge with vertex v as the output must be mapped onto one of the data paths with data exported to node n.

$$\forall v, e, n | G(v, e), \ \Sigma_{l|H(n,l)} \ M_{el}(e, l) = M_{vn}(v, n) \tag{4.7}$$

$$\forall v, e, n | G(e, v), \ \Sigma_{l|H(l,n)} \ M_{el}(e, l) = M_{vn}(v, n) \tag{4.8}$$

In addition, the scheduler needs to ensure that each directed edge is mapped onto a contiguous path of links. This constraint is reflected in the expression that for any given directed edge e, any routing node r either has no input or output mapped onto that directed edge, or has exactly one input and one output mapped onto that directed edge.

$$\forall e \in E, \quad r \in R, \sum_{l|H(l,r)} M_{el}(e, l) = \sum_{l|H(r,l)} M_{el}(e, l) \tag{4.9}$$

$$\forall e \in E, \quad r \in R, \quad \sum_{l|H(l,r)}, \quad M_{el}(e, l) \leq 1 \tag{4.10}$$

As shown in Fig. 4.30, the above constraints reflected in this example are $M_{el}(e_2, l_1) = 1$, $M_{el}(e_3, l_{24}) = 0$, $M_{el}(e_1, l_7) = 1$, and $M_{el}(e_3, l_{25}) = 0$. Note that while the mapping scheme on link l_7 in Fig. 4.30c is clearly not reasonable, it does not violate Eq. (4.10) because it holds for directed edges e_1 and e_2, which will need to be clarified later in the managing utilization part.

Some hardware structures have special requirements. For example, a signal propagating in the x direction can continue to propagate in the y direction, but the signals propagating in the y direction cannot continue to propagate in the x direction. To characterize this, matrix $I(L, L)$ can be added to the variables describing the hardware, representing all pairs of links that cannot be mapped onto the same directed edge. For example, $I(l, l') = 1$ indicates that for some reason link l and link l' cannot be mapped onto any directed edge simultaneously. Then, the above constraint can be written as:

$$\forall l, \quad l' | I(l, l'), \quad e \in E, \quad M_{el}(e, l) + M_{el}(e, l') \leq 1 \tag{4.11}$$

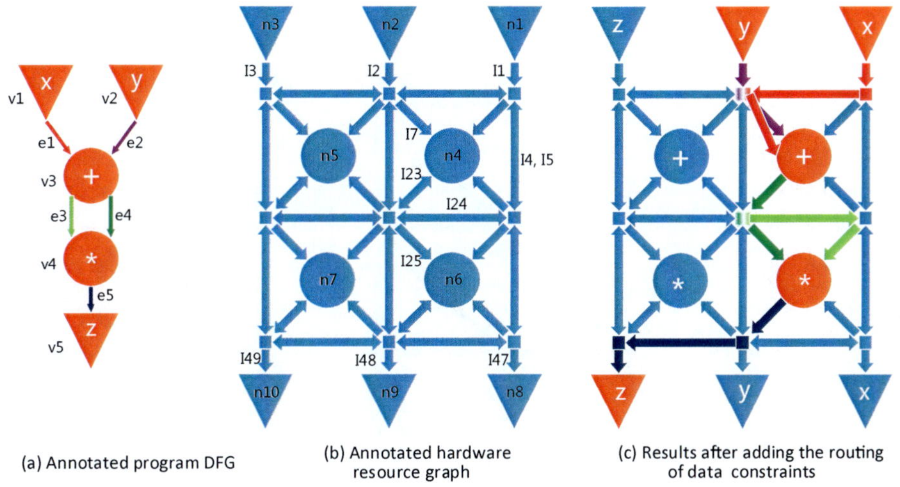

Fig. 4.30 Results after adding the routing of data constraints (see the color illustration)

There are various other expressions similar to the constraint of Eq. (4.11) in terms of routing, but they are similar and can be added or deleted as required.

3. Managing Event Timing

A new set of variables $T(V)$ characterizing time is introduced to represent the time when vertex $v \in V$ is reached, and its dimension is the number of clock cycles. Calculating the time required to route all directed edges from vertex v_{src} to vertex v_{dst} in the DFG can be expressed by $T(v_{\text{dst}}) - T(v_{\text{src}})$. In logical order, this value consists of three parts: (1) the number of clock cycles $\Delta(E)$ needed from the start of timing until the data of v_{src} is ready, (2) the routing delay needed for propagation on the directed edge, whose value is the sum of the number of clock cycles needed for data to propagate on the link corresponding to the mapping of these edges, and (3) the time deviation, which represents the difference in the number of clock cycles needed to reach the vertex v_{src} for different directed edge paths, expressed as $X(E)$. Then, the expression of the timing is as follows:

$$\forall v_{\text{src}}, e, v_{\text{dst}} | G(v_{\text{src}}, e) \& G(e, v_{\text{dst}}), T(v_{\text{src}}) + \Delta(e) + \sum_{l \in L} M_{el}(e, l) + X(e) = T(v_{\text{dst}}) \tag{4.12}$$

However, the above calculation does not circumvent the case of excessive computation of delay as shown in Fig. 4.31b, which does not violate the constraint of Eq. (4.10) for the same reason as in Fig. 4.30c. The above case can be circumvented in some routing constraints similar to Eq. (4.11), and an alternative approach will be used here: a new variable $O(L)$ that indicates the order in which the links are activated is introduced, as indicated by the black numbers in Fig. 4.31b and c. If a certain directed edge is mapped onto two connected links, the following constraint

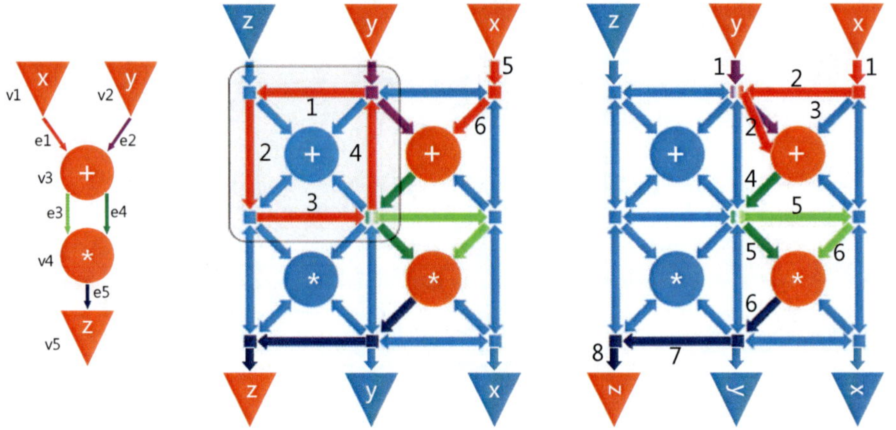

(a) DFG of the program (b) Situations to be handled by timing constraints (c) Situations that cannot be handled by timing constraints

Fig. 4.31 Results after the addition of event timing management constraints (see the color illustration)

will ensure that the downstream link comes after the upstream link:

$$\forall l,\quad l'|I(l,l'),\quad e\in E|H(l,l'),\quad M_{el}(e,l)+M_{el}(e,l')+O(l)-1\le O(l') \tag{4.13}$$

4. Managing Utilization

In simple terms, the number of clock cycles required for completing a single task can be used to quantify the utilization of hardware resources, since this value characterizes how long the hardware resource cannot accept a new task. However, the above simple statement is actually consistent with the idea of the event timing section. The following is to evaluate the utilization of different types of hardware resources.

First, the utilization of links $U(L)$ is considered:

$$\forall l\in L,\quad U(L)=\sum_{e\in E}M_{el}(e,l) \tag{4.14}$$

Equation (4.14) represents the number of times link l is mapped onto by different directed edges, which is valid when and only when the resources occupied by these directed edges do not conflict. Possible cases include but are not limited to: different directed edges occupy different control steps; different directed edges occupy different byte bits to transfer data; these directed edges transfer the same data; there is a conflict between the directed edges and it is impossible for them to be simultaneously valid. Many hardware architectures allow reusing the data path, and in this example the directed edges e_3 and e_4 actually share the data path l_{23} all the time (Fig. 4.30b), which is reasonable. To represent the constraint on data path reuse, a new variable B_e is introduced to represent the directed edge-bundle, as shown in Fig. 4.32a, which represents those directed edges that can be mapped onto the same link without additional overhead. $B(E, B_e)$ is used to describe the relationship between the directed edge-bundle and the directed edge. Both $B(E, B_e)$ and $C(V, N)$ can be determined before scheduling. The following three constraints ensure that the mapping results have no data conflicts between the directed edge-bundle and the link, or between the directed edge and the link, and the link's utilization is presented:

$$\forall e, \quad b_e | B(e, b_e), \quad l \in L, \quad M_{bl}(b_e, l) \geq M_{el}(e, l) \tag{4.15}$$

$$\forall b_e \in B_e, \quad l \in L, \quad \sum_{e \in B(e, b_e)} M_{el}(e, l) \geq M_{bl}(b_e, l) \tag{4.16}$$

$$\forall l \in L, \quad U(L) = \sum_{b_e \in B} M_{bl}(b_e, l) \tag{4.17}$$

As shown in Fig. 4.32, the case in Fig. 4.32b can now be determined to be illegal according to the above three equations. The directed edges e_1 and e_2 belong to different directed edge-bundles, but $\sum_{b_e \in B} M_{bl}(b_e, l_7) > 1$, which indicates that different directed edge-bundles are mapped onto the same link. The mapping of the directed edge l_{23} does not have this problem, and $M_{el}(e_3, l_{23}) + M_{el}(e_4, l_{23}) = 2 > M_{bl}(b_3, l_{23}) = 1$, which does not violate the link reuse rules.

Then, the utilization of PEs $U(N)$ is considered. It is necessary to consider the number of clock cycles $\Delta(V)$ for a node to fully occupy its mapped computational resources. This value is usually equal to 1 when the overall architecture is fully pipelining, but it increases when the non-fully pipelining limits the use of node n in subsequent cycles.

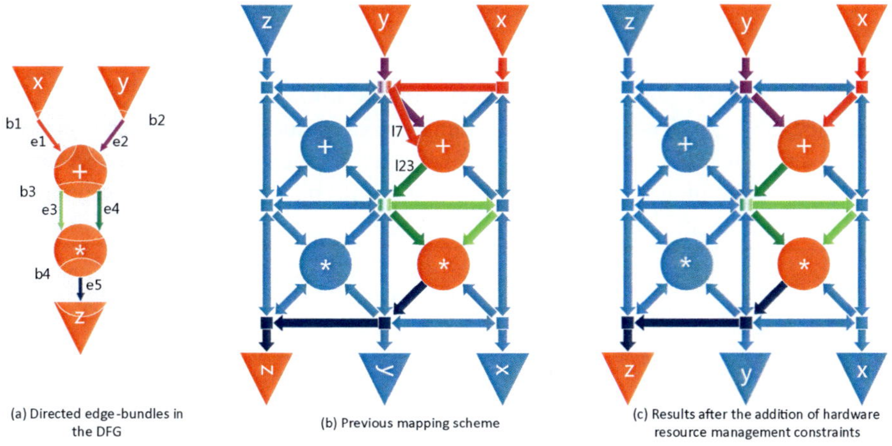

Fig. 4.32 Results after the addition of hardware resource management constraints (see the color illustration)

$$\forall n \in N, \quad U(n) = \sum_{v \in V} \Delta(V) M_{vn}(v, n) \tag{4.18}$$

Typically, different CGRA architectures have different upper limits on hardware resource usage, expressed by constraints as follows:

$$\forall l \in L, \quad U(l) \leq \mathrm{MAX}_L \tag{4.19}$$

$$\forall n \in N, \quad U(n) \leq \mathrm{MAX}_N \tag{4.20}$$

MAX_L and MAX_N indicate the parallel processing capability of links and computational resources for a particular CGRA structure, and usually $\mathrm{MAX}_L = \mathrm{MAX}_N = 1$ for the DySER in this example as well as for most CGRA hardware structures.

5. Optimization Objective

All of the previous constraints focus only on a separate part only, but the ultimate responsibility of the scheduler is still to provide an optimized mapping as a whole while ensuring the mapping is error-free. This means that, in practice, the scheduler needs to balance the performance in terms of throughput and latency, and to make an appropriate trade-off if there is a need for both. The quantitative expressions of the above two aspects are given below.

In the calculation of the latency on the critical path, the time of the input node is written as 0 and the maximum latency for the output node is agreed to be LAT. The lower limit of this value can be used by the scheduler to estimate how long it will take to complete the code block.

$$\forall v \in V_{\text{in}}, \quad T(v) = 0 \tag{4.21}$$

$$\forall v \in V_{\text{out}}, \quad T(v) \leq \mathrm{LAT} \tag{4.22}$$

In addition, to quantify the constraint on the throughput, the service interval is defined as SVC to represent the minimum number of clock cycles between successive invocations without data dependences between the loop bodies of each invacation. For fully pipelining mapping results, there should be $\mathrm{SVC} = 1$, so essentially it is both an optimization objective and a constraint.

$$\forall n \in N, \quad U(n) \leq \mathrm{SVC} \tag{4.23}$$

$$\forall l \in L, \quad U(l) \leq \mathrm{SVC} \tag{4.24}$$

In addition to throughput and latency, minimizing the time difference MIS of data arriving at the PE is also one of the optimization objectives. The optimal mapping obtained subject to all the above constraints in this example is shown in Fig. 4.33c,

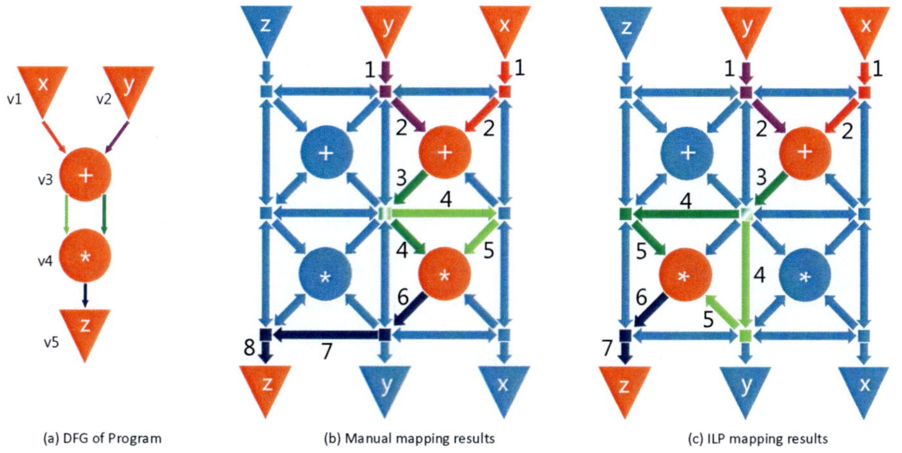

Fig. 4.33 Final results after the addition of all constraints (see the color illustration)

and both LAT = 7 and MIS = 0 reach the minimum value. In the manual mapping results LAT = 8 (Fig. 4.33b), MIS = 0 also exists when the data arrives at the multiplication PE, which is obviously not as good as the result obtained by the ILP method.

So far, through the description of five aspects, a very well-developed constraint system has been given, and many other additional constraints can be summarized into these five categories.

4.2.4.2 Application of Integer Linear Programming

Since the ILP method becomes the mainstream processing method for CGRA automatic compilation, most of the barriers between the top-level C code block and the bottom-level CGRA hardware configurations have been removed. A more complete architecture for CGRA simulation and the ILP constraint system, i.e., the CGRA-ME system described earlier, is presented below, and the framework of CGRA-ME is shown in Fig. 4.34.

For CGRA-Modeling and Exploration (CGRA-ME), the software interface is high-level languages, such as a piece of C code, and the hardware interface is an XML-based custom language to describe the specific CGRA structure. The open-source compiler infrastructure LLVM is used to obtain the DFG of the loop statement blocks in the top-level code block. The DFG and the modulo routing resource graph (MRRG) transformed from the hardware structure description are used for mapping. Then, the compiler uses the ILP processing method to obtain the mapping results which are then simulated to evaluate the performance, energy consumption and area. The main focus here is the mapper, which is characterized by its hardware-insensitive mapping method, i.e., various CGRA architectures can be described in their designed XML-like language. However, the support for object code is not satisfactory, because

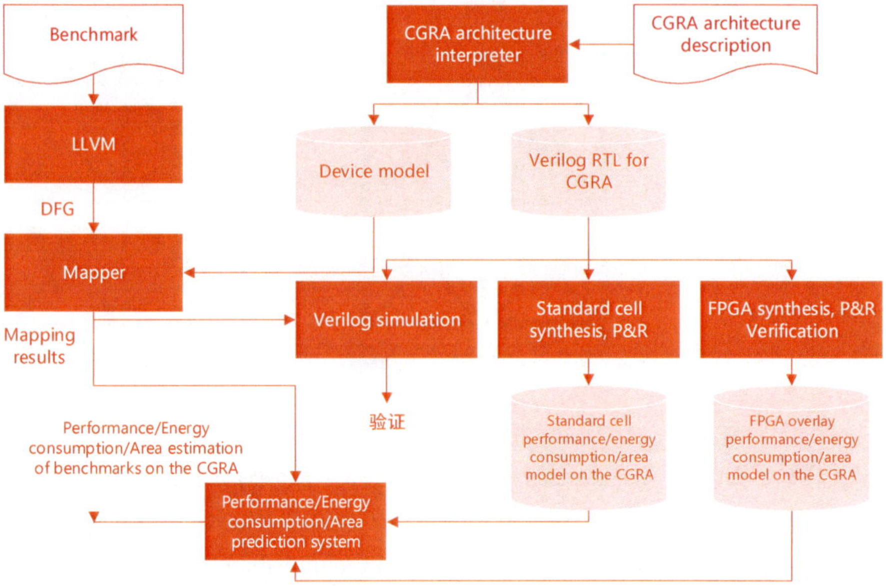

Fig. 4.34 Framework structure of CGRA-ME

although LLVM can get the SSA of the whole C language code block, CGRA-ME can only complete the mapping of the internal loop bodies, rather than the branches and loops, which is limited by the DFG format. This drawback is not so obvious in practice, because as described at the beginning of this book, the main advantage of CGRA application is to process loop bodies with high degree of parallelism.

The mapper uses two solvers, GUROBI [52] and SCIP [53], to perform mapping operations of the kernel, and the constraint scheme it uses differs significantly from the previous descriptions due to its customization and some modifications for the hardware description of the MRRG. The following is a detailed description of the constraint scheme of CGRA-ME, which is of great help to understand the writing of the constraint scheme.

CGRA-ME uses three sets of key variables to abstract the mapping, which are referred to as follows:

$F_{p,q}$: When the value is 1, it means that function node p in the MRRG is used as the mapping of operation node q in the DFG.
$R_{i,j}$: When the value is 1, it means that routing node i in the MRRG is used as the mapping of numerical node j in the DFG.
$R_{i,j,k}$: When the value is 1, it means that routing node i in the MRRG is used as a mapping from numerical node j in the DFG to its receiving node k.

It is easy to see that the set of variables used in this constraint scheme is a subset of the previous scheme Node-to-node mapping types are differentiated and edge-to-edge mapping in the graph is replaced by the form of kth downstream node. This is actually reasonable because numerical nodes can be added anywhere in the

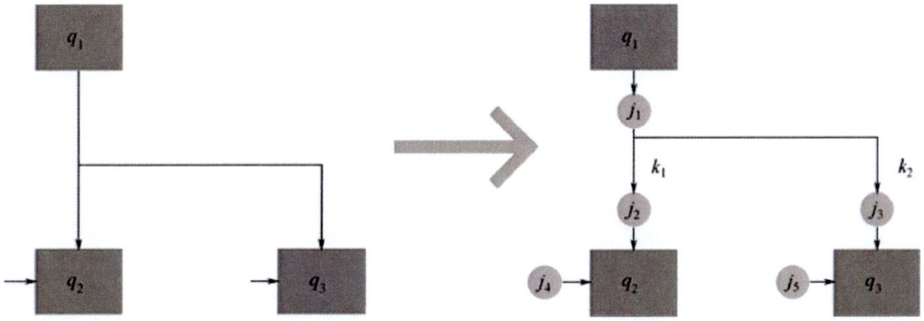

Fig. 4.35 Description of the variables used by the MRRG

DFG without affecting the actual dataflow. In the meanwhile, all numerical nodes except the input and output nodes on the dataflow path of DFG can be deleted. Therefore, the expression using downstream node is equivalent to that using directed edges. One numerical node is by default added before and after every operation node respectively, and the added nodes can be mapped together as part of that operation node to a function node of the MRRG (Fig. 4.35). The constraint scheme is as follows:

(1) Placement of computation: each operation node in the DFG needs to be mapped onto a function node of the MRRG.

$$\sum_{p \in \text{FuncUnits}} F_{p,q} = 1, \quad \forall q \in \text{Ops} \tag{4.25}$$

(2) Exclusivity of function nodes: each function node of the MRRG is mapped by at most one operation node of the DFG.

$$\sum_{q \in \text{Ops}} F_{p,q} \leq 1, \quad \forall p \in \text{FuncUnits} \tag{4.26}$$

(3) Validity of function node mappings: each operation node in the DFG is ensured to be mapped to a function node that can support that operation.

$$\begin{aligned} &F_{p,q} = 0 \\ &\forall p \in \text{FuncUnits}, \quad q \in \text{Ops} \\ &\text{where} : q \notin \text{SupportedOps}(p) \end{aligned} \tag{4.27}$$

(4) Exclusivity of routing nodes: each routing node in the MRRG is ensured to be mapped at most one numerical node of the DFG.

$$\sum_{i \in \text{Vals}} R_{i,j} \leq 1, \quad \forall i \in \text{RouteRes} \tag{4.28}$$

(5) Fan-out routing: is used to ensure data coherence of adjacent routing nodes. For each fan-out of a routing resource that is mapped with a numerical node, there must be at least one downstream routing resource that is used by the same value, whether it is used as another routing resource or as a receiving point of the fan-out. Therefore, at least one output node of a routing node is driven by the same numerical node, or that routing node has at least one output that is an input to some downstream function node.

$$R_{i,j,k} \leq \sum_{m \in \text{fanouts}(i)} R_{m,j,k}$$
$$\forall i \in \text{RouteRes}, \quad \forall j \in \text{Vals}, \quad \forall k \in \text{sinks}(j) \tag{4.29}$$

(6) Implicit routing: the fan-out of a route is ensured to be used as an input to a function node, and thus a mapping from an operation node on the DFG to that function node will be implied. Here, the $\rightarrow$ is used to indicate the presence of a directed edge between two nodes in the DFG. Subject to this constraint, if the mapping of the receiving point k of numerical node j is one of the inputs of function node p, k is also an input of operation node q and q is mapped onto p.

$$\begin{aligned}
&F_{p,q} \geq R_{i,j,k} \\
&\forall p \in \text{FuncUnits}, \quad \forall q \in \text{Ops}, \quad \forall i \in \text{RouteRes}, \quad \forall j \in \text{Vals} \\
&\text{where} : \exists (j \rightarrow q) \wedge (i \rightarrow p) \\
&\forall k \in \text{sinks}(j)
\end{aligned} \tag{4.30}$$

(7) Initial fan-out: the routing resource that is the output of the function node is ensured to be the mapping of the output of the operation node on the DFG.

$$\begin{aligned}
&F_{p,q} = R_{i,j,k} \\
&\forall i \in \text{RouteRes}, \quad \forall j \in \text{Vals}, \quad \forall p \in \text{FuncUnits}, \quad \forall q \in \text{Ops} \\
&\text{where} : \exists (q \rightarrow j) \wedge (p \rightarrow i) \\
&\forall k \in \text{sinks}(j)
\end{aligned} \tag{4.31}$$

(8) Routing resource usage: the routing resources must be mapped by the corresponding values.

$$\begin{aligned}
&R_{i,j} \geq R_{i,j,k} \\
&\forall i \in \text{RouteRes} \\
&\forall j \in \text{Vals} \\
&\forall k \in \text{sinks}(j)
\end{aligned} \tag{4.32}$$

(9) Exclusivity of multiplexers inputs: self-reinforcing routing loops may result in fan-outs within the loop being terminated rather than exported to the desired

receiver, which should be prevented.

$$
\begin{aligned}
&R_{i,j} = \sum_{m \in \text{fanins}(i)} R_{m,j} \\
&\forall i \in \{\text{RouteRes} \,|\, |\text{fanins(i)}| > 1\} \\
&\forall j \in \text{Vals}
\end{aligned}
\tag{4.33}
$$

It is easy to see that the CGRA-ME constraint system focuses on data routing, while default settings are used for timing synchronization, resource management, and optimization objectives. Meanwhile, compared to the previously described constraints on data routing, these nine constraints are more complicated because the CGRA-ME imposes more tight constraints. For example, no more than one operator can be mapped onto the same FU.

ILP processing schemes used in practice are given above. So far, the ILP method is still the best solution to deal with the mapping problem, but as mentioned above, its biggest drawback is that it is too slow for large-scale graph mapping problems and takes too long to compile, requiring decomposition of large-scale graphs into smaller ones and then processing them separately, which is undoubtedly a major obstacle for the automatic compilation process.

4.2.5 Irregular Task Mapping

SDCs usually act as the coprocessors of GPPs to accelerate the data-intensive part of the program, while the control part of the program is handled by GPPs. Data communication is required between GPPs and SDCs, and, for example, MorphoSys [54] and FLORA [55] use shared registers or dedicated buses to connect GPPs to SDCs. When dealing with control-intensive algorithms, the SDC communicates with the GPP frequently. This large communication cost cancels out the performance gain from the SDC. In this case, if the part containing simple control flow can be mapped to the SDC, the latency and performance loss caused by the intermediate variable transfer of the program can be effectively eliminated, and the overall performance of the program can be substantially improved by working out suitable scheduling and mapping methods to further exploit the parallelism of the kernel with control flow.

ADRES is an SDC architecture that can handle both control-intensive and data-intensive algorithms, which allows efficient modulo scheduling of the innermost loop and can be used as a VLIW processor to execute non-loop and out-of-loop code. ADRES uses predicate operations to handle the control flow within the loop, and it contains four lines of FUs, with the first line of FUs and a central register file providing VLIW functions. When executing non-inner loop code, the SDC activates only the first line of FUs, with the remaining three lines deactivated. The traditional SDC does not execute non-inner loop code, which is executed on the GPP, when the entire array of FUs of the SDC is idle. ADRES has higher performance compared to a heterogeneous platform such as the traditional SDC and GPP, as it eliminates the slow transfer of intermediate variables between the GPP and the SDC, and provides more hardware resources for non-loop code than a GPP. However, ADRES simply divides the FUs into two categories, processing non-inner loop code and inner loop code. For non-inner loop code, it is still scheduled and executed in a sequential manner, and out-of-loop and in-loop code are executed alternatively. The possible parallelism of inner loop code and out-of-loop code as a whole has not been well exploited, which is essentially not much different from a heterogeneous platform including the GPP and SDC. The performance gains for control-intensive algorithms are therefore quite limited.

Tasks containing simple control flow, which need to be mapped onto an SDC for execution, are referred to as irregular tasks. A distinctive feature of an irregular task is that it contains multiple basic blocks in its intermediate representation. A simple DFG cannot represent the data dependences and control dependences between individual operations, but a CDFG with basic blocks as nodes is required to abstract this type of task. Irregular tasks are broadly divided into nested loops and branches. Nested loops can be roughly divided into perfect nested loops and imperfect nested loops. Imperfect nested loops are nested loops with statements outside the innermost loop (i.e., odd statement) or nested loops with multiple inner loops at the same nesting level, i.e., nested loops containing multiple inner loops of the same level. Branches can be further divided into if–then-else (ITE) branches and nested if–then-else (NITE) branches. For a program containing a nested loop or branch, the corresponding CDFG consists of multiple basic blocks, and the data transfer and control jumps between the basic blocks are dynamically determined during execution, which is the major difference between the mappings of irregular tasks and regular tasks. A regular task contains only one basic block and the execution time of each operation within the basic block can be determined statically. The mapping only needs to consider the DFG within the basic block, while it is difficult to determine whether and when some operations in an irregular task are executed during compilation. Moreover, an error may occur when executing a simple static dataflow configuration, and the mapping needs to consider the whole CDFG containing explicit control flow. Therefore, it is a big challenge to implement static mapping and scheduling of irregular tasks using compilation techniques. It is a bigger challenge to break the execution order of basic blocks limited by the control flow, exploit more parallelism, and improve the performance of mappings for irregular tasks.

Currently, the SDC handles irregular tasks mainly by converting control flow into data flow. The inefficiency of architecture and mapping for control flow limits the scope of target code that can be accelerated by SDCs. Therefore, solving the mapping problem for irregular tasks requires not only finding suitable mapping algorithms, but also making certain modifications to the SDC architecture to efficiently support compilation and execution of control flow. The co-design of hardware and software is the main idea to implement high-performance irregular task mapping. The common way is to extend the hardware architecture according to the needs of the mapping algorithm, or to develop related algorithms based on a specific hardware.

As mentioned earlier that irregular tasks can be abstracted by the CDFG, it is a relatively simple and intuitive solution to implement irregular task mapping by mapping the CDFG onto a modulo routing resource graph that models hardware resources. In addition, there are many mapping algorithms designed for two types of irregular tasks, namely, loops and branches. The mapping techniques and CDFG mapping algorithms for branches and nested loops are discussed below respectively.

1. Branch Mapping

Research shows that more than 40% of the compute-intensive loops in SPEC CPU 2006 and the benchmark program of some digital signal processing applications contain branch structures. Another study further shows that more than 70% of the conditional branches in compute-intensive loops of the SPEC CPU 2006 benchmark test suite are nested branch structures with higher complexity. The control flow limits the execution order of some basic blocks, and according to the well-known Amdahl's Law, the control-intensive part of the loop will become the performance bottleneck of the entire program when the acceleration of the compute-intensive part of the loop reaches its relative limit. ITE and NITE structures are widely found in compute-intensive loops and become the main constraint on the performance of SDCs.

For branch structures, typical processing methods can be divided into predicated execution and speculative execution. Speculative execution combines branch prediction and speculation to predict the paths with higher execution probabilities by using branch predictors. The operations of the path prediction can be prefetched or even executed before branch instructions, thus shortening the length of the dependency chain and effectively reducing the II of the software pipeline. The costs of prediction errors include refreshing and repopulating the processor pipeline, using additional buffers to store the modifications to the system state due to the speculative instructions and implementing rollbacks of these modifications when necessary. The significant advantage of speculative execution is improving the parallelism; however, it requires the assistance of a branch predictor. The prediction accuracy of the branch predictor directly affects the performance of the entire program; currently, it is generally used in GPPs. The SDC does not support this branch processing mechanism at present, due to the low accuracy of branch prediction during configuration-level execution and the relatively high cost of branch prediction errors. In the case of a prediction error, the PE needs to be reconfigured and switching configuration packets introduces considerable time overhead [56]. Predicated execution is an important part of the explicit parallel technique, which adds a source operand (i.e., predicate) to each instruction as a condition for instruction execution. When the predicate is true, it executes the operation in the instruction, otherwise transforms the instruction into an NOP instruction. The advantage of predicated execution is that it converts the control flow into data flow, which can combine the basic blocks of the original branches into a hyperblock and increase the granularity of compilation and scheduling, thus increasing the ILP in the basic blocks and effectively improving the performance of the software pipeline or modulo scheduling. For the new hardware architectures of SDC, predicated execution has been evolved into various versions, which can be approximately categorized into three types, the partial predication (PP) [57], the full predication (FP) [58], and the dual-issue single execution (DISE) [59]. The processing methods of these three branches are introduced as follows.

1) FP

FP is a type of techniques that map the operation of updating the same variables in a branch to different control steps of the same PE, so that only the operation that is conditionally selected is actually executed at program runtime. Although the actual performed operation is dynamically variable from iteration to iteration and the moment when the variable value changes is not fixed, the computation result of that PE after the longest time must be the correct output of that variable. Since the final value of this variable comes from a fixed PE, no additional selection operation is required. In the case of a mapping to an ITE with n operations on any path, in the worst case, the number of operation nodes of the FP DFG is $2n$, but at the same time the placement constraint on these $2n$ nodes needs to be added. FP mainly has the following disadvantages: (1) although only operations on one path are executed at runtime as per the conditional results, unnecessary instruction prefetch is wasted because the operations of both branches are statically mapped onto the corresponding PEs before runtime; (2) the operations of modifying variables occupy PE resources at the corresponding time regardless of whether they are executed or not, resulting in a waste of PE resources; (3) the variable can be used only after the longest time, and the II of the software pipeline is large if the program has an iterative dependence on the variable. Figure 4.36a to d show the source code, the code after FP processing, the corresponding CDFG, and the mapping results of the loop body containing the ITE structure.

The state-based full predication (SFP) approach [60] introduces sleep and wake-up mechanisms to change the PE state and decide whether to execute instructions on a specific path. A 1-bit status register is embedded in each PE to indicate the current state. The pseudo branch SFP (PSFP) approach [61] uses predicate registers to mimic the branch behaviors and control the wake-up timing of each PE. PSFP introduces an additional wake-up operation. The counter-based SFP (CSFP) approach uses counters to implement automatic wake-up, thus eliminating the wake-up operation. Both PSFP and CSFP require an unconditional sleep operation to mimic the unconditional jump instruction in traditional GPPs. The SFP can handle both ITE and NITE, but there are redundancies in the additional operations due to state switching.

The tag-based full predication (TFP) approach [62] improves the SFP. The redundant operations in SFP are mainly caused by the conditional judgments and indirect predicate passing between operations on each path. The TFP eliminates the state transfer operation and uses tags to directly transfer predicate information for instruction invalidation, thus eliminating the performance overhead due to state switching while enabling the invalidation of distributed instructions and parallel rewriting of tag registers to improve performance. Similar to the SFP, the TFP does not execute branch operations that do not meet the conditions, thus saving energy. Unlike PP that strives for a minimal architecture modification, the TFP requires the support of more additional hardware and instructions: (1) each PE needs to a multi-bit tag register

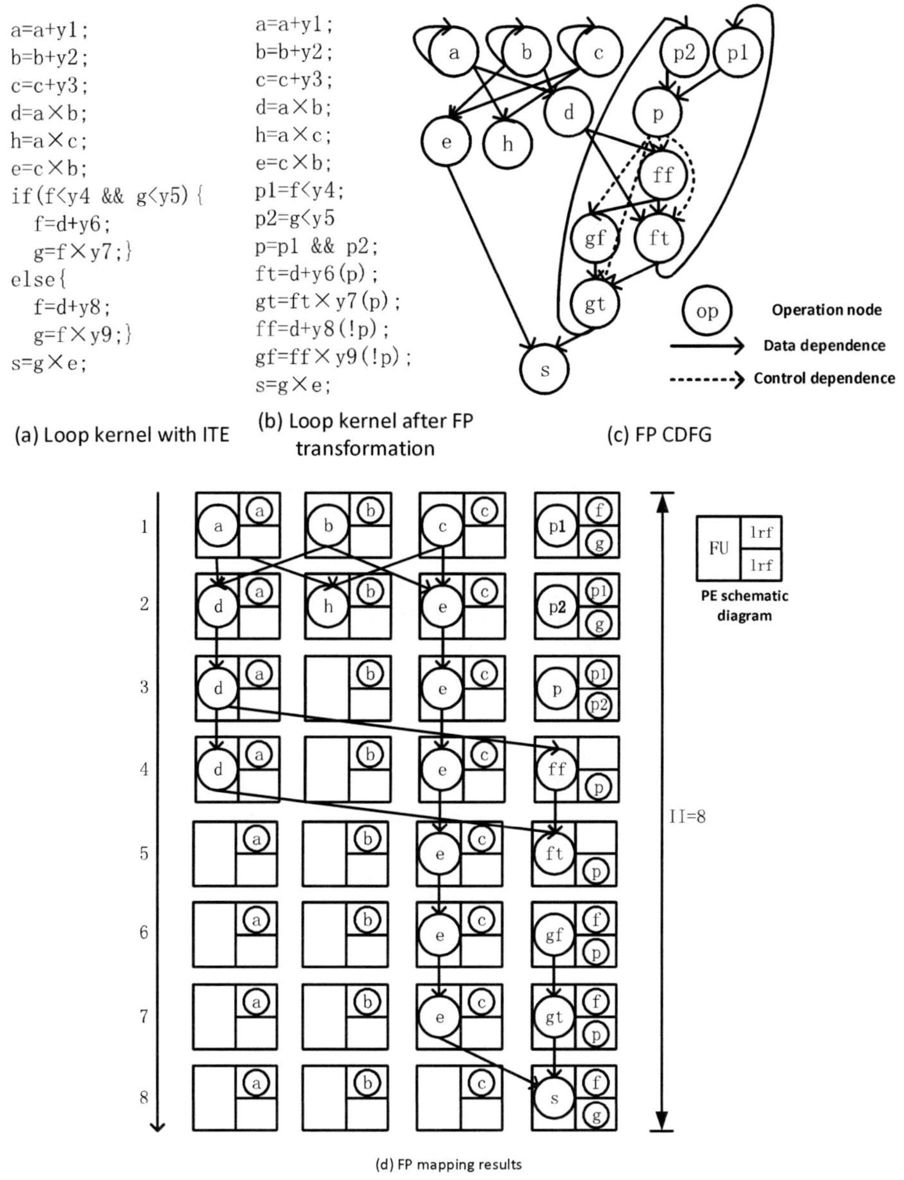

Fig. 4.36 Example of FP branch processing

TReg to indicate which path the PE needs to cancel; (2) additional tagged fields (CT and NT) will be added to each instruction word; (3) additional comparators are required to determine whether TReg and CT have the same value, deciding whether to disable the TReg to hold the current value; (4) the register DReg is required to invalidate the results of the general-purpose computations. Overall, the TFP has more requirements on hardware resulting in a poor generality, which may not be applicable to general SDCs.

2) PP

PP refers to the techniques that map operations in two paths of a branch structure to different PEs so that they can be executed simultaneously. All variables updated by the branch structure need select operations before they are used, and the correct value of a variable is obtained as per the conditional execution result. Obviously, when the same variable is updated in if and else parts, the selection needs to be made from the operational results of both branches. However, when the variable is updated only in one the if and else part, the selection needs to be made between the result of this computation and the old value of the variable (i.e., the variable value before entering the branch structure or the value updated in the last iteration of this branch). Figure 4.37a to c show the code, the CDFG, and the mapping result on the PEA obtained by using PP processing on the source code shown in Fig. 4.36a. With sufficient PE resources, both paths can be executed in parallel with the condition, or even before the condition is executed. Thus, PP can achieve a software pipeline with a smaller II. In the case of a large number of operations in a branch, this approach requires the construction of a predicate network that propagates the conditional computation results to the selection operations corresponding to each variable that may be modified by the branch. In the worst case, a total of $3n$ nodes are required in order to map an ITE structure with n operations on each path. PP has the following significant disadvantages: (1) additional selection operations consume extra time, PE resources, and energy, which is bad for performance; (2) if the two operations to update the same variable in the if and else branches have different scheduling time, i.e., the time to reach the final selection operation is not equal, registers need to be used to preserve the intermediate results or extend the II to avoid the next computation of the pipeline overwriting the computation result of the previous iteration. Using registers to balance the timing will increase the register pressure, and register overflow problems may occur when there are too many unbalanced operations. Extending the II will decrease the throughput and hence the performance gains. The specific method used to solve the problem of unbalanced input timing between the two selection operations depends on the performance indicators concerned by the application scenarios. A tradeoff is made according to the more focused targets. PP can handle both ITE and NITE structures.

3) DISE

DISE indicates the techniques that combine and pack the instructions of two branches into one node and issue them to the same control step of the same PE, and which of these two instructions is executed at program runtime depends on the result of the conditional computation. In the case of a mapping to an ITE with n operations on any path, in the worst case, the CDFG of DISE has n operation nodes, with no placement constraints. The implementation of DISE requires the support from the compiler and it only applicable to dynamic-scheduling SDCs. How the operation nodes are packed affects not only the correctness but also the performance. BRMap [63] is

```
a=a+y1;
b=b+y2;
c=c+y3;
d=a×b;
h=a×c;
e=c×b;
p1=f<y4;
p2=g<y5
p=p1 && p2;
ft=d+y6;
gt=ft×y7;
ff=d+y8;
gf=ff×y9;
f=select(ft,ff,p);
g=select(gt,gf,p)
s=g×e;
```

(a) Loop kernel after SSA transformation

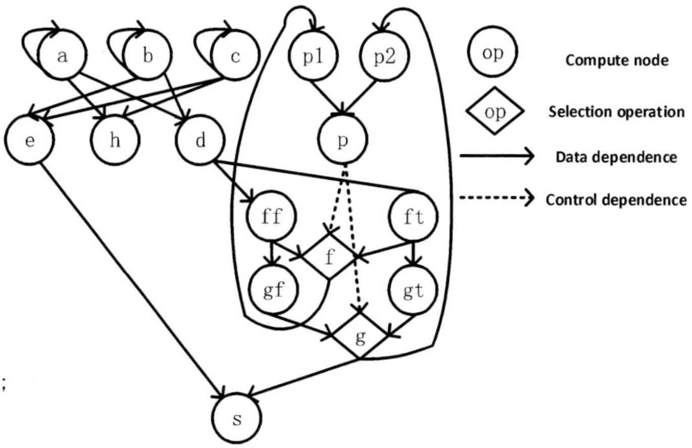

(b) PP CDFG

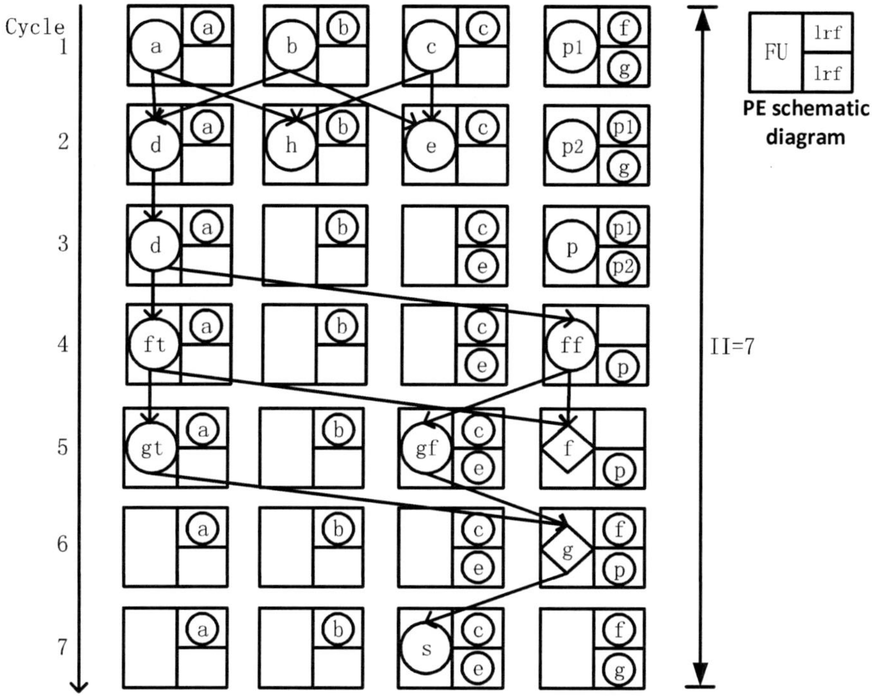

(c) PP mapping results

Fig. 4.37 Example of PP branch processing

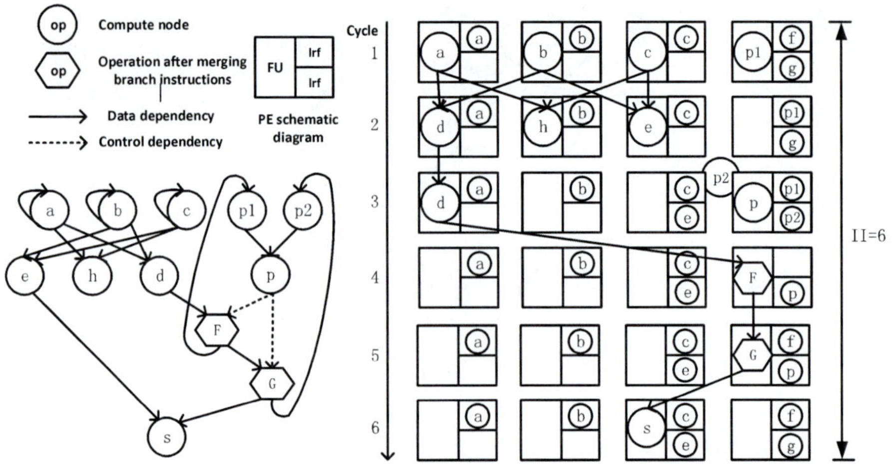

Fig. 4.38 Example of BRMap branch processing

a mapping algorithm based on node packing. It initially schedules the DFG after PP transformation, packs and combines the operations with overlapping scheduling windows within ITE structures as much as possible to minimize the number of DFG nodes while ensuring the correctness. Thus, BRMap effectively reduces the resource-constrained II and achieving an effective mapping of the DISE mechanism on the SDC. DISE can reduce the pressure on PP registers and accelerate ITE execution, whereas the DISE mechanism cannot support NITE, i.e., it does not work for the branch structures that require multi-bit predicate control. Figure 4.38a and b show the CDFG and the mapping results of the node packed PP CDFG shown in Fig. 4.37b by using BRMap.

TRMap [64] is an improvement of the DISE scheme. The algorithm is proposed based on a new control paradigm of the SDC denoted as triggered instruction architecture (TIA). The TIA instruction set provides control instructions to handle branches.

The PE structure of TIA has been described in detail in Chap. 3, which contains an ALU, some registers and a scheduler. The scheduler consists of a trigger resolution and a priority encoder. The trigger resolution accepts the tags, channel status, internal predicate registers, and programmer-specified triggers as the inputs to evaluate whether the trigger condition is true. If it is true, the rigger resolution sends the instruction-ready signal to the priority encoder. The priority encoder selects the instruction with the highest priority from the set of available instructions ready for execution and sends it to the PE for execution. The final PE executable instruction generated by the compiler back end is a trigger instruction; it consists of a trigger condition and a general instruction, and each instruction can be executed only when the trigger condition is satisfied. The trigger is a Boolean expression that consists

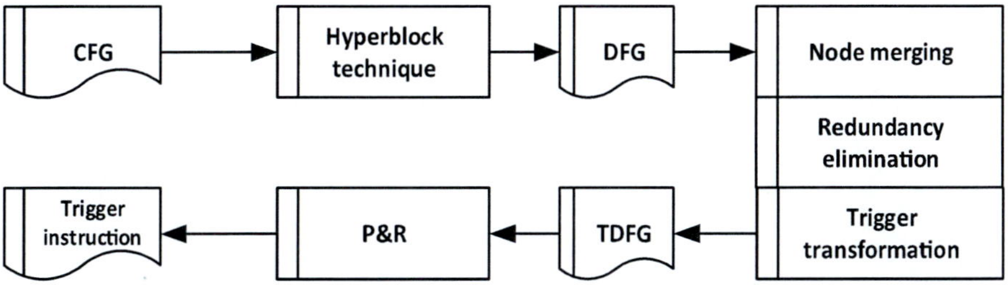

Fig. 4.39 Schematic diagram of TRMap processing flow

of a set of logic operations of the architecture state of the PE to control instruction execution.

TRMap strives to reduce the number of redundant nodes in the DFG for the purpose of reducing the ResMII of the application software pipeline under time and resource constraints. TRMap reduces the size of the DFG in two ways: (1) similar to the DISE mechanism, it merges the operations on different branch paths. Unlike DISE that only allows merging two operations, TRMap can merge more than two operations. TRMap can also support the NITE structure where whether the innermost branch is executed is jointly constrained by multiple conditions and the path selection is cooperatively controlled by multi-bit predicates. (2) It considers the Boolean operations that take the source operands in the DFG as the results of the conditional computation and are used to control flow decisions as redundant nodes, transfers them to the scheduler for execution, and removes these nodes from the original DFG. The processing flow of TRMap is shown in Fig. 4.39. First, TRMap inputs a CFG and constructs a DFG from multiple basic blocks of the CFG via SSA transformation using the hyperblock front-end technique. Then, three optimization methods, operation merging, redundant node effective elimination and triggered instruction transformation, are applied to minimize the DFG and generate the triggered DFG (TDFG). Finally, TRMap uses the P&R algorithm based on modulo scheduling to map the TDFG to TIA and generate trigger instructions.

TRMap combines the two methods of merging branch path operations and offloading the Boolean operations for the control flow decision to the scheduler in TIA for execution, which largely reduces the size of the DFG and improves the throughput of the software pipeline. Figure 4.40 shows the DFG and the corresponding mapping results processed by using TRMap. Compared to Fig. 4.38a, Fig. 4.40a further reduces the number of nodes in the DFG by eliminating the Boolean operation node p where the source operand is a predicate. Comparing between Figs. 4.36d, 4.37c, 4.38b, and 4.40b, it can be seen that, with the same source code, the IIs of the software pipeline implemented by using FP, PP, BRMap, and TRMap are 8, 7, 6, and 5, respectively. TRMap's reduction and optimization in the DFG size results in a significant reduction in the II.

Note that the II is affected by both the number of DFG nodes and the number of PEs. When the size of the PEA is large, reducing the number of DFG nodes may not be enough to change the II, and TRMap has little effect in this case. Therefore, it can

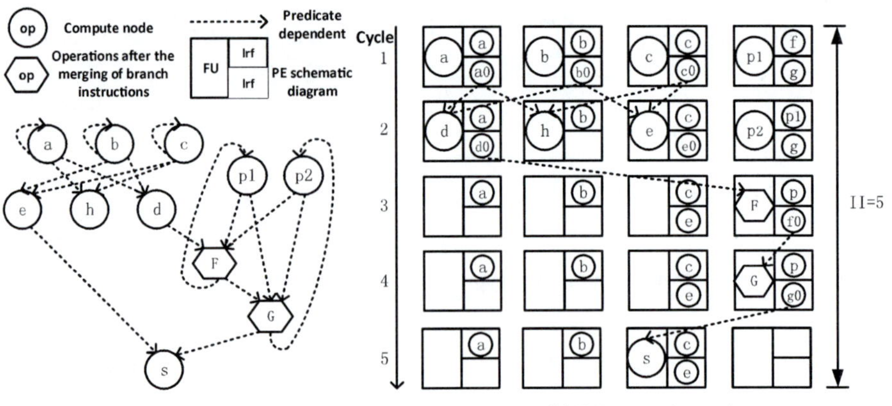

(a) TRMap TDFG (b) TRMap mapping results

Fig. 4.40 Example of TRMap branch processing

be concluded that the smaller the number of PEs in the SDC structure, the larger the number of branch operations and predicate logic operations in the source program (i.e., the larger the number of nodes that can be merged and deleted), and the more significant the performance improvement of TRMap. Neither FP nor DISE can be used for accelerating the loops with NITE. TRMap is not limited by the number of branch paths and can legitimately merge the operations from more than two branch paths, effectively eliminating the connections of predicates. Figure 4.41 shows the whole process of processing a loop with the NITE structure.

(e) BRMap 映射结果	(e) BRMap mapping results
合并节点	Merge node
消除冗余操作	Operations of eliminating redundancies
(f) TRMap 打包后的触发数据流图	(f) TDFG after TRMap packing
周期	Cycle
(g) TRMap 映射结果	(g) TRMap mapping results

As mentioned above, TRMap is based on the SDC architectures that support the control paradigm, such as TIA. The strong hardware support makes it possible to map NITE branches efficiently onto an SDC. However, the data-driven tag-based dynamic scheduling execution mode results in a significant high overhead in the hardware implementation. If some hardware functions are transferred to the compiler, the cost will be compromised. Therefore, the co-design of hardware and software with a reasonable distribution of control and mapping tasks to the compiler and hardware can be a future trend to address control flow mapping problems.

2. Nested Loop Mapping

The compute-intensive tasks of multimedia applications contain a large number of nested loops. As the execution time of the program is mainly consumed by the large

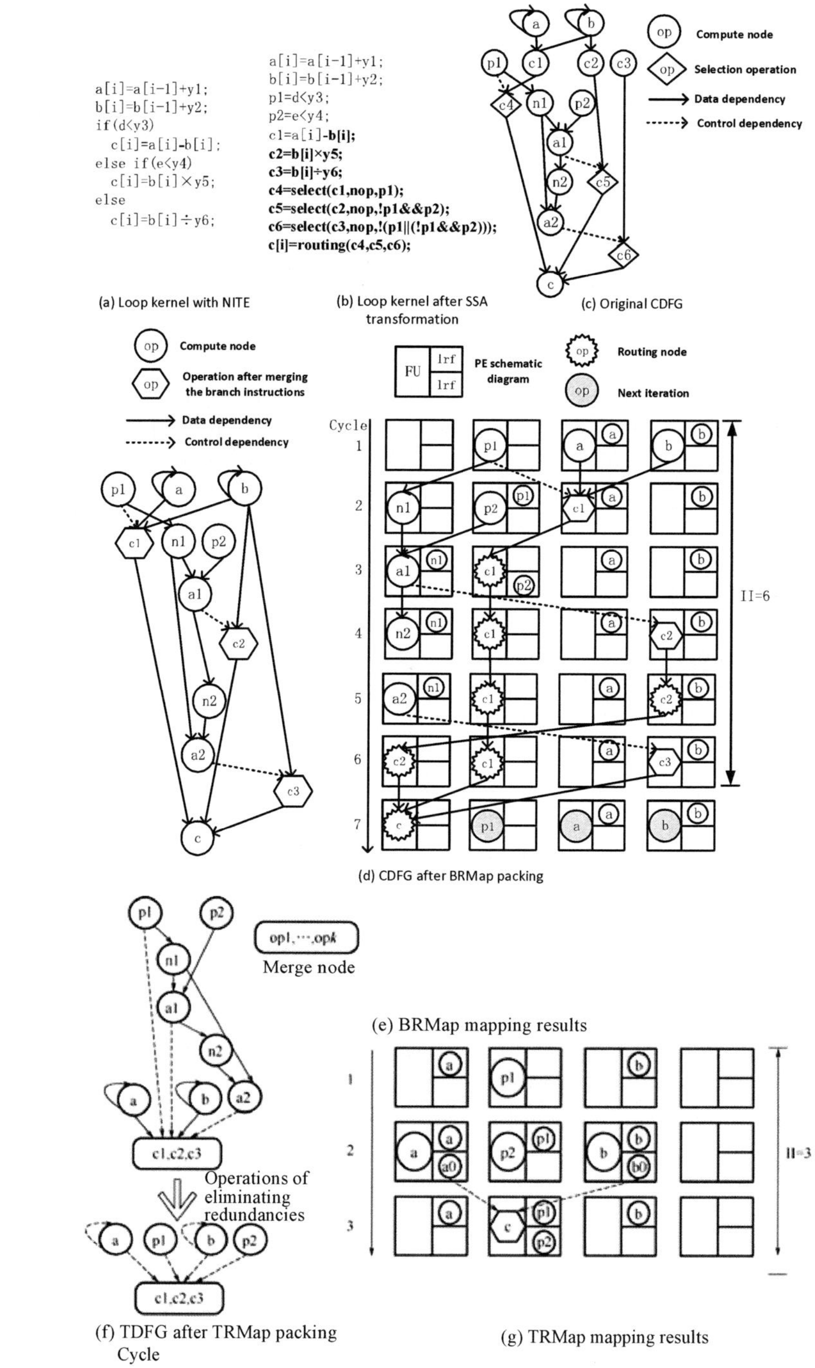

Fig. 4.41 Example of TRMap processing NITE

number of nested loops, the mapping optimization of nested loops is important to improve the performance of compute-intensive applications. Optimizing the loop structure in the algorithm according to the target hardware structure and obtaining the maximum ILP is the core idea of loop optimization. There are many mature techniques for loop optimization on other processors (e.g., CPU and GPU). Loops are one of the most important objective codes of SDCs, and loop optimization for SDCs has received much attention in the past decade. The goal of loop optimization is to exploit the ILP, DLP, and LLP in the loop code, including the typical techniques such as loop unrolling, software pipelining and polyhedral model.

1) Loop Unrolling

Loop unrolling is a simple and effective loop optimization technique that is widely used in compilers for processors with various architectures; it is equally applicable to SDC compilers. It reduces the control code of a loop (e.g., iteration counting and end-of-loop condition judgment) by copying the body code of the loop. It improves the execution performance of a loop by mapping the same instructions to different hardware for parallel execution (similar to SIMD) in a manner similar to vectorization. The loop unrolling factor depends on the size of the operators within the loop and the number of PEs in the hardware structure. One of the characteristics of SDC computing structures is the abundance of hardware resources, which requires the compiler to exploit more computational parallelism to reach its full potential. Meanwhile, the control implementation of this type of structure is costly, which makes the loop unrolling technique well suited for SDC structures. However, loop unrolling is applicable to a limited number of source programs. In particular, only loops without data dependencies between iterations can be unrolled. Also, the loops with branches cannot be unrolled. Loop unrolling is widely used in single-layer loops and can also be applied in some perfect nested loops to improve performance. Note that the data corresponding to the unrolled copy of each loop body should be properly organized to ensure correctness. Taking a single-layer loop as an example, Fig. 4.42 shows the transcoding process of loop unrolling.

Since configurations can be reused by different PEs performing the same operation, the data only needs to be allocated and stored properly and accessed correctly. The maximum loop unrolling factor is $\lfloor N_{\mathrm{PE}}/N_{\mathrm{op}} \rfloor$.

```
int a[20];
for(int i=0;i<20;i++){
    sum+=a[i];
}
```

(a) Source code

```
int a[20];
for(int i=0;i<10;i++){
    sum+=a[2i];
    sum+=a[2i+1];
}
```

(b) Partial loop unrolling

```
int a[20];
for(int i=0;i<10;i++){
    sum+=a[0];
    sum+=a[1];
    ...
    sum+=a[19]
}
```

(c) Full loop unrolling

Fig. 4.42 Example of loop unrolling

2) Multi-level Software Pipelining

Software pipelining techniques are used to accelerate loops on SDC architectures, which can improve the system performance on the basis of almost no increase in the amount of code and configurations. In addition, it can be applied to a certain extent to optimize infinite loops. The software pipelining algorithm for single-layer loops has been detailed in the previous section, and here we focus on techniques for implementing software pipelining for nested loops, i.e., multi-level software pipelining. Perfect nested loops can be implemented by using the software pipelining technique for single-layer loops because two-layer perfect nested loops with m and n cycles respectively can be converted to single-layer loops with $m \times n$ cycles. No additional operations is required for alternating the inner and outer loops, and only the scheduling of the innermost loop body operations needs to be considered. However, software pipelining of the imperfect nested loops on an SDC is a thorny problem. Since imperfect nested loops may contain multiple inner loops of the same level as well as operations outside the inner loop, it is not possible to directly apply modulo scheduling to generate a single-level software pipelining containing all statements because different statements may occur at different frequencies. How to fully utilize the parallelism of the inner and outer loops has a significant impact on the performance. The two-level software pipelining approach [65] solves this problem with the objective code of two-layer imperfect nested loops. When this approach is extended to handle more than two layers of nested loops, the two-level software pipelining technique can be used layer by layer from the inside out, and can be referred to as a multi-level software pipelining.

The following describes the technical details of the two-level software pipelining. The core idea is to first perform Level 1 pipelining operation in the inner loop that may be multiple, where the odd statements (i.e., non-inner loop statements within the outer loop) are viewed as the inner loops with one iteration, fully exploiting the loop parallelism of each inner loop. The Level 2 pipelining operation is then performed in the outer loop to exploit the parallelism between inner loops at the same level. The following problems need to be considered.

(1) Since pipelining operations are performed in both inner and outer loops, each inner or outer loop has a different II. How to apply a global strategy to optimize these IIs uniformly rather than minimizing them individually, so that we can achieve an overall optimum rather than just a local optimum?
(2) The two-level pipelining may lead to oversized pipeline kernels. How to compress the kernels to avoid the long time spent on mapping oversized kernels and the oversized configuration packages (which may exceed the capacity of 'the configuration memory)?

To solve the problem (1), [65] expresses a set of inner and outer loop pipelining IIs, $\mathrm{IIT} = (\mathrm{II}_{i0}, \mathrm{II}_{i1}, \ldots, \mathrm{II}_{im\square 1}, \mathrm{II}_{o})$, limited by resource constraints and data dependencies between iterations as a set of equations or inequalities related to the pipeline execution time L_{ix}, the kernel width W_{ix}, and the loop iterations TC_{ix}, as shown in Eqs. (4.34) to (4.39) (the corresponding meanings of the notations involved in the equations are shown in Table 4.2); the objective function is then constructed by minimizing the total execution time of the outer loop. Thus, the problem of optimizing the IIT can be expressed as Eq. (4.40). This is a typical integer nonlinear programming (INLP) problem. As the search space is $(m+1)$-dimensional and the constraints are relatively complex, the time complexity is high if to solve it by using the general INLP solving method. The analysis and derivation of the original inequality or equation indicate that: (1) the minimum inner loop execution time L_i leads to the minimum outer loop II_o; (2) the total outer loop execution time L_o is positively related to each II, and the minimum II_{ix} and II_o lead to the minimum L_o. + Therefore, the complete chain of relationships is shown in Eq. (4.41). The minimum II_{ix} is computed based on W_i, and the minimum II_o can be computed based on II_{ix}. These IIs constitute the optimal IIT, which results in the minimum L_o. Assuming that the number of PEs in the SDC is N_{PE}, the range of W_i is 1 to N_{PE}, the time complexity of the algorithm is $O(N_{\mathrm{PE}})$.

$$L_{ix} = L_{dx} + \mathrm{II}_{ix} \times (\mathrm{TC}_{ix} - 1) \tag{4.34}$$

$$L_i = \sum \{L_{ix}\} \tag{4.35}$$

$$\begin{aligned} L_o &= L_i + \mathrm{II}_o \times (\mathrm{TC}_o - 1) \\ &= \sum \{L_{dx}\} + \sum \{\mathrm{II}_{ix} \times (\mathrm{TC}_{ix} - 1)\} + \mathrm{II}_o \times (\mathrm{TC}_o - 1) \end{aligned} \tag{4.36}$$

$$W_{ix} = \left\lceil \frac{L_{dx}}{\mathrm{II}_{ix}} \right\rceil \times W_{dx} \tag{4.37}$$

$$W_i = \max\{W_{ix}\} \tag{4.38}$$

$$W_o = \left\lceil \frac{L_i}{\mathrm{II}_o} \right\rceil \times W_i \tag{4.39}$$

$$\min L_o \ \text{s.t.} \mathrm{II}_{ix} \geq \mathrm{RecMII}_{ix}, \quad \mathrm{II}_o \geq \mathrm{RecMII}_o, \quad W_o \leq N_{\mathrm{PE}} \tag{4.40}$$

$$\left. \begin{array}{c} W_i \rightarrow \mathrm{II}_{ix,\min} \rightarrow L_{i,\min} \rightarrow \mathrm{II}_{o,\min} \\ \mathrm{II}_{ix,\min} \end{array} \right\} \tag{4.41}$$

To solve the problem (2), a multi-step compression method is designed in [65] to compress the very large kernel into several smaller parts, which can effectively control the size of the configuration package and the compilation time. The specific approach is shown in Fig. 4.43d to e, where the outer loop kernel (OLK) is first divided into several segments as per the boundary of the inner loop. A segment kernel (SK) is then extracted from each segment. The SK contains several iterations of several inner loops, and the height of the SK, H_{SK}, is determined by the height of the segment H_s and the minimum value of the least common multiple of the IIs of all

Table 4.2 Meanings of notations

Notation	Meaning
L	Computation latency
W	Kernel width
II	Initialization interval
TC	Times of cycles
The first subscript i	Inner loop
The first subscript o	Outer loop
The first subscript d	DFG of the inner loop
The second subscript x	Index of the inner loop

innermost loops contained in the segment, i.e., $H_{\text{SK}} = \min(H_s, \text{LCM}\{\text{II}_{ix}\})$. If the SK contains incomplete inner loop kernels, the missing operations (gray operations in Fig. 4.43e) should be added to the SK to ensure the completeness of the mapping. Finally, the SK is further divided into several segment kernel elements (SKEs) by column. It can be seen that, after compression, each operation or kernel is represented by a hierarchical multi-level index.

In summary, the two-level software pipelining can better exploit the parallelism of nested loops and make full use of the rich computing resources of the SDC.

3) Polyhedral Model

The polyhedral model is a convenient, flexible and expressible alternative representation of nested loops; it is a mathematical framework for optimizing the transformations of nested loop code. The polyhedral model provides an abstraction that uses integral points of a polyhedron to represent the computations and data dependencies of a nested loop. In particular, each iteration entity of the nested loop is viewed as a grid point of a virtual polyhedron. Taking the execution performance of the loop code as the optimization objective, the loop code is then transformed into an equivalent form but with a higher-performance on this polyhedral structure using affine transformations or general non-affine transformations. The polyhedral model is based on mature mathematical theories and is suitable for structural transformations of complex loops. The polyhedral model can also be used to compile the perfect nested loop mappings in SDCs. Research shows that it is an effective method to complete the mappings for the perfect nested loops by analyzing the constraints on the SDC and applying the polyhedral model to optimize the parallelism of the loops. PolyMAP [66] is a genetic algorithm-based heuristic loop transformation and mapping algorithm, which converts the loop mapping problem into a nonlinear optimization problem based on the polyhedral model; it uses the total execution time (TET) of the loop as a performance indicator to guide the mapping. As shown in Fig. 4.44, for a perfect two-layer nested loop, the original iteration domain (i.e., the set of all execution instances of each statement) is a rectangle. A parallelogram iteration space is then obtained after affine transformations. Tiling this new iteration domain onto the PEA can yield better performance than that of the original rectangular iteration domain.

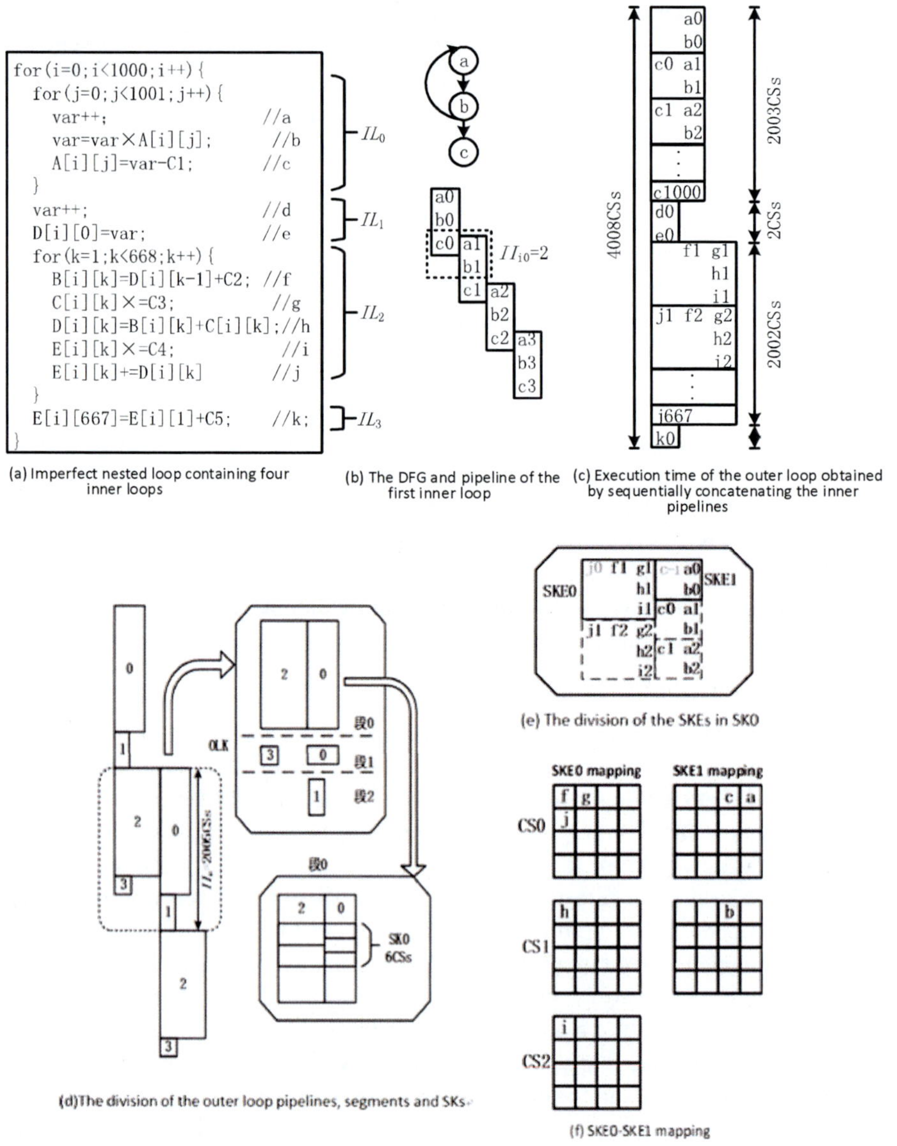

Fig. 4.43 Example of a two-level software pipeline

Imperfect nested loops can be converted into perfect nested loops using approaches such as merging transformation, and then the algorithms that are suitable for perfect nested loops such as PolyMAP can be applied to achieve mappings with high parallelism.

The PolyMAP method is further improved in [67] aiming at solving the problems of poor hardware utilization and low execution performance of existing loop pipelines in dealing with nested loops. The affine transformation is used to simplify

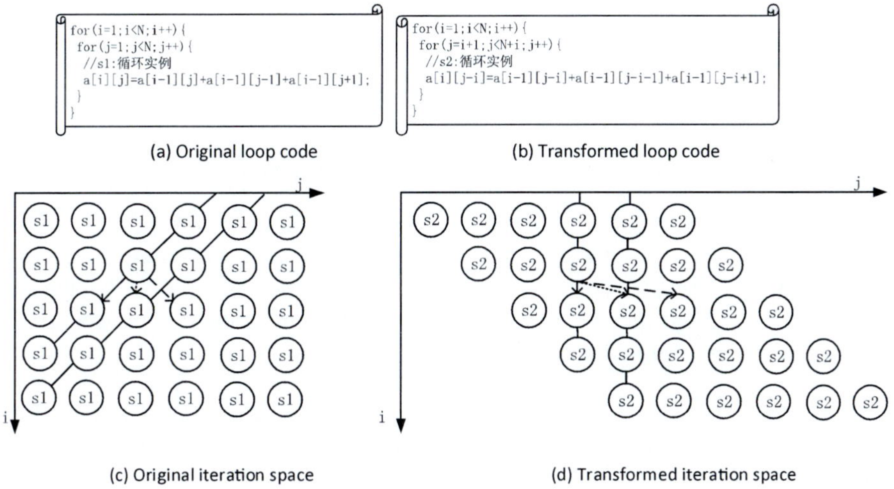

Fig. 4.44 Transformation example of the polyhedral model

the nested loop pipeline. Combined with the multi-pipeline merging method, it can make full use of the parallelism in the inner loop to improve the PE utilization, and reduce the data memory dependency due to the outer loop, thus reducing the access overhead during the alternation of the inner and outer loops. The affine transformation parameters and pipeline II are incorporated as the variable parameters into the performance model based on the polyhedral model for iterative optimization. The specific implementation framework is shown in Fig. 4.45. First, all valid pipelining-beneficial loop transformation parameter sets are generated based on the polyhedral model. According to each parameter set, the affine transformation is performed on the nested loops to generate feasible merging factors that are used to merge the inner loop pipelines. The detailed implementation algorithm is described in [67]. Then, the performance of a series of candidate transformed and merged nested loops is evaluated by calculating the total execution time, and ranking all candidates as per the execution time in ascending order. Finally, starting from the initial position of the candidate list, the P&R is performed in sequence to find the effective mapping with the best performance. The key to this method is the polyhedral modeling of the execution time of the nested loops using the affine transformation parameters and the merging factor as parameters. It can be seen that the polyhedral model has advantages in terms of flexibility and effectiveness in optimizing nested loop pipelines. This model can adaptively mine the optimal affine transformation parameters for its source program based on the target loop code.

4) Other Loop Transformation Methods

(1) Non-basic-to-Basic-Loop Transformation [68]

The non-basic-to-basic-loop transformation method moves all statements to the same innermost loop, creating an ideal nested loop. Since the location of some statements

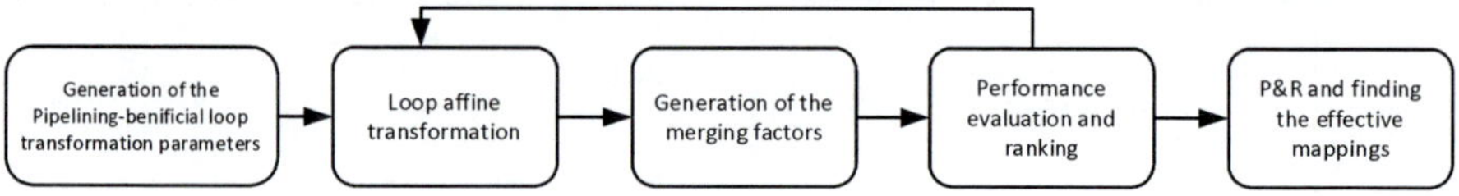

Fig. 4.45 Iterative optimization framework for affine parameters based on the polyhedral model

in the source code is changed, it may be necessary to add if statements to ensure that the program is executed correctly. This method is simple and intuitive, but the added operations may significantly reduce the performance, with the loss outweighing the gain. Thus, it has limited application.

(2) Loop Fission

The loop fission breaks an imperfect loop into several perfect nested loops, which is invalid when the original loop has iterative dependencies.

(3) Loop Tiling [69]

As an important loop optimization technique, the loop titling maps loops based on the iteration-level granularity. However, SDCs map source programs based on the operation-level granularity. Thus, it is not suitable for SDCs.

(4) Flattening-Based Mapping of Imperfect Loop Nests [70]

This method moves the odd statements to the extended hardware for execution without affecting the inner pipeline. However, the outer loops still use the method of merging the Epilog-Prologs of multiple inner loop pipelines followed by a sequential execution. As a result, the parallelism of the outer loops is not fully exploited. The PE utilization is still limited by the kernel of the innermost loop; the memory access overhead is not considered. In addition, this approach cannot use non-fissionable inner loops of the same level to handle imperfect loops.

(5) Single-Dimension Software Pipelining (SSP) [71]

This is an outer-loop pipelining approach, which selects the most profitable loop level of the nested loops and fully unrolls it into a single loop for pipelining so that a sufficient overlap can be achieved beyond the innermost level. It can support imperfect nested loops with odd statements; however, the imperfect nested loops with inner loops at the same level are not supported. In addition, the approach is not applicable to SDCs due to the complexity of resource allocation, data transmission, and loop control.

3. CDFG Mapping Algorithms

While keeping the CDFG of the original program unchanged, an efficient way to handle the mapping of irregular tasks is mapping the complete CDFG structure containing control flow and dataflow to an SDC that supports jump and conditional jump instructions, and a lightweight global synchronization mechanism [4]. The input to the mapping algorithm is a high-level abstraction of the application characterized

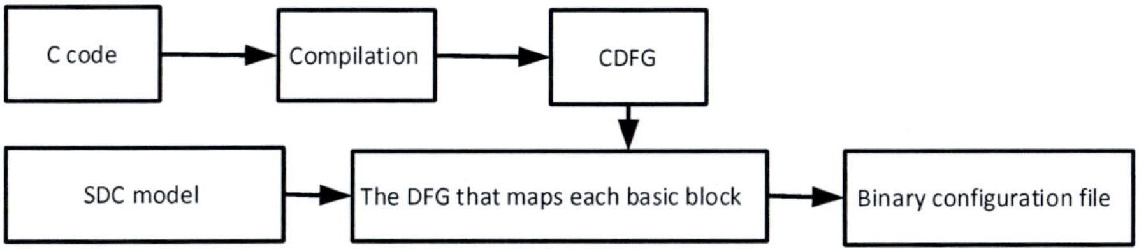

Fig. 4.46 Schematic diagram of the CDFG mapping process

by a CDFG and an SDC model, as shown in Fig. 4.46, where the CDFG is denoted by $G = (V, E)$, V is the set of the basic blocks, and $E \subseteq V \times V$ is the set of the directed edges of the control flow. Each vertex of the CDFG is a basic block (BB), which is represented as a DFG, and the edges in the basic blocks represent the data dependencies between operations. The basic block consists of a series of consecutive statements, where the control flow enters at the beginning of the basic block and ends at the end. The control flow at the end of the basic block is supported by the jump (jmp) and conditional jump (cjmp) instructions. The source program, the corresponding CDFG, and the mapping result are shown in Fig. 4.47. The specific mapping process is that each basic block in the CDFG is scheduled sequentially in a certain order. There is no competition among the basic blocks for the hardware resources at the same moment. Thus, each basic block can be independently mapped onto the whole PEA using any of the previously mentioned algorithms for regular task mapping.

The following problems need to be considered:

(1) Synchronization. When the execution process jumps from one basic block to another, all PEs in the SDC must switch to the configuration of the next basic block simultaneously.
(2) Scheduling order of the basic blocks. The traversal order of the CDFG affects the memory allocation and lifecycle of the variables, and an appropriate scheduling order will improve the mapping performance. The common traversal schemes include the forward breadth-first traversal, the backward breadth-first traversal, the depth-first traversal, and the random traversal. Usually, the forward breadth-first traversal follows the execution order of the program and is a relatively good traversal method.
(3) Data transfer between basic blocks. Three storage medium, memory, global registers and local registers, can be used to route intermediate variables. Using memory to route intermediate variables requires a high bandwidth and long access time. In comparison, using local registers results in the least cost. Since the same variable appearing in different basic blocks may occupy the same registers, the allocation of registers needs to be further considered.

Compared with the typical processing techniques for PP and FP, this mapping method avoids the prefetch and execution of unnecessary instructions. As listed in Table 4.3, it shows significant advantages in energy consumption, which is suitable for the application areas with more stringent power requirements, such as the IoT.

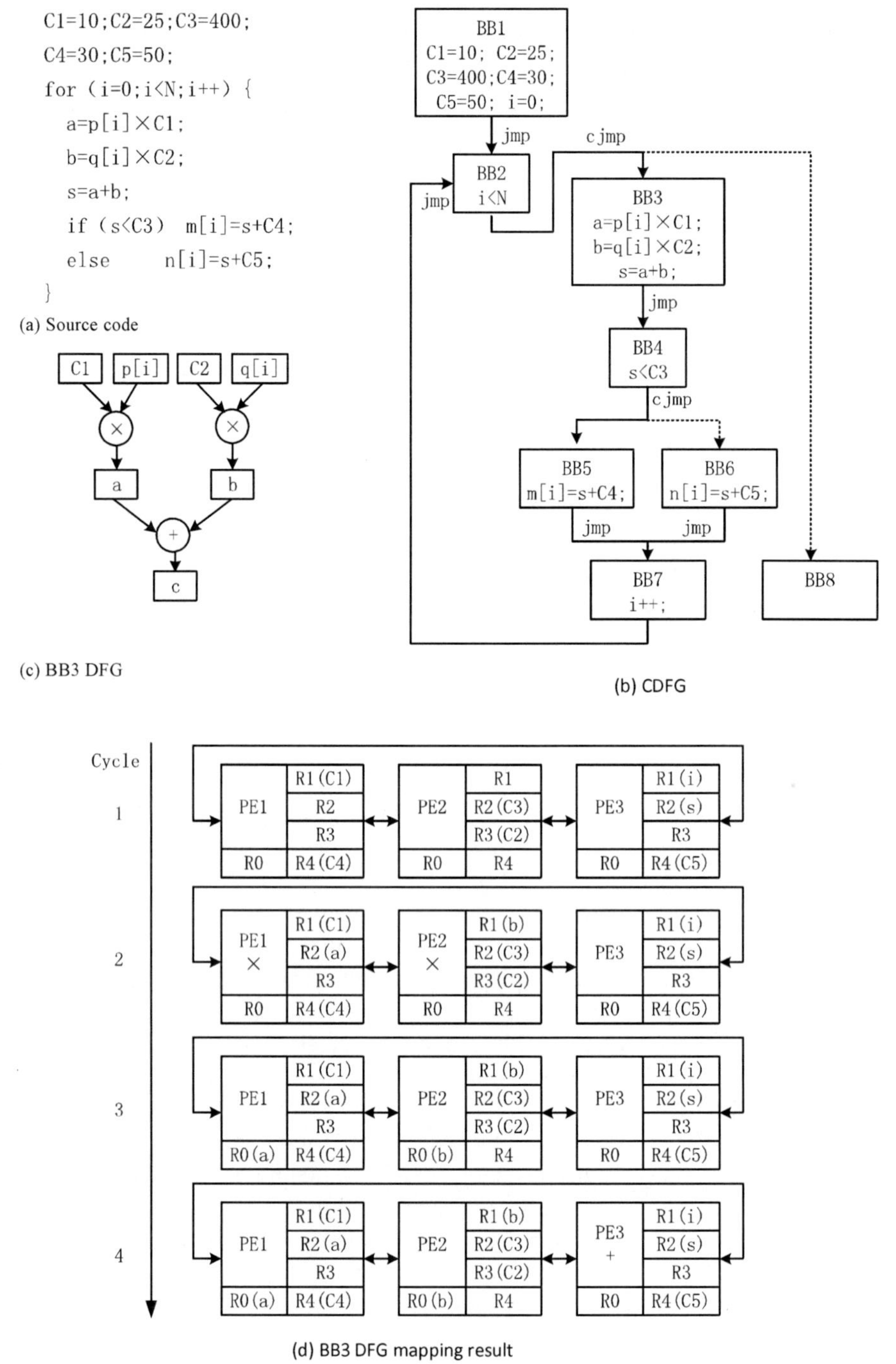

Fig. 4.47 Example of the CDFG mapping algorithm

Table 4.3 Comparison for the times of the prefetch and execution of the branch instructions

Instruction prefetch				Instruction execution			
PP	FP	DISE	CDFG	PP	FP	DISE	CDFG
$3n$	$2n$	n	n	$3n$	n	n	n

Meanwhile, the CDFG mapping algorithm has no constraints on the type of the control flow; it can handle both loops and branches, support complex jumps, and has a high flexibility in mapping the complete applications. This method treats the execution of the basic blocks as independent, and the mapping of each basic block is done independently and can occupy the whole PEA. Moreover, each basic block can use the maximum hardware resources, increasing the mapping flexibility. However, this kind of scheduling based on the basic block granularity can make the PE resources less utilized because the operations within one basic block are generally not enough to cover the whole PEA. Also, the basic block-by-block scheduling has a strict limitation on the execution order of the basic blocks without considering the ILP between them. In addition, when intermediate variables are routed using local registers, the operations and their source operands may be laid out to different PEs.

Then, additional routes are added to send the source operand to the PE where the destination operation is located, thus increasing the latency and causing performance loss. The hardware supporting the CDFG mapping algorithm needs to support two additional instructions, jmp and cjmp. Each PE can compute the address of the first instruction in the next basic block based on the offset. The hardware architecture is more complex compared to the common PEs in SDCs. Thus, this approach achieves flexible control flow mapping at the cost of an increased hardware complexity.

In summary, the mapping optimization for irregular tasks like nested loops and branches on SDCs, has attracted extensive and intensive research in many domestic and international research institutions in the last decade. This is because data-intensive code in real applications is typically nested loops, and branch structures inevitably exist inside the loops. When the entire data-intensive part is offloaded to the SDC for execution, the dynamic jumps triggered by branches and nested loops can limit the overall performance of the program, making it necessary to address the problem of irregular task mapping. The key to improving the performance of the irregular task mapping is to develop more ILPs, which requires the mapping algorithm to make full use of the structural characteristics of the SDC: (1) rich PE resources, allowing large pipeline kernels and supporting high parallelism; (2) direct interconnects between PEs, allowing data to be passed directly between producers and consumers without going through intermediate registers or memory, and (3) one configuration for multiple executions, and the size of the configuration package does not increase with the number of the loop iterations. Implementing an irregular task mapping containing complex control flow requires the co-design at the compiler and hardware levels, i.e., optimizing the scheduling mapping algorithm to maximize the exploitation of parallelism and maintain a correct program execution subject to control flow constraints, and extending the structure of the PE to provide the resources required by the storage and routing of some fine-grained control signals in the algorithm. Co-optimization of hardware and software allows for a good tradeoff between the compilation time and hardware complexity, resulting in the best performance.

4.3 Dynamic Compilation Methods

The increasing scale of the integrated circuits and the demand for computing power of data-intensive applications (e.g., AI, bioinformatics, data centers, and IoT) drive the integration of increasingly rich computing resources on a chip, resulting in an exponential increase in the size of the mapping problem in static compilation of SDCs. To address this problem, clustering algorithms and analytical placement techniques have been proposed to reduce the time overhead of the compilation algorithms with the "divide-and-rule" idea, to improve the scalability. Therefore, the target hardware of a static compilation must be limited to an affordable size, and the part larger than this size will be divided into multiple subproblems to be solved. In order to improve the utilization of the hardware resources and implement multithreading, SDCs can be dynamically compiled by taking advantage of their dynamic reconfigurability. Based on the hardware virtualization, the dynamic compilation uses the configuration flow or instruction flow that has been compiled offline to convert the static compilation result into a dynamic configuration flow subject to the dynamic constraints at runtime, to avoid an excessive time overhead of the dynamic conversion process. Therefore, dynamic compilation can be classified into two categories, the instruction flow-based configuration dynamic generation method and the configuration flow-based dynamic configuration transformation method. This section will first introduce the hardware virtualization technique of SDCs, followed by the introduction of the two dynamic compilation methods.

4.3.1 Hardware Resource Virtualization

The concept of hardware virtualization started in 1960 originally to describe virtual machines (IBM M44/44X); it is a way to logically divide computer system resources. Virtualization gives users an abstract computing platform that allows computing elements to run on virtual hardware, to achieve dynamic resource allocation and flexible scheduling at the software level. For example, NVIDIA offers the virtual GPUs [72] that can be flexibly deployed in the data center with the features of real-time migration, optimized management and monitoring, and the use of multiple GPUs in a single virtual machine. Theoretically, virtualization has been proven to be effective in improving the system utilization and can accommodate the dynamic load characteristics and resource variability presented in most big data applications. Today, there is an increasing demand for virtualization, where the cloud computing is a typical example.

Cloud computing is an Internet-based compute mode that has now become a new computing paradigm. For cloud computing, virtualization is an essential technique that enables the isolation among users, the flexible and scalable for systems, and improved security by virtualizing hardware resources. Thus, the hardware resources can be fully utilized. The most commonly used virtualization technique in cloud computing is the server virtualization. Server virtualization virtualizes a computer into multiple virtual machines (VMs), and the hardware and OS can be separated through a virtualization layer between hardware and software; this virtualization layer is referred to as a hypervisor or virtual machine manager (VMM). By using dynamic partitioning and scheduling, the virtualization layer allows multiple OS instances to share the physical server resources, and each VM has an independent set of virtual hardware devices and an independent guest OS that can run user applications.

Docker container [73] is another technique commonly used in cloud computing. Docker is an open-source application container engine that allows users to package the applications in the OS and the dependent packages into an independent container, which can improve the efficiency of delivering applications. Docker is similar to the VM. However, Docker is an OS-level virtualization, while the VM is a hardware-level virtualization. In comparison, Docker is more flexible and efficient. The following introduces several main hardware virtualization technologies.

1. GPU Virtualization

With thousands of compute cores, GPU can process the workloads in parallel and efficiently; with many parallel PEs, they are much faster than GPPs. GPU virtualization allows multiple client VM instances to share the same GPU or multiple GPUs for computation. However, because GPU architectures are not standardized and the GPU vendors offer architectures with widely varying degrees of support for virtualization, traditional virtualization cannot be applied directly. Due to the complex GPU architecture and more limitations compared to GPPs, it was not until 2014 that two full virtualization strategies for mainstream GPU platforms implementing hardware-assisted virtualization emerged, GPUvm based on NVIDIA GPUs and gVirt based on Intel GPUs. GPU virtualization remains a challenging problem, and several types of GPU virtualization techniques are briefly described below [74, 75].

1) API Remoting

This approach virtualizes the GPU at a higher level. Since GPU vendors do not provide source code for their GPU drivers, it is difficult to virtualize GPUs at the driver level as it is done for the hardware such as virtualized disks. API remoting provides the client OS with a GPU wrapper library that has the same API as the original GPU library. This wrapper library intercepts the calls to GPUs from applications (e.g., the calls to OpenGL, Direct3D, CUDA, and OpenCL), remotes the intercepted calls to the host OS for execution, and returns the processed result to the client OS. However, API remoting is limited by the platform. For example, a Linux host cannot execute DirectX commands remoted from a Windows client. At the same time, API remoting can cause a lot of context switching, which can lead to a relatively large performance loss.

2) Para/Full Virtualization

Full virtualization completely mimics the entire hardware, providing the same environment for the client OS as the actual hardware, while para virtualization modifies some elements of the client OS on top of the full virtualization, adding an extra API to optimize the instructions issued by the client OS. GPU virtualization can be implemented at the driver level based on the architectural documentations for some GPU models that are now publicly provided by vendors or obtained via reverse engineering. Para virtualization modifies the custom driver in the client to deliver sensitive operations directly to the host to improve performance, while full virtualization eliminates this step because it mimics a complete GPU.

3) Hardware-Supported Virtualization

This approach allows the guest OS to access the GPU directly through hardware extensions provided by the vendor. This direct access to GPUs is achieved by remapping DMA and interrupts to each client OS. Both Intel VT-d and AMD-Vi support this mechanism, but they do not allow sharing a GPU between multiple client OSs. NVIDIA GRID lifts this restriction and allows multiplexing NVIDIA GPUs designed for cloud environments.

2. FPGA Virtualization

Due to the advantages of the programmability and high performance, many cloud service providers have been using FPGAs to provide cloud services in recent years, such as Amazon, Alibaba, Microsoft and among other companies. Compared to GPU, FPGA accelerators provide energy-efficient and high-performance schemes for DNN inference tasks. In the traditional FPGA development model, users typically use HDLs for modeling and map the hardware model to the FPGA using development tools to eventually generate an executable FPGA mapping. Thus, FPGA can only be developed and used by a single user. In the cases where an application requires fewer resources and does not need to run continuously, most of the hardware resources in the FPGA are left idle, which prevents the FPGA from being fully utilized. Therefore, virtualization-based FPGAs are needed to meet multi-client and dynamic load scenarios in practical applications. Stuart Byma, et al. from the University of Toronto provided partially reconfigurable regions among multiple FPGAs as cloud computing resources via OpenStack, allowing users to start user-designed or pre-defined hardware accelerators connected by a network as if they were VMs. Chen Fei, et al. from IBM China Research Institute proposed a general framework for integrating FPGAs into data centers and established a prototype system based on OpenStack, LinuxKVM, and Xilinx FPGAs, thus achieving the isolation of multiple processes in multiple VMs, precisely quantified accelerator resource allocation, and priority-based task scheduling.

The logical abstraction for FPGAs through virtualization can to some extent improve their development efficiency, better utilize FPGA logic resources, facilitate the large-scale deployment and application of FPGAs, and save the top-level users from having to pay too much attention to the implementation and details of the FPGA hardware. The research on the virtualization of FPGAs are focusing on addressing two issues, abstraction and standardization; they are the key to integrating FPGAs into OSs. The following is a brief description of several typical virtualization implementations for FPGA.

1) Virtualization Layer

The virtualization layer (overlay) is a layer of virtual programmable architecture that connects the FPGA hardware to the top-level application. The virtualization layer aims to provide a programming architecture and interface familiar to the upper-layer users, so that they can use high-level languages like C in programming without caring about the specific circuit structure. As a result, the abstraction and virtualization of the underlying hardware resources are realized. Using CGRA as a virtualization layer may be an effective solution for FPGA abstraction [76, 77], and the resultant agile development can be several orders of magnitude faster than that of HLS. Moreover, due to the dynamic reconfiguration of CGRA, the CGRA virtualization layer offers a higher degree of flexibility than the vendor-provided CAD tools. In addition, since the virtualization layer provides logic processing units or soft-core processors that are typically independent of the underlying FPGA hardware, upper-layer designs can be easily ported across different FPGA architectures.

2) Partial Reconfiguration and Virtualization Manager

Partial reconfiguration means to carve out one or more regions in an FPGA and program and configure them separately at runtime without affecting the operation of the other parts of the FPGA. Partial reconfiguration allows the FPGA to switch among multiple tasks directly at the hardware level. In general, the FPGA virtualization through partial reconfiguration requires the introduction of additional management layers. Similar to the VM monitor, the management layer manages and schedules the various resources of the virtualized FPGA in a unified manner, such as the Shell layer in the catapult project [78].

3) Virtual Hardware Process

The virtual hardware process contains the abstract concept of managing the execution and scheduling of hardware resources. The virtual hardware process is assigned to the appropriate execution units depending on the resources, which requires standard software APIs, hardware interfaces and protocols, and a unified execution model. FPGAs have two execution abstractions depending on the functions. If the FPGA acts as an accelerator, it can be controlled by the processor as a slave device with drivers. References [79] and [80] discussed the standard interfaces and library designs for such FPGA accelerators. This type of application is very common and many suppliers of commercial FPGAs provide a variety of supporting tool chains for supporting this type of applications. If the FPGA acts as a processor equivalent to the GPP, the architecture can be abstracted to a series of hardware applications that are software-independent and can interact with the software applications by means of communication and synchronization. If the communication between the software and hardware applications uses a message passing mechanism, this hardware application is referred to as a hardware process [81]. If a coupled memory sharing mechanism is used, this hardware application is referred to as a hardware thread.

In the resource management of virtual hardware processes, since most FPGAs use an island architecture, a task/process/thread can only be placed on one island, which allows easy suspension and recovery. Reference [82] suggests that a more fine-grained grid structure can be used to improve the hardware utilization, allowing a thread to use multiple idle grids at the same time.

4) Standardization

Standardization requires a unified solution that is generally accepted by the industry and can achieve better portability, reusability, and development efficiency for reconfigurable computing. Several commercial companies, such as ARM, AMD, Huawei, Qualcomm, and Xilinx, have already involved in the development of general standards. The resultant achievements include CCIX [83] and HSA Foundation [84].

Overall, FPGA virtualization is still in the early stage of development; it is a hot topic of research in both industry and academia.

3. SDC Virtualization

Since the CGRA is the best implementation of SDCs, the following will discuss SDC virtualization with the CGRA as an example. The reconfigurable computing adopted by the CGRA is an emerging computing paradigm that is reflected in both research and practical applications, making new flexible computing architectures possible, yet with some challenges. The first is the productivity gap between hardware design and software programming. Compared to software programming, hardware design shows significant differences in both the underlying methodology and the time required to perform the design iteration. Software compilation is very quick, whereas hardware synthesis is a very time-consuming process. This gap hinders the rapid scaling of the reconfigurable computing. The underlying architecture of the CGRA has very diverse designs and development environments, which restricts the ease use of the CGRA. Virtualization, however, is an effective solution to this problem. Virtualization can make the instruction scheduling and mapping processes of the CGRA transparent to the programmer, so that the programmer does not have to consider the specific hardware architecture.

1) Challenges of Virtualization

CGRA virtualization is very similar to the relatively mature FPGA virtualization, but much less studied. This is because the hardware and software designs of the CGRA lack the widely available commercial products, recognized underlying research platforms and compilation systems, common evaluation benchmarks, and acknowledged architectural templates and hardware abstractions. For example, some CGRAs, such as ADRES, support only PEA-based reconfiguration, while some CGRAs such as PipeRench [85] support PE line-based reconfiguration, where each PE line is a stage of the pipeline; some CGRAs like TIA support PE unit-based reconfiguration, where many communication channels exist between PEs, such as the shared register files, the crossbar switches, and the FIFOs. Therefore, it is difficult to make an objective comparison for different architectures (e.g., external interfaces, memory systems, system control methods, PE functions, and interconnects). The architectural designs are diverse and controversial, which is a major challenge for CGRA virtualization.

Another challenge lies in the compilation. The CGRA mainly uses static compilation to determine the usage of the compute and interconnect resources; thus, it is difficult to dynamically schedule the configurations with determined execution order and synchronous communication.

2) Virtualization Methods and Possible Directions

As shown in Fig. 4.48, CGRA virtualization provides a unified model that includes standardized interfaces, communication protocols, and execution abstractions. Based on this model and the input application, the static compiler generates a virtual configuration suitable for this family of CGRAs. Then, the virtual configuration is optimized and mapped online or offline to a specific physical CGRA. In addition, the generated configuration is sent to the system scheduler to determine the placement and eviction of runtime tasks based on the resource utilization and status. Virtualization facilitates the use of the CGRA through a unified model, which allows for an easy incorporation of the CGRA into the OS. Virtualization also benefits the CGRA designs, where designers can generate any application-dependent CGRA that fits into a co-development environment.

The implementations of CGRA virtualization have been controversial. Several possible directions of CGRA virtualization are listed below.

(1) Abstraction and Standardization

One of the main challenges for the future CGRA is to provide users unfamiliar with the underlying concepts with higher design productivity and simpler ways to use

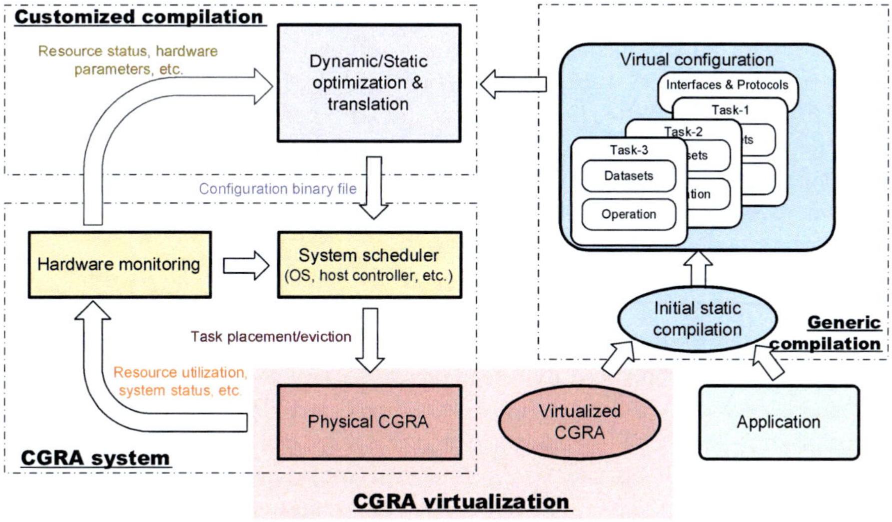

Fig. 4.48 CGRA virtualization and its supported systems

reconfigurable computing systems. One way to achieve this goal is to increase the level of abstraction and standardization. For example, the HLS is used to increase the level of abstractions, i.e., designing the hardware at a high level. This is analogous to introducing a high-level language to replace the assembly language in software development, which greatly simplifies the design process. Standardization refers to the provision of a set of well-defined interfaces and protocols that enable developers to design hardware based on these unified models. Standardization can be done at different levels: either at the hardware level by defining a set of standardized interfaces and protocols (see CCIX consortium [83]), or at a high level by providing standardized interfaces at the OS level [86], allowing both users and developers to use a unified view regardless of the hardware details.

The following provides the cases to be considered and implemented in integrating the CGRA into the OS. The two main tasks of an OS are to abstract and manage resources. Abstraction is a powerful mechanism to handle complex and disparate hardware tasks in a generic way. As one of the most fundamental abstractions of an OS, a process is like an application running on the VM, where the threads allow different tasks to run concurrently on this VM, enabling task-level parallelism. To allow different processes and threads to work in coordination, the OS must provide the methods for communication and synchronization. In addition to abstraction, resource management of the underlying hardware components is also necessary because the virtual computing resources provided by the OS to processes and threads need to share available physical resources (processors, memory, and devices) in both the spatial and time domains. Brebner [87], based on the Xilinx XC6200, is one of the first implementations. Brebner introduced the concept of "virtual hardware" and named the reconfigurable part as a swappable logic unit. One practice is to provide an accelerator API framework for reconfigurable computing. For example, HybridOS (2007) [88] extends the Linux with some standardized APIs to provide an accelerator virtualization; Leap (2011) [79] provides a basic device abstraction and a set of standardized I/O and memory management services for FPGAs. Another practice is to provide a real OS, such as BORPH (2007) [81], FUSE (2011) [89], SPREAD (2012) [90], RTSM (2015) [91], etc. Basically, all systems use island-based hardware architectures. However, there is a lack of some standardized test suite to evaluate the reconfigurable computing OSs in a complete way. Also, there is a considerable space for improving the security and standardization (portability) of OS, which is the same in dynamic and partial reconfigurations.

In addition, since the standardized software APIs, hardware interfaces and protocols of CGRAs are very similar to those of FPGAs, it is not difficult to implement CGRA standardization. However, it is a big challenge to make the standard universally accepted and followed by researchers and users. Some CGRAs already have standard software APIs, such as the special drivers and library files for configuration expressions. Most CGRAs belong to this category, such as MorphoSys, ADRES, and XPP. Some CGRAs, such as TIA and TRIPS, can be used as parallel computing processors, equivalent to GPPs in OSs. They can be abstracted to the hardware applications, which can interact with GPPs through standard hardware/software communication and synchronization. There are also CGRAs that serve as the alternative data paths of GPPs, such as DySER and DynaSpAM [9], which can be abstracted at a lower level to the extensions of the instruction set architecture.

(2) Virtual Hardware Process

Virtual hardware processes are relatively easy to implement on CGRAs because CGRAs have coarse-grained resources that facilitate dynamic scheduling.

Virtualization can mask the architectural differences of different CGRAs. Some CGRAs, such as MorphoSys and ADRES, are executed statically, i.e., the location and execution time of operations are determined at compile time. Some other CGRAs are executed dynamically. For example, KressArray uses tokens to control execution; BilRC uses an execution triggered model, which triggers the PEs at runtime for static configuration and dynamic compilation. In [92], it is stated that dynamic scheduling can improve the CGRA performance by 30% compared to static scheduling, with an area overhead of about 8.6% of that of the static scheduling. Resilient CGRAs [93] uses the idea of resilient circuits, i.e., the scheduling of the operations on the data path is not fixed, but is dynamically determined by the readiness of the operands. It is a synchronous circuit that operates like an asynchronous circuit. There are two types of interconnected circuits, one for processing data and the other for controlling the flow of operations. These two circuits are seamlessly mapped onto the resilient CGRA. Only a small amount of information, such as the latency and topology of the components, is sufficient to create the configurations for the resilient CGRA. It enables the code with binary compatibility when mapping onto different resilient CGRAs. PipeRench and Tartan [94] provide the virtualized execution models for their compilers. In particular, the PipeRench architecture completes the virtualized pipeline computation through pipeline reconfiguration, which allows executing deep pipeline computations even if the configuration is never completely present in the architecture. Tartan can use a considerable amount of hardware to execute the entire generic application through the virtualization model. TFlex [95] is a composable CGRA with PE units that can be arbitrarily aggregated together to form a larger single-threaded processor. The thread can be placed on an optimal number of PEs and run at the most energy-efficient node, providing a dynamic and efficient PE-level resource management solution for the OS. Pager et al. [96] proposed a software scheme for multithreaded CGRAs, which uses a traditional single-threaded CGRA as a multithreaded coprocessor for a multithreaded GPP. This approach converts the configuration binary file at runtime, which makes the thread occupy fewer pages (the pages are similar to the islands in FPGAs), providing a dynamic page-level resource management scheme and allowing CGRAs to be integrated into multi-threaded embedded systems as multi-threaded accelerators. Park et al. [97], on the other hand, proposed a virtualized execution technique on a specific CGRA that can transform static modular scheduling kernels at runtime based on different resources and boot times.

(3) Simple and Effective Virtualization Layer With Powerful Programmability

Although CGRA can be used as a virtualization layer for FPGAs and improve their configuration performance, an effective virtualization layer for CGRA has not been studied. We expect that the virtualization layer of the CGRA have simple but powerful programmability.

Overall, current CGRA virtualization techniques are mainly used to facilitate static compilation, and more CGRA virtualization techniques for OSs and dynamic resource management are needed in the future. Virtualization is the first and most important step for CGRAs to be widely used. Further developments of virtualization urgently requires continued exploration and cooperation from the CGRA research communities.

4.3.2 Instruction Flow-Based Dynamic Compilation

As mentioned in Sect. 4.3.1, there are various hardware architectures for SDCs, and the industry does not have a unified standard. When the hardware architecture of an SDC changes, or when it does not change physically but the hardware virtualization virtualizes various hardware architectures by means of software, programmers have to redesign the compiler back end or change the source code to accommodate the new SDC hardware architecture, in order to better use the SDC. However, both approaches face the problem that whenever the target SDC architecture is improved by the developers, existing applications will not be able to be adapted to the new underlying hardware. Therefore, for each iterative upgrade of the SDC hardware architecture, the engineers need to redesign or recompile the machine code for the new architecture. As a result, new hardware features cannot be used quickly and efficiently by users. The binary translation (BT) [98] can be a good solution to the above problem. Several examples of early BT applications are reviewed below. Rosetta [99] is a binary translation software developed by Apple to allow programs running on the PowerPC platform to run on the McIntosh computer on the Intel platform. FX!32 [100] is a program compiled for the original $\times$ 86 system to run on the Alpha processor. Transmeta Crusoe [101] is a processor specifically designed to translate $\times$ 86 code into VLIW machine languages. DAISY [102] is a system designed by IBM to make VLIW compatible with popular architectures; it performs BT at runtime.

1. Basic Concepts

Depending on the execution time, BT can be divided into static binary translation (SBT) and dynamic binary translation (DBT). SBT translates one machine code into another before code execution, while DBT dynamically analyzes the running machine code at runtime and translates it into the machine code supported by the target machine. Hardware virtualization is actually a new hardware architecture virtualized by the resource manager in the SDC at runtime, based on the target application characteristics and its resource consumption. Therefore, compared with SBT, DBT is more suitable for SDCs. DBT is usually represented as a system that monitors, analyzes and transforms part of the binary code, so that it can be executed on another architecture. Its main advantage is that it does not require extra effort

Table 4.4 Examples of some DBT techniques

DBT Name	Release Time	Target Machine
DIF [103]	1997	VLIW
DAISY [102]	1997	VLIW
Transmeta [104]	2000	VLIW
CCA [105]	2004	One-dimensional CGRA
Warp [106]	2004	FPGA
DIM [107]	2008	One-dimensional CGRA
GAP [108]	2010	One-dimensional CGRA
MS DBT [109]	2014	Crossbar switch CGRA
DynaSPAM [9]	2015	Two-dimensional CGRA
DORA [10]	2016	Two-dimensional CGRA

from the programmer and does not break the standard tool flow used in the software development process.

DBT has a long history and has been applied in many hardware architectures since its birth. Some researchers, in order to further increase the running speed of the application, have dynamically translated parts of the code originally executed on the GPP to run on the coprocessor or accelerator via DBT. Table 4.4 lists the applications of some DBT techniques on different hardware architectures.

The principle of all current mainstream DBTs is to identify and accelerate a program trace. A trace is a program block consisting of one or more basic blocks, where can contain branches but not loops. The trace to be accelerated should be selected from a program path with high execution frequency in order to maximize the performance gain of DBT. In general, DBT can be divided into three phases, trace detection, trace mapping, and trace offloading. Figure 4.49 illustrates the operation mechanism of DBT. The trace detection phase identifies the hot traces that need to be accelerated, usually based on the evaluation of the program running on the GPP. A counter is commonly used to count the execution frequency of the program traces. If the execution frequency of a trace is higher than a preset threshold, it is considered as a hot trace and will be cached in the T-Cache. In this case, the program can quickly retrieve all instructions and map them to the PEA the next time it runs the trace on the PEA. The trace mapping phase continuously detects the runtime PCs of the GPP, and if the PCs match the addresses already cached in the T-Cache and the branch predictors also match the hot traces, all instructions in the hot traces will be dynamically mapped to the PEA and the generated configuration will be cached. The trace offloading phase means that the GPP offloads the workload to the PEA after the dynamic mapping is complete. The input variables of the trace are passed to the PEA via registers and accelerated by the PEA for execution, and finally the output is returned to the GPP.

As the advantage of the CGRA lies in the ability to quickly complete configuration switching, which provides the possibility of fast workload switching. DBT turns this possibility into reality, and such application scenarios are very common nowadays;

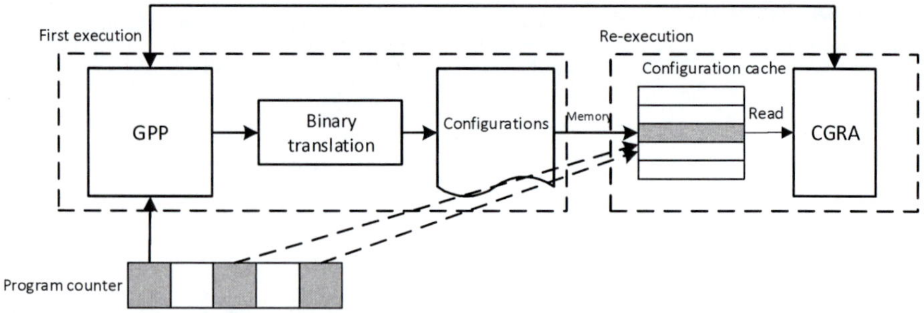

Fig. 4.49 Operation mechanism of DBT

see the application part in later chapters of this book for more details. In addition, SBT is not compatible across hardware generations, while DBT can achieve this goal. In order to get the above advantages, DBT needs to make certain trade-offs between mapping optimization and operation cycle, or adopts more complex optimization strategies to get closer to the compilation performance of SBT.

The most important feature of dynamic compilation is the ability to quickly obtain configurations, or the ability to convert the high-level language code in the candidate region into a bit stream that conforms to the hardware architecture interface in real time, and then complete the CGRA configuration and perform data operations. The previous sections have detailed some methods to obtain optimized mappings. However, in dynamic mapping, as the goal is to obtain mapping results quickly, it is clear that the time-consuming algorithms like ILP are not suitable. When faced with this mapping problem, the simplest dynamic compilation strategy is to map instructions in instruction order, i.e., directly mapping any candidate instruction onto a free hardware resource adjacent to the previous instruction. This seemingly rough scheme actually has many benefits, especially for large CGRA structures with a large number of PEs.

2. Implementation Methods of Dynamic Compilation

The following is a practical use of the greedy way to dynamic mapping, which is equivalent to reducing the mapping time in the compilation process from the algorithm level. For a given instruction, the PE closest to the center of each input port is first selected as the initial point, and an attempt is made to find a feasible route between the PE and the input. If no such route exists, the mapping is switched to the adjacent PE, and the search for a feasible route is conducted again until all feasible PEs have been traversed or a suitable route has been found. The algorithm does not consider the output interface, because the output is treated as the input to the next mapping when it needs to be considered.

As shown in Fig. 4.50a, for a 4×4 CGRA PEA, when the mapping of a 2-input operator is traversal handled by using the greedy algorithm, the midpoint unit is first found, and then the nearby PEs are sorted according to their distances from the midpoint unit. In the figure, only five PEs are marked from 1 to 5; in fact, all feasible PEs need to be detected. Figure 4.50b shows that the feasible route is found when the

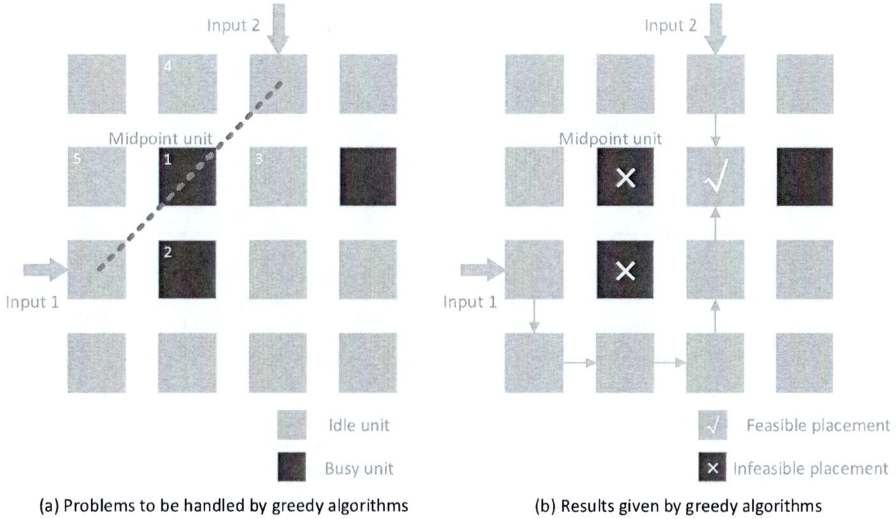

Fig. 4.50 Example of greedy algorithm processing

PE3 is tested, which is then used as the actual mapping scheme. It is easy to see that the routing resource calls would have been more reasonable if the computation had been performed at PE4 or PE5. However, these better results are ignored in order to get the mapping results faster.

Another approach to optimize the hardware resource topology from the hardware level and thus reduce the complexity of mapping is presented below. Ferreira et al. [109] proposed a DBT mechanism for CGRA dynamic modulo scheduling. The operation mechanism of dynamic modulo scheduling is shown in Fig. 4.51, where the CGRA coprocessor copies data from the register file and writes the return results back to the registers after the hot code block is executed. The monitor module detects the hot loop code blocks during the operation of the GPP. The binary conversion module is implemented by software and dynamically generates the CGRA configuration at runtime. The monitor is implemented based on an FSM; its goal is to dynamically identify the loop code blocks in the instructions. Upon detecting a loop, the method first estimates whether the instructions in this loop are compatible with the CGRA. Then, it converts the code block into a CGRA configuration and stores it in the configuration memory. When this piece of code is subsequently executed again, the CGRA can be directly called for execution to achieve acceleration. This mechanism simplifies the structure of the CGRA, and it is only applicable to the CGRAs that are interconnected by crossbar switches. This is because in such CGRAs, any PE can access the output of any other PE, resulting in lower requirements in the placement of the operators. Therefore, the DBT algorithms can achieve very low complexity. The method traverses each vertex in the DFG in topological order. For an unbalanced DFG, the dynamic modulo scheduling algorithm inserts registers at runtime to balance the branch paths.

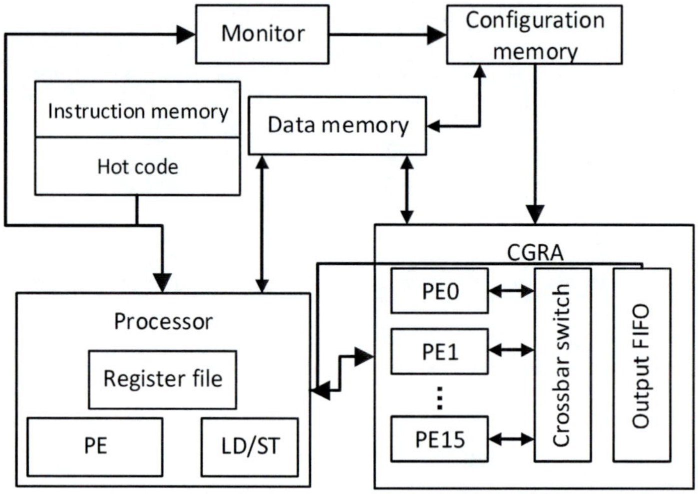

Fig. 4.51 Operation mechanism of dynamic modulo scheduling (LD/ST stands for load/store)

During the implementation of DBT, in addition to the optimizations in scheduling, mapping and other compilation aspects, the dynamic optimizations can also be implemented at the hardware level. A very typical example is the dynamic voltage and frequency scaling (DVFS), the fundamental of which is to collect the load-related signals during the real-time operation of the chip to compute the real-time load parameters of the system, and then speculate the hardware resources required for the next possible system load. When it is concluded that there is redundancy in the current utilized resources, it will turn off some PEs, or reduce the load voltage and frequency of the system, to achieve the purpose of reducing the power consumption.

3. Examples of Dynamic Compilation Systems

The following uses the DORA [10] system as an example to analyze its efforts in dynamic compilation. Applying the idea of dual-issue OoO processors, DORA first evaluates some potentially optimizable code from each source in the pending area, and then optimizes corresponding schemes before and after mapping depending on the selected hardware architecture. The relatively suitable structures for DORA are DySER and other hardware structures that use dynamic registers for special instructions. Note that DORA's optimization scheme for DySER is not necessarily optimal for other structures. For example, many structures require modulo scheduling, but some structures represented by DySER imply the step of modulo scheduling due to its OoO kernel structure and integration scheme supporting pipelining.

Table 4.5 lists the optimization names, the performance comparisons, and the reductions in the translation time of the DORA dynamic compilation scheme compared to the DySER static compilation system. It can be seen that they have many schemes with similar performance to the original compilation system; however, DORA introduces some schemes that are not adopted by the DySER static compilation system, some of which are even better. At the same time, some of the optimization

Table 4.5 Comparison of optimization strategies

Optimization name	Performance comparison	Translation time reduction/%
Loop unrolling	Similar	64
Loop store-and-forward	Similar	0
Loop deepening	Similar	5
Load/Store vectorization	Similar	9
Accumulator extraction	Similar	−4
Useless code elimination	Already have	0
Operation merging	Already have	0
Real-time constant insertion	Not implemented	0
Dynamic loop transformation	Not implemented	0
CGRA placement	Better	–

schemes have significantly reduced the translation time. The data in the table shows the time reduction due to a particular optimization alone, and the corresponding reduction varies when a combination of optimizations is used.

Taking the source code in Fig. 4.52a as an example, the schemes listed in Table 4.5 are briefly introduced next. As each scheme differs in terms of time and space used, they can be briefly classified into the following categories.

1) Pre-mapping Processing

When the coprocessor receives an interrupt and finds a new executable program trace, it reads the content in the trace cache into local memory. In most cases, the program traces that are heavily reused will be the looping basic blocks. If there are multiple basic blocks in the trace cache at the same time, the compiler will sort them according to the possibility of reuse and process them sequentially. After getting the basic blocks to be processed, the compiler first identifies the data dependencies and control dependencies, and then adopts the corresponding optimization strategies for each instruction before mapping.

Optimization Scheme 1: Identification and Exclusion of Reduction Variables

Generally speaking, a certain reduction variable (or register) will only be updated after a certain number of loop iterations, so that the loop execution process can be easily monitored. It is better not to update the reduction variables on the CGRA. At the same time, the code that regularly modifies the reduction variables can be removed. In this example, the instruction corresponding to "add \$0 × 1, %rax" is deleted.

```
begin: //原码
    movss (%rcx,%rax,4),%xmm0
    subss (%r8,%rax,4),%xmm0        •
    add $0x1,%rax                   •
    cmp %eax,%ebx
    mulss %xmm0,%xmm0               •
    addss %xmm0,%xmm1               •
    jg begin
```

(a) Source code

```
begin:
    movss (%rcx,%rax,4),%xmm0
    movss (%r8,%rax,4), %tmp
    subss %tmp,%xmm0                •
    mulss %xmm0,%xmm0               •
    movss %xmm0,%atmp               •
    ...
    mulss %xmm0,%xmm0               •
    addss %xmm0,%atmp               •
    addss %atmp,%xmm1
    jge begin
```

(b) Loop unrolling

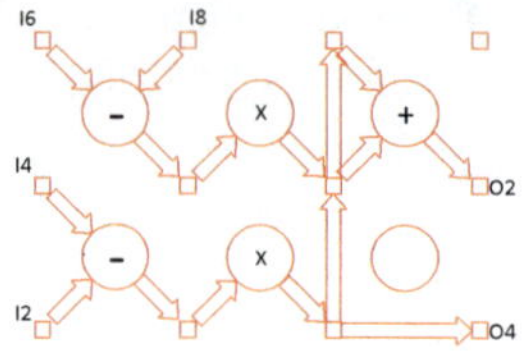

(c) Accumulator conversion

```
begin:
    movss (%rcx,%rax,4),%xmm0
    movss (%r8,%rax,4), %tmp
    subss %tmp,%xmm0                •
    mulss %xmm0,%xmm0               •
    addss %xmm0,%xmm1               •
    movss 1(%rcx,%rax,4),%xmm0
    movss 1(%r8,%rax,4), %tmp
    subss %tmp,%xmm0                •
    add $0x2,%rax
    mov %ebx, %tmp
    sub %eax, %tmp
    cmp $2, %tmp
    mulss %xmm0,%xmm0               •
    addss %xmm0,%xmm1               •
    jge begin
```

(d) Code form of the mapping result

```
begin:
    dmovss (%rcx,%rax,4),I2
    dmovss (%r8,%rax,4), I4
    dmovss 1(%rcx,%rax,4),I6
    dmovss 1(%r8,%rax,4), I8
    add $0x2,%rax
    mov %ebx, %tmp
    sub %eax, %tmp
    cmp $2, %tmp
    drecv O1, %xmm0
    drecv O2, %atmp
    addss %atmp,%xmm1
    jge begin
```

(e) Hardware representation of the mapping result

```
begin:
    dmovss (%rcx,%rax,4),I2
    dmovss (%r8,%rax,4), I4
    dmovss 1(%rcx,%rax,4),I6
    dmovss 1(%r8,%rax,4), I8
    drecv O1, %xmm0
    drecv O2, %atmp
    addss %atmp,%xmm1
    dmovss 2(%rcx,%rax,4),I2
    dmovss 2(%r8,%rax,4), I4
    dmovss 3(%rcx,%rax,4),I6
    dmovss 3(%r8,%rax,4), I8
    add $0x4,%rax
    ...
```

(f) Loop deepening result

Fig. 4.52 Processing results of each stage

Optimization Scheme 2: Loop Unrolling

As many loop body codes do not use all the hardware resources, the source-code is loop unrolled in order to increase the computing speed. In this example, there are two addition/subtraction instructions and one multiplication/division instruction that can be mapped onto PEs. Thus, if there are four mappable PEs on the hardware structure, one of them can be used for loop unrolling, i.e., executing the operations in the next iteration cycle. Figure 4.52b shows the code after loop unrolling. In many cases, it is not possible to get the number of iterations the loop needs from the source code alone. Hence, in order to get the correct loop results, all the expanded data during loop unrolling needs to be preserved, and the code corresponding to the end of the loop needs to be modified appropriately to get the correct number of iterations.

Optimization Scheme 3: Accumulator Identification

Accumulation operations are very common in loops, which introduce the register data dependencies that often prevent the arrays from iterative pipelining. The solution is to enable the array to output only the sum of the internal values and accumulate that value into the total sum. DORA looks for the accumulator by finding a register that is both the input and the output of an addition PE, and the register is not used by the other instructions. The instructions identified as accumulators are accumulated into a temporary register, and an unmapped instruction is inserted to add the value in that temporary register to the accumulation register. In this example, register %xmm1 is identified as an accumulator, so the modified code is shown in Fig. 4.52(c).

Optimization Scheme 4: Store-and-Forward

Inside some loops, some values may be first stored and then loaded in the subsequent loops, which has a large impact on the time performance. Thus, such storage operations can be identified during the optimization and data associated with the subsequent operations calling that register. The method of identifying these load/store pairs is also simple. As long as the address registers are the same and there is no operation written to the register between the two instructions, the data can be associated.

2) Real-Time Register Monitoring

In contrast to static compilation, dynamic compilation can monitor the register data in real time in order to achieve real-time dynamic optimization. For the given code blocks in the pending area, the compiler selects some instructions of interest to monitor. For example, when the entire program dataflow needs to be tracked, it can filter out all the registers that are read without store operations. These registers must have been computed before the execution of the code block; they can be considered as constants. When monitoring the real constants, the register is obviously characterized by only read operations, with no store operations.

Optimization Scheme 5: Hardware Monitoring

First, the feasibility of implementing real-time monitoring from the hardware level is discussed. The monitoring at the hardware-level is to compare whether the value of a register has changed from the previous one in real time. Thus, after a monitoring instruction is selected in the pre-mapping phase, the information about the instruction or the register is stored in a special register. When the register is read, the relevant operation is activated. A very practical activation operation is to initiate an interrupt when detecting the execution of an instruction, which is equivalent to the implementation of a controllable interrupt program. To implement this function, additional registers are needed to store the information used to match the instructions of interest. Six 66-bit registers are used in DORA to describe the interested instruction, where 64 bits are used to store the PC, 1 bit to select the register (typically two registers for an instruction), and one valid bit.

Optimization Scheme 6: Dynamic Register Optimization

An application of the register monitoring is to process constants or constants-like numbers. For registers that have been recognized as constants, repeatedly using the load instructions will obviously degrade the performance; so embedding the constant in the hardware structure is a good optimization strategy. During program runtime, although some registers are related to both read and store instructions during preprocessing, it is found that no storage has been performed in actual use. Such memory address is also likely to be a constant, and the compiler will embed it as a constant in the field registers of the hardware at runtime.

Another application of the register monitoring is to observe the changes of the value in a register bound to a loop, to indicate the number of times that the loop is unrolled. If a large number of iterations have very small boundaries, loop unrolling will prevent many iterations from using the code in the optimized region. If the monitoring results indicate that the unoptimized code will be executed when the unrolled number of times of a loop exceeds the threshold number of the iterations, the code will rollback to the original state and execute the previous unoptimized scheme.

There are many other applications of the register monitoring, and there may be new usages for different applications. However, regardless of the functionality, the register monitoring should be executed as little as possible for satisfying the energy requirements. For example, when monitoring constants, once the "constant" is found to be written, it is obvious that the value no longer needs to be monitored.

3) Dynamic Mapping

Subject to the requirements for the mapping speed, methods such as ILP and modulo scheduling algorithms are not applicable to dynamic mappings. However, the greedy algorithms are used here because they can get relatively good mapping results in a short time. Examples of specific dynamic mapping algorithms have been given in the previous section, so we will not repeat them here. Figure 4.52d and e show the dynamic mapping results for this example.

Optimization Scheme 7: Instruction Set Specialization

The control flow of the x86 instruction set usually uses the EFLAGS register to store the information of conditional instructions. As DySER does not have such register, DORA tracks the most recently-used branch instructions and stores their source information during the BT process, to replace the corresponding function in the × 86 architecture.

4) Post-mapping Optimizations

Further optimizations can be performed after the scheduling is complete. These optimization schemes will improve the parallelism and eliminate the redundant code.

Optimization Scheme 8: Loop Deepening

Loop unrolling is to fully utilize the computing resources on the reconfigurable PEA, while loop deepening is to fully utilize the input and output bandwidths on the reconfigurable hardware. This optimization strategy analyzes each group of associated load/store code. If not all the bandwidths are fully used by the group of code in the current mapping, the compiler will compound the data required by the subsequent load/store code into this data path in advance as far as possible, to achieve the path reuse. In this example, there are two sets of related load instructions, (%rcx, %rax, 4) and (%r8, %rcx, 4), each requiring only a 64-bit bandwidth. As DySER allows 128-bit transfer, i.e., executing four load instructions at the same time, loop deepening is a good strategy to be employed. The optimized result is shown in Fig. 4.52f.

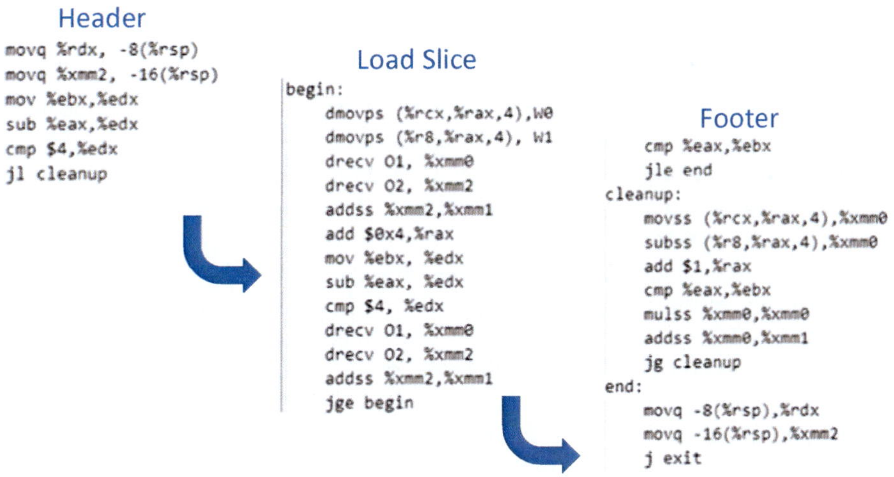

(a) Header (b) Body (c) Footer

Fig. 4.53 Parts of the final code

Optimization Scheme 9: Load/Store Vectorization

The vectorization strategy will combine the load and store operations with only different offsets into a single operation with large bit-width. As long as it can be confirmed that the load/store does not read or write to the locations written or read by intermediate memory operations, these associated load/store instructions can be candidates for vectorization strategies. In this example, when there are only load or store instructions in the code region, or when load and store are not interleaved, these instructions can always be vectorized. Figure 4.53 shows the vectorized code, where the dmovps instructions are the load instructions after the vectorized combination.

In the case of interleaved load and store instructions, the situation is a bit more complicated. The compiler will use mark check to trace the load/store instructions of the loop body, which are usually placed at the beginning of the code block, as shown in Fig. 4.53a. When the associated load/store instructions do not change the address computation registers or only change a fixed value, the vectorization strategy can be adopted for optimization.

Optimization Scheme 10: Code Finishing

As the transformed loop must perform the same number of iterations as the source code, additional code is often needed to determine the correct number of iterations after operations like loop unrolling and loop deepening. As shown in Fig. 4.53a and c, for a register-determined loop, a header is needed to ensure that the same number of iterations is executed, and a footer is needed to complete the additional content after loop unrolling.

4.3.3 Configuration Flow-Based Dynamic Compilation

Configuration flow-based dynamic compilation techniques typically use the method of dynamic transformation of configurations. Compared with the instruction flow-based dynamic compilation techniques for SDCs, since the ILP in the source code has been fully exploited by static compilation of the configuration flow before transformation, the static compilation results can be used to generate the configuration flow without dynamically analyzing the dependencies in the code. Therefore, the performance of the configuration flow generated by this method is better. Since the oriented hardware architectures before and after the transformation are similar (hardware virtualization techniques usually only change the size of the available hardware resources), the recoding process is not required in the transformation. Therefore, if a configuration transformation algorithm with low time complexity can be found, a low dynamic transformation overhead can be achieved. The following details how to solve the problem of the dynamic transformation of configurations, the problems in the current mainstream approaches, and a general template-based approach to dynamic transformation of configurations.

1. Abstraction and Modeling of the Dynamic Configuration Transformation Problems

Unlike the configuration dynamic generation technique that transforms the instruction flow of the GPP into the configuration flow of the CGRA, the dynamic configuration transformation in the SDC transforms the configuration flow into a new configuration flow that is subject to the constraints generated at runtime. Figure 4.54 compares the configuration dynamic generation and transformation techniques. Figure 4.54a shows the DFG IR of the application. Figure 4.54d shows the instruction flow of the GPP in the SDC system executing the application, and Fig. 4.54b is the configuration flow of the CGRA in the SDC executing the application using all the resources. Figure 4.54c shows the configuration flow of the application executed by the virtualized CGRA when 50% of the resources for the running CGRA are occupied by other applications. As described in Sect. 4.2.2, the DFG mapping problem can be modeled as a graph homomorphism problem. After the graph homomorphism problem is solved, if all the vertices and edges of the DFG are labeled on the vertices and edges of the corresponding TEC, all the labeled vertices and edges will form the subgraphs of the TEC, which can be referred to as the time extended CGRA subgraph (TECS). In fact, the TECS can represent both the kind of operations performed by PEs at a given moment and the usage of interconnect resources of the CGRA. Thus, the TECS can be considered as an equivalent expression of the CGRA configurations. EPIMap gives the conclusion that the DFG mapping problem can be transformed to the problem of finding an epimorphism mapping F that satisfies the following relation.

$$F : \mathrm{TEC}(V, E) \rightarrow G(V, E) \tag{4.42}$$

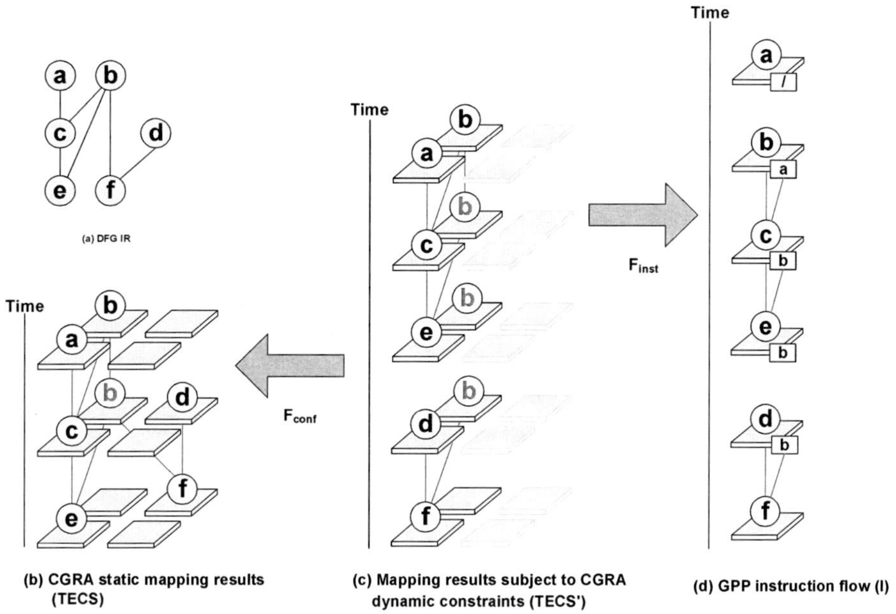

Fig. 4.54 Dynamic compilation modeling

However, since F is surjective, not all vertices or edges on the TEC can actually have mappings found in $G(V, E)$. If the vertices or edges without mappings in the TEC are removed, the problem can be modeled as finding an epimorphism mapping F_s that satisfies the following relation.

$$F_s : \mathrm{TECS}(V, E) \rightarrow G(V, E) \tag{4.43}$$

Dynamic generation of configurations can be described as finding an epimorphism mapping F_i that satisfies the following relation.

$$F_i : \mathrm{TECS}'(V, E) \rightarrow I(V, E) \tag{4.44}$$

where TECS' refers to the instruction flow executed by the CGRA subjected to dynamic resource constraints. This dynamic compilation can provide good software transparency, so that users cannot feel that it is using hardware other than GPPs. However, the instruction flow-based configuration generation technique has to analyze the complex dependencies (including data dependencies and control dependencies) in the program during dynamic operation; it tries to execute the operations without dependencies in parallel. Without sacrificing the hardware resources (with static scheduling only, rather than scheduling using hardware), trying to maximize the instruction parallelism would result in considerable time overhead, which is however not tolerated by the dynamic compilation. Although partial parallelism can be extracted by using a processor that supports OoO execution, the extracted parallelism by using this method is very limited subject to the size of the instruction window. Dynamic transformation of configurations can be described as finding an epimorphism mapping F_c that satisfies the following relation.

$$F_c : \mathrm{TECS}'(V, E) \rightarrow \mathrm{TECS}(V, E) \quad (4.45)$$

Since the configuration dynamic generation technique can use the static mapping results that are themselves capable of representing the vast majority of dependencies in the program (there are almost no dependencies between operations scheduled at the same control step, as will be discussed in detail next), it is possible to avoid analyzing the dependencies in the instruction flow. Therefore, the dynamic configuration transformation technique is theoretically able to generate better quality (higher performance) configurations compared to the instruction flow-based configurations. Nevertheless, hardware virtualization makes the hardware resources in the dynamic case change compared to the static one; thus, all operations in the static configuration need to be rescheduled and rerouted. However, in general, the time overhead for solving these two problems remains significant. The existing configuration flow-based dynamic compilation of SDCs is solved in several ways as follows.

1) Model Simplification

Pager et al. [96] proposed an approach to enable CGRAs to support multithreading, which in essence uses the virtualization techniques to virtualize CGRAs into multiple sub-arrays and employs dynamic configuration transformations to schedule statically generated configurations to the idle resources in CGRAs. In order to achieve dynamic transformation of configurations on CGRAs, this method simplifies both the hardware architecture and the compilation scheme. As shown in Fig. 4.55, in terms of the hardware architecture, the PEA must be able to be partitioned into multiple identical pages that consists of a set of PEs, where all pages must form a ring topology. Any virtualized PEA must satisfy the above hardware architecture constraints. Once the partitioning is complete, the configurations can be transformed only at the page granularity. In addition, the PEA must have sufficient global registers for use in the dynamic configuration transformation. In terms of compilation, this approach requires that each page in the static compilation result can only communicate with at most two pages adjacent to it, and that only global registers rather than local registers can be used during static compilation. While making the dynamic configuration transformation easier and faster, these model simplification approaches can degrade the quality of the static mapping. This can slow down the application when it has exclusive access to the entire CGRA. Moreover, as not all architectures can be divided into pages with a ring topology, the simplification approach is not applicable to general CGRA architectures.

2) Switching Multiple Configurations

The polymorphic pipeline architecture (PPA), proposed by Park et al. [97], is essentially a CGRA. The PPA consists of multiple cores interconnected by a mesh, where

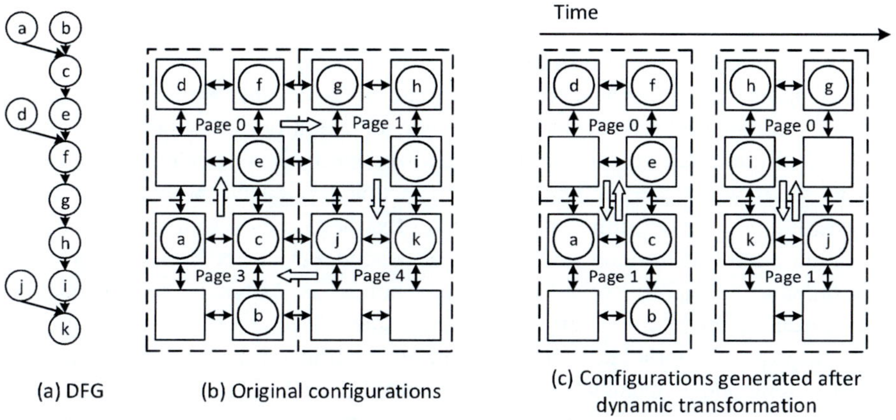

Fig. 4.55 Dynamic configuration transformation under the simplified model

each core contains four mesh interconnected PEs. During the hardware virtualization, taking a core as the granularity, it generates configurations suitable for different sizes of virtual hardware during the static compilation time, and then transforms them to configurations oriented to different resources or IIs based on runtime information. However, in practice, there are still simplifications in this method for modeling dynamic transformation of configurations: (1) the granularity of the hardware virtualization is the core rather than the base unit PE, which may lead to a lower utilization of hardware resources; (2) the PPA only supports traditional folding and unfolding methods for the configuration transformation, which is less flexible.

3) Static Template-Based Transformation

The static template-based dynamic configuration transformation is to first statically generate configurations and a set of templates, and then transform static configurations online through the templates into configurations subject to the dynamic constraints, thus achieving dynamic transformation algorithms with low complexity. Reference [2] proposed a template-based dynamic configuration transformation method for general-purpose SDC architectures and arbitrary transformation targets. The following describes in detail how to construct a problem model for the template-based dynamic configuration transformation.

If TEC and TEC′ are respectively used to denote the hardware architecture before and after transformation, all configuration transformations should be considered on the premise of $\text{TEC}' \subseteq \text{TEC}$. There are two main reasons for this: (1) if $\text{TEC} \subseteq \text{TEC}'$, i.e., the hardware architecture after transformation is a superset of the one before transformation, this situation is similar to the instruction flow-based configuration dynamic generation. It requires a dynamic exploitation of parallelism in the original configuration, which is difficult to ensure the quality of the transformed configuration; (2) in an SDC that supports multithreading, the application should try to use the idle hardware resources to improve its utilization rate, and thus the performance.

Therefore, it is more important and meaningful to transform the configurations to the cases orienting fewer resources.

Dynamic configuration transformation can be divided into two steps, rescheduling and rerouting. Rescheduling is to reschedule all operations in the original configuration to different control steps of the virtualized hardware resources, while rerouting determines which routing resources are occupied when the producer in the DFG passes data to the consumer. Regardless of the method used to implement dynamic configuration transformation, the original dependencies should be kept unchanged in order to ensure functional consistency before and after the transformation. If the rescheduling and rerouting can be done statically, the result can be represented in the form of a template. The configuration is transformed at runtime according to the template to achieve lower time complexity. Is it possible to find such a template? The answer is Yes, because the two previously mentioned methods can essentially be attributed to the template-based configuration transformation method. The following section describes in detail what computational templates and routing templates represent and where they come from.

(1) Computational Template

Rescheduling is essentially to remap all operations (operators) in configurations to hardware resources on different control steps, and the operations can be abstracted as the vertices on the TECS. Thus, the rescheduling process is essentially to find a mapping $P{:}V(\mathrm{TECS}') \rightarrow V(\mathrm{TEC})$. Moreover, in order to ensure the functional consistency before and after the transformation, every vertex in the TECS must have at least one preimage, which means that P must be surjective. In order to cover all possible cases of configurations (so that different configurations can be transformed), the most complex case must be considered at static time, i.e., $V(\mathrm{TECS}) = V(\mathrm{TEC})$. Therefore, if a mapping $\mathrm{PT}{:}V(\mathrm{TECS}') \rightarrow V(\mathrm{TEC})$ can be found, followed by removing some definition and value domains of the PT at dynamic time depending on the specific configurations, the desired mapping P can be obtained. The computational template is an expression of PT. Figure 4.56a shows a folded template using a matrix, where each line of the template represents a control step, each column represents a PE in the virtualized architecture, and the elements of the matrix represents the operators executed by the PEs in the original architecture (here directly by the PEs in the original architecture).

(2) Routing Template

Since the operators in the original configurations have been rescheduled, the communication between operators must be rerouted. In order to enable each edge in the TECS to be routed and to reduce access conflicts, the use of the global registers is minimized. Therefore, a constraint is imposed here on the computational template: $\forall(w, v) \in E(\mathrm{TECS})$, $P^{-1}(w)$ and $P^{-1}(v)$ are the same PE or adjacent PEs. This constraint ensures that a piece of data routed from one PE to another does not go through a third PE or global register. The rerouting problem can be described as finding a mapping $Q{:}R \rightarrow 2^{B}$, where R is all registers in the TEC' (including output

Time step \ Resource	PE_1'	PE_2'
1	PE_1	PE_3
2	PE_2	PE_4

(a) Computational template

Original route	Assigned registers
PE_1.OR	PE_1'.Reg1
PE_2.OR	PE_1'.Reg2
PE_3.OR	PE_2'.Reg1
PE_4.OR	PE_2'.Reg2 PE_1'.Reg3

(b) Routing template

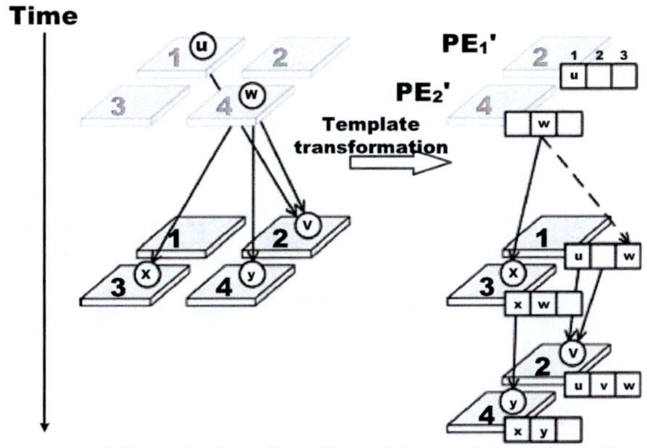

(c) Example of configurations via template transformation

Fig. 4.56 Example of computational templates and routing templates

registers and local registers of each PE), and B is the set of all cache units of the virtualized architecture. Why do we need to do this? See the example shown in Fig. 4.56, where the two operators x and y executed in parallel on the original architecture have to be executed sequentially on the virtualized architecture due to the resource constraints. If the data resulting from the computation of x is not cached, but only temporarily stored in the output register of the PE, it will flushed out by y. Therefore, the transformation algorithm must allocate a new cache for each PE's output register in the original architecture. In addition, if an edge in the TECS satisfies Eq. (4.46), a new register r (e.g., edge (w, v) in Fig. 4.56c) should also be assigned to it. Meanwhile, the edges satisfying Eq. (4.46) introduce additional time overhead for routing, which will affect the quality of the transformed configuration. This problem will be dealt with in detail in a later section. Of course, since the transformation algorithm is to assign the computational tasks of multiple PEs to fewer PEs to accomplish the same function, new caches should also be allocated for the PE's local registers in the original architecture. In summary, generating a routing template is actually to allocate registers for routing to registers in the original architecture.

$$\forall e = (w, v) \in E(\text{TECS}), \quad \text{if } \text{PE}(\text{PT}^{-1}(w)) \neq \text{PE}(\text{PT}^{-1}(v)), \\ \text{then } \exists r \in \text{PE}(\text{PT}^{-1}(v)) \tag{4.46}$$

The same process of rescheduling and rerouting exists in the two methods of dynamic configuration transformation [96, 97] described earlier. They can actually be summarized as template-based transformation, except that they use some deterministic templates (e.g., folded) without trying to acquire more efficient templates.

2. Framework of the Template-based Dynamic Configuration Transformation

The framework of the template-based dynamic configuration transformation aims to use as much information as possible from static compilation results to generate

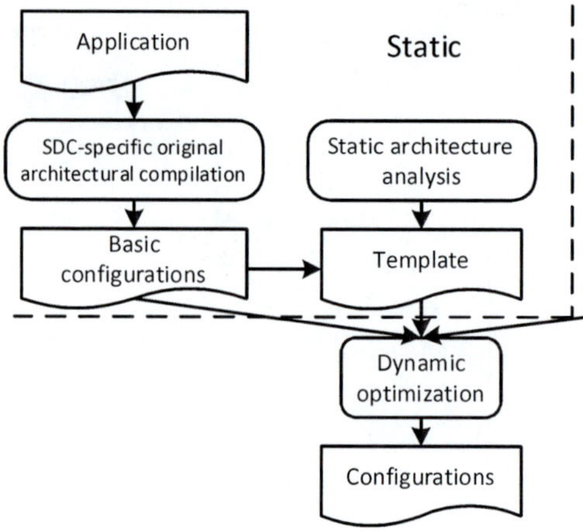

Fig. 4.57 Dynamic compilation framework

templates that can guide the dynamic transformation process, thus greatly reducing the time complexity of the dynamic process. Therefore, the framework can be divided into static and dynamic parts. As shown in Fig. 4.57, the static part first generates the compilation results of the original architecture of the application-specific SDC. The template generator then generates efficient templates based on the characteristics of the basic configuration generated statically and the analysis for the architecture of the SDC. Note that since the static process cannot predict the actual dynamic operation, it is important to generate as many templates as possible for different virtual hardware structures on the one hand, and to consider the quality of the templates on the other hand. The templates that do not provide higher performance for the SDC system should be discarded. The dynamic part generates the configuration flow satisfying the constraints generated online by the dynamic manager in the SDC, by using the dynamic optimization algorithms on the GPP.

The dynamic configuration transformation also requires hardware support. As shown in Fig. 4.58, the part inside the dashed-line box is the execution process of the static basic configurations on the SDC. The PEA configuration controller will obtain the corresponding configurations through the configuration counter based on the operation (e.g., the cycle index). The PEA will execute the corresponding operations based on the current configurations. When the SDC generates a constraint on the mapping of new tasks due to that the other applications occupy the resources or the power requirements, and the template corresponding to the constraint is available, the SDC will start the configuration transformation mechanism to store the newly generated configuration in the new configuration cache. After the transformation, one of the static basis configurations (on one control step) will be transformed into multiple configurations in the idle virtualized resources based on the template.

In summary, the dynamic configuration transformation algorithm for SDCs can be described using a generic framework based on templates, and all other similar approaches are actually the special cases described by this framework (using fixed

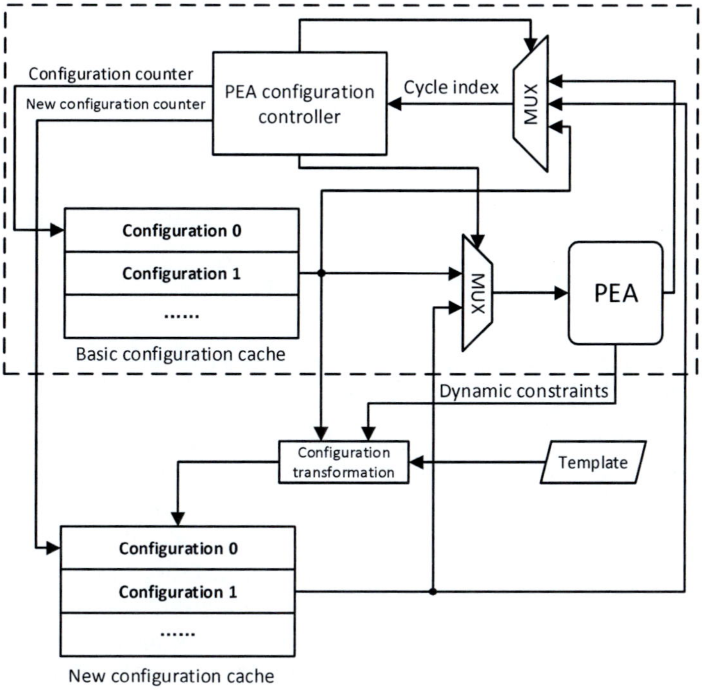

Fig. 4.58 Dynamic compilation system

templates). This framework uses statically generated basic configuration as well as templates to dynamically generate configurations subject to dynamic constraints at runtime. The generation and optimization of the templates and the dynamic transformation algorithms are important to the framework; they are introduced in detail below.

1) Template Generation and Optimization

Template generation and optimization is to generate valid and efficient templates, i.e., the transformed configuration can retain all data dependencies in the original configuration with the same functions, and deliver good performance. Two rules should be followed in the template generation and optimization:

(1) Rerouting the dependencies in the original DFG. If this is done within a PE (using local registers inside the PE), except for increasing the register pressure of the PE, it does not introduce additional cycles for routing, as the edges (w, x) in Fig. 4.56.
(2) Because the output register of the PE in the original configuration is mapped onto a secure local register in the new configuration, and the template may require the data in that register to be routed to the local register of another PE (as the edge (w, v) in Fig. 4.56), the probability of occurring an extra cycle for routing is equal to the that of the inter-PE communication in the newly generated

$$
\begin{array}{c|cccc}
\text{PE} & 1 & 2 & 3 & 4 \\
\hline
1 & 0 & 0.4 & 1 & 0 \\
2 & 0.3 & 0 & 0.5 & 0 \\
3 & 0 & 0.1 & 0 & 0.5 \\
4 & 0 & 0 & 0.5 & 0
\end{array}
\xleftrightarrow{\text{Transform}}
\begin{bmatrix} 0 & 0.4 & 1 & 0 & 0 & 0.3 & \ldots & 0.5 & 0 \end{bmatrix}_{1\times 16}
$$

(a) Communication characteristic matrix (b) Communication characteristic vector

Fig. 4.59 Two expressions of PE communication characteristics

configuration, and the mathematical expectation of the additional cycles due to routing should also be equal to this.

The above two rules show that the communication characteristics in the static basic configuration affect the quality of the template. The communication characteristics are the frequency of the communication between each PE pair, which can be expressed by the utilization of interconnects between PEs. Figure 4.59a shows a matrix representation of the communication characteristics, where the element in row i and column j represents the utilization of the interconnects from PE_i to PE_j, i.e., the average number of times the interconnect between PEs is used to transfer data in each cycle. The communication characteristics can be obtained by parsing the static basic configuration.

The algorithm for the generation and optimization of the computational template is shown in Fig. 4.60. The algorithm is essentially a branch and bound algorithm with four inputs, TEC (the description of the original architecture), TEC' (the description of the target hardware architecture, i.e., the number of columns of the matrix when a matrix is used to describe the computational template), D (the estimated template depth, i.e., the number of rows of the matrix when a matrix is used to describe the computational template), and FM (the communication characteristic matrix extracted from the underlying configuration). Its output is the PT, i.e., the computational template that can be represented by the matrix in Fig. 4.59(a). The algorithm searches for the feasible templates by using branch and bound method; it records the expected value of the additional routing overhead generated by that template after finding a feasible template. In the subsequent search process, if the additional routing overhead generated by the obtained template exceeds the best template currently found, the algorithm cuts all subsequent branches even if the contents in the template are not yet fully populated, in order to reduce the useless search space, speed up the search, and finally obtain a template optimized for specific communication characteristics. If a legitimate template is not found after a long time, the template depth will be increased and the algorithm will rerun.

In general, more than one application is supported by the SDC, and the basic configurations compiled and generated for these applications reflect different communication characteristics. As a result, the template generated and optimized for the communication characteristics of one application can lead to significant routing overhead when used by another application. Moreover, the applications supported

Computational template generation and optimization algorithms

```
Inputs: TEC,TEC',D,FM
Outputs: PT
1   i ← 1, bflag ← 0, PT ← ϕ;
2   Cost ← 0, Threshold ← +∞;
3   M_0 ← NullMatrix(TEC'.length, D);
4   S_i1 ← { (x,t)|M_{i-1}(x,t) = null};
5   T ← {(x,t)|(M_{i-1}(x,t), PE_i) ∈ E(TEC)};
6   if T ≠ ϕ then
7     S_i2 ← {(x,t)|∀(x_0,t_0) ∈ T, (PE'_{x_0}, PE'_x) ∈ E(TEC')};
8   else
9     S_i2 ← S_i1;
10  S_i ← S_i1 ∩ S_i2;
11  M_i ← M_(i − 1);
12  bflag ← 1;
13  if S_i ≠ ϕ then
14      M_i(S_i(1)) ← PE_i;  // Generate one branch
15    Cost ← AC_Ex}(M_i,FM);
16      if Cost < Threshold then
17          delete S_i(1)
18          bflag ← 0;
19  if i = TEC.length then
20      PT ← M_i; // Find a valid pattern
21      Threshold ← Cost;
22  if bflag = 0 then
23      i ← i + 1;
24      goto line4;
25  else
26      i ← i − 1
27      if i = 0 then
28          break;  // All branches have been traversed
29      goto line11;
```

Fig. 4.60 Computational template generation and optimization algorithms by the SDC are expanding, and if the SDC system generates a template for every supported application, it will incur high memory costs. Therefore, this is unrealistic. It is worth discussing how to find the templates within a certain number and enabling high transformation efficiency for all the algorithms supported by the SDC.

Considering the principles of the template generation and optimization algorithms, it can be seen that the template generated by the algorithm are actually only related to the "direction" of the communication characteristic matrix. To better explain the problem, the communication characteristic matrix is first transformed into a communication characteristic vector, as shown in Fig. 4.59b. According to the law of the template optimization, if the communication characteristic vectors of two applications A and B are collinear, the optimal template for application A must also be the optimal template for application B. Therefore, in order to support dynamic configuration transformation for a larger number of applications with fewer templates, the template can be shared between applications whose communication characteristic vectors are close in direction. The framework for dynamic configuration transformation should select a limited number of representatives among the many communication characteristics and make the templates generated by them relatively efficient for all applications.

To address this problem, a template selection method based on the K-means algorithm is proposed in [2]. K-means is a data clustering algorithm aiming at dividing n multidimensional data into k clusters, where the data is generally called the clustered object, n and k are determined before the algorithm executes. The process of the K-means algorithm is as follows: (1) select k data from all data as the initial cluster centroids; (2) calculate the distances from each data to these cluster centroids and assign the data to the nearest cluster; (3) calculate the centroids of all newly generated clusters; (4) repeat steps (2) to (3) until the cluster to which each data belongs no longer changes. In the template selection problem, the communication characteristic vectors of all applications can be viewed as the clustered objects, and k cluster centroids obtained by the clustering algorithm can be viewed as the representatives of the communication characteristics of all applications in the clusters. An important step of the K-means algorithm is the definition of the distance, and the more commonly used ways to define the distance are Euclidean distance and cosine distance. In particular, Euclidean distance is the most familiar way of distance definition to us, which refers to the distance between two clustered objects in Euclidean space; it can be described as Eq. (4.47). Cosine distance is used to measure the difference between the vectors represented by two clustered objects in direction, which can be described as Eq. (4.48). If two vectors are collinear, the distance computed by this formula should be zero. In the template selection problem, since the generation of the template is only related to the direction of the applied characteristic vectors, the cosine distance should be used as the distance measure of the clustered objects when using the K-means algorithm for the template selection. Of course, a prerequisite for using the K-means algorithm to solve the template selection problem is considering that all applications run at the same frequency on the SDC. If not, different weights should be assigned to different applications to reduce the system expectation of the additional overhead due to the configuration transformation.

$$\forall x = [x_1, x_2, \cdots, x_n], \quad \forall y = [y_1, y_2, \cdots, y_n],$$
$$\text{distance}(x, y) = \sqrt{(x_1 - y_1)^2 + \cdots + (x_n - y_n)^2} \tag{4.47}$$

$$\forall x = [x_1, x_2, \cdots, x_n], \quad \forall y = [y_1, y_2, \cdots, y_n], \quad \text{distance}(x, y) = 1 - \cos\langle x, y\rangle \tag{4.48}$$

2) Methods of Template-Based Dynamic Configuration Transformation

To ensure the functional consistency of the configuration before and after transformation, the data dependencies of all instructions in the configuration before transformation must be maintained. Three kinds of dependencies can exist in all computing systems, read after write (RAW), write after read (WAR), and write after write (WAW). Since the template-based dynamic configuration transformation is to sequentially transform one control step in the original configuration into configurations with multiple control steps (Fig. 4.61c), the dependencies between operations belonging to two different control steps in the original configuration must be maintained in the new configuration. Therefore, only the dependencies within the same control step need to be considered. In the basic configuration, operations within the same control step are executed concurrently, and there will be no RAW dependencies between operations. Also, to avoid the output uncertainty, there is no WAW dependencies between operations, and only WAR dependencies may occur. WAR dependency indicates that the original two concurrently executed instructions i and j are now executed sequentially, and instruction i may modify the operands of instruction j in advance, resulting in wrong operation results of instruction j. Therefore, the dynamic configuration transformation needs to allocate more registers to avoid inconsistent functions before and after transformation due to the violation of WAR dependencies. Figure 4.61a shows the basic configuration with WAR dependency. Assume that the dynamic transformation method only allocates two registers on $PE_1^{'}$, Reg1 and Reg2, for the output register of PE_1 and the local register Reg1 of PE_2 in the original configuration; the transformed configuration is shown in Fig. 4.61b. Since the addition operation modifies the input operands of the multiplication operation in advance, the transformed and original configurations cannot maintain functional consistency. To solve the possible problems caused by the WAR dependency, the dynamic transformation method allocates two registers Reg1 and Reg2 for the output registers of PE_1 on $PE_1^{'}$, as the mappings of the output registers that act as input and output in the original configuration, respectively, and allocates Reg3 for the local register Reg1 of PE_2.

The transformed configuration is shown in Fig. 4.61c. As can be seen from the figure, Reg1 and Reg2 in $PE_1^{'}$ constantly exchange functions (output registers as input or output) to ensure the functional consistency of the configuration.

During the dynamic configuration transformation, the registers can be allocated in advance at static time by using routing templates, i.e., the allocation considering the worst case. However, this may lead to considerable allocation pressure on local

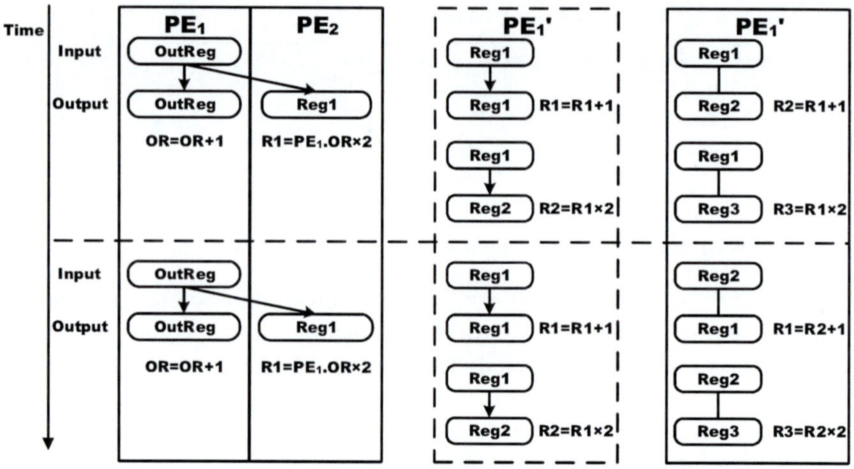

Fig. 4.61 Correlation in configuration transformation

registers within the PE, exceeding the local register resources and making the transformation impossible. Similar to the computational templates, some applications need to share the same routing template. Another way for the register allocation is to dynamically allocate registers as needed and withdraw registers that exceed their lifecycles in time. The static register allocation can effectively reduce the runtime of the dynamic transformation algorithm, while the dynamic register allocation algorithm requires constantly maintaining the data structure that records the allocation information of each register, resulting in a higher time overhead than that of the static allocation method. The template-based dynamic configuration transformation algorithm is shown in Fig. 4.62, which transforms the configuration TECS of a control step into a new configuration TECS' via a template. Due to the limited space, only the pseudocode using dynamic register allocation is listed here. In the algorithm, Table_1 is the data structure that records the new registers allocated for all local and output registers in the original configuration. Table_2 is the data structure that records the intermediate registers used for the inter-PE routing in the newly generated configuration.

Note that lines 9 and 15 of the algorithm may generate additional configuration for routing, which may increase the newly generated configuration. To maintain the linear complexity of the algorithm, the generation of this part of the configuration should also an algorithm with at most linear complexity.

3) Transformation Examples

The advantages of the template-based dynamic configuration transformation are presented via an example shown below. Figure 4.63a shows the interconnect structure of the original architecture of the SDC, where the communication characteristics

Template-based dynamic configuration transformation algorithm

Inputs: TECS, Pattern

Outputs: TECS'

for $j \leftarrow 1: Pattern.depth$ do
    for $k \leftarrow 1 : Pattern.length \times Pattern.width$ do
        $p_{kj} \leftarrow Pattern(k,j);$
        $TECS'(k,j).op \leftarrow TECS(p_{kj}.op);$
        if $TECS(p_{kj}).in\ depends\ on\ LReg\ or\ OutReg$ then
            $TECS'(k,j).in \leftarrow Table_1(p_{kj})(TECS(p_{kj}).in);$
        elseif $TECS(p_{kj}).in\ depends\ on\ OutReg\ of\ other\ PEs$ then
            $TECS'(k,j).in \leftarrow Table_2(p_{kj})(TECS(p_{kj}).in);$
        else
            $TECS'(k,j).in \leftarrow TECS(p_{kj}).in;$
        if $TECS(p_{kj}).out\ outputs\ to\ LReg$ then
            $TECS'(k,j).out \leftarrow Allocate\ RF(new)\ in\ TECS'(k);$
            $Update\ Table_1;$
        elseif $TESC(p_{kj}).out\ outputs\ to\ OutReg$ then
            $TECS'(k,j).out \leftarrow Allocate\ Buffer(new);$
            $Update\ Table_1\ and\ Table_2;$
        else
            $TECS'(k,j).out \leftarrow TECS(p_{kj}).out;$

Fig. 4.62 Template-based dynamic configuration transformation algorithm

of its PEs are also marked on the directed edges between PEs (if there are no edges between two PEs, it means that there is no interconnect between these two PEs or the probability of communication between these two PEs is zero in the static compilation result). For the convenience of later descriptions, the probability of communication between PE_i and PE_j is expressed as $cr(PE_i, PE_j)$. The following five data routing paths exist in the original configuration: (PE_1, PE_2), (PE_2, PE_1), (PE_2, PE_3), (PE_3, PE_4), and (PE_4, PE_2). Figure 4.63b shows the routing configuration based on a folded template that assigns the operators of PE_1 and PE_4 in the original architecture to PE'_1 f the virtual architecture, and the operators of PE_2 and PE_3 to PE'_2 f the virtual architecture. Thus, only (PE_2, PE_3) in the five paths in the original configuration is mapped to the new configuration without incurring additional routing overhead. Therefore, the mathematical expectation of the routing overhead due to the folded template is $E_F = cr(PE_3,PE_4) + 2 \times cr(PE_2, PE_1) = 3$. Figure 4.63(c) shows the route configuration based on an optimized template that assigns the operators of PE_1 and PE_2 in the original architecture to PE'_1 if the virtual architecture, and the operators of PE_3 and PE_4 to PE'_2 of the virtual architecture. Therefore, only (PE_2, PE_3) and (PE_4, PE_2) of the five paths in the original configuration incur additional routing

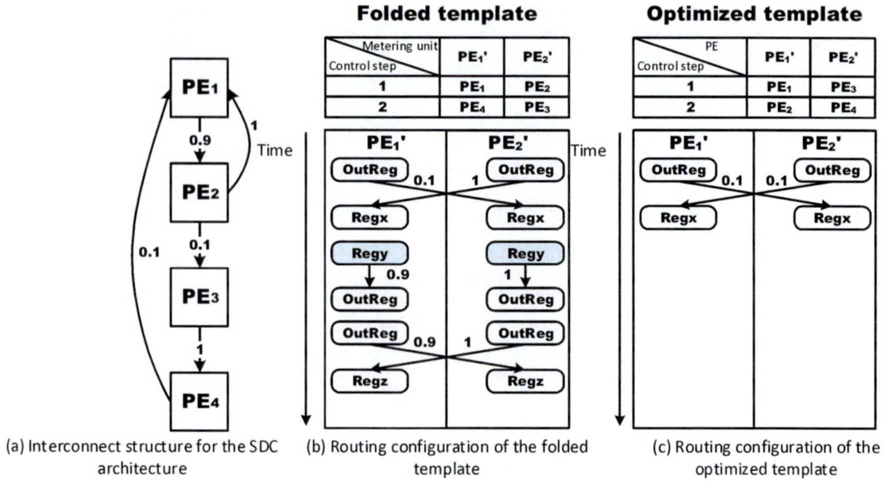

Fig. 4.63 Example of the template-based dynamic configuration transformation

overhead. As a result, the mathematical expectation of the routing overhead due to the optimized template is $E_O = \text{cr}(\text{PE}_2, \text{PE}_4) = 0.1$.

In summary, thanks to the dynamic reconfiguration nature, the SDC is able to support dynamic compilation by hardware virtualization. In the design of a dynamic compilation system for SDCs, in order to generate the configurations online, the most important indicator is the time overhead. Currently, the dynamic compilation techniques of the SDC are commonly include the instruction flow-based configuration dynamic generation and the configuration flow-based configuration dynamic transformation techniques. The instruction flow-based dynamic compilation technique can achieve the purpose of software transparency and high ease of use. In order to meet the real-time requirements of the dynamic compilation, the instruction flow-based dynamic compilation technique often adopts a greedy algorithm with low complexity. However, this approach can hardly analyze the dependencies between operators in the original application online, which can seriously sacrifice the compilation quality. The configuration flow-based dynamic compilation techniques, in essence, can all be summed up as the template-based dynamic compilation. These techniques simply employ some feasible and easily accessible templates, but these templates are not necessarily efficient. While satisfying the condition of low time overhead, it is also important to balance the quality of the dynamic compilation (the performance of the transformed configuration). Therefore, it is recommended to select a template with low communication overhead.

This chapter discusses how to design a compilation system for SDCs from the aspects of static and dynamic compilations, and details the complete compilation process from high-level languages to the configurations of SDC. Starting from the static compilation problem, this chapter first introduces the abstract representation (IR) of hardware and software, then abstracts the mapping into a mathematical model from the perspective of IRs, theoretically analyzes how to solve this mathematical model, and finally elaborates the mapping problem of irregular tasks across basic blocks in detail. Then, this chapter provides an in-depth discussion on the hardware mechanism of the dynamic compilation problem and two approaches to the dynamic compilation of SDCs. As the compilation problem for SDCs can essentially be reduced to an NP-complete problem, it is inevitable to trade off indicators for large compilation problems, making the development of the compilation systems for SDCs difficult. The authors believe that the future research on SDC compilation systems may fall into the following three directions: (1) efficient software and hardware IRs and compact model representations to provide optimization space for compilation, (2) comprehensive quality model for compilation that can accurately measure the quality of static and dynamic compilations, to provide optimization objectives, and (3) co-design of software and hardware to enable software and hardware designs to guide and compromise with each other, and thus optimize the SDC system.

References

1. Aho AV, Lam MS, Sethi R, et al. Compilers: principles, techniques, and tools. 2nd ed. Boston: Addison-Wesley Longman Publishing; 2006.
2. Liu L, Man X, Zhu J, et al. Pattern-based dynamic compilation system for CGRAs with online configuration transformation. IEEE Trans Parallel Distrib Syst. 2020;31(12):2981–94.
3. Rau BR. Iterative modulo scheduling. Int J Parallel Prog. 1996;24(1):3–64.
4. Das S, Martin KJM, Coussy P, et al. Efficient mapping of CDFG onto coarse-grained reconfigurable array architectures. In: Asia and South Pacific design automation conference, 2017. p. 127–32.
5. Mahlke SA, Lin DC, Chen WY, et al. Effective compiler support for predicated execution using the hyperblock. In: International symposium on microarchitecture, 1992. p. 45–54.
6. Lam MS. Software pipelining: an effective scheduling technique for VLIW machines. ACM Sigplan Notices. 1988;23(7):318–28.
7. Hamzeh M, Shrivastava A, Vrudhula S. EPIMap: using epimorphism to map applications on CGRAs. In: Design automation conference, 2012. p. 1284–91.
8. Zhao Z, Sheng W, Wang Q, et al. Towards higher performance and robust compilation for CGRA modulo scheduling. IEEE Trans Parallel Distrib Syst. 2020;31(9):2201–19.
9. Liu F, Ahn H, Beard SR, et al. DynaSpAM: dynamic spatial architecture mapping using out of order instruction schedules. In: International symposium on computer architecture, 2015. p. 541–53.
10. Watkins MA, Nowatzki T, Carno A. Software transparent dynamic binary translation for coarse-grain reconfigurable architectures. In: IEEE International symposium on high performance computer architecture. IEEE; 2016. p. 138–50.
11. Govindaraju V, Ho C, Sankaralingam K. Dynamically specialized datapaths for energy efficient computing. In: International symposium on high performance computer architecture, 2011. p. 503–14.
12. Dave S, Balasubramanian M, Shrivastava A. RAMP: resource-aware mapping for CGRAs. In: Design automation conference, 2018. p. 1–6.
13. Canis A, Choi J, Fort B, et al. From software to accelerators with LegUp high-level synthesis. In: International conference on compilers, architecture and synthesis for embedded systems, 2013. p. 1–9.
14. Budiu M, Goldstein SC. Pegasus: an efficient intermediate representation. Pittsburgh: Carnegie Mellon University; 2002.

15. Izraelevitz A, Koenig J, Li P, et al. Reusability is FIRRTL ground: hardware construction languages, compiler frameworks, and transformations. In: International conference on computer- aided design, 2017. p. 209–16.
16. Wang S, Possignolo RT, Skinner HB, et al. LiveHD: a productive live hardware development flow. IEEE Micro. 2020;40(4):67–75.
17. Bachrach J, Vo H, Richards B, et al. Chisel: Constructing hardware in a Scala embedded language. In: Design automation conference, 2012. P. 1212–21.
18. Sharifian A, Hojabr R, Rahimi N, et al. μIR—An intermediate representation for transforming and optimizing the microarchitecture of application accelerators. In: IEEE/ACM International symposium on microarchitecture, 2019. p. 940–953.
19. Hamzeh M, Shrivastava A, Vrudhula S. REGIMap: Register-aware application mapping on coarse-grained reconfigurable architectures (CGRAs). In: Design automation conference, 2013. p. 1–10.
20. Koeplinger D, Feldman M, Prabhakar R, et al. Spatial: a language and compiler for application accelerators. In: ACM SIGPLAN conference on programming language design and implementation, 2018. p. 296–311.
21. Sujeeth AK, Brown KJ, Lee H, et al. Delite: A compiler architecture for performance-oriented embedded domain-specific languages. ACM Trans Embed Comput Syst. 2014;13(4s):1–25.
22. Wilson RP, French RS, Wilson CS, et al. SUIF: an infrastructure for research on parallelizing and optimizing compilers. ACM SIGPLAN Not. 1994;29(12):31–7.
23. Yin C, Yin S, Liu L, et al. Compiler framework for reconfigurable computing system. In: International conference on communications, circuits and systems, 2009. p. 991–5.
24. Baumgarte V, Ehlers G, May F, et al. PACT XPP: a self-reconfigurable data processing architecture. J Supercomput. 2003;26(2):167–84.
25. Callahan TJ, Hauser JR, Wawrzynek J. The Garp architecture and C compiler. Computer. 2000;33(4):62–9.
26. Lattner C, Adve V. LLVM: a compilation framework for lifelong program analysis & transformation. In: International symposium on code generation and optimization. IEEE Computer Society; 2004. p. 75–86.
27. Chin SA, Sakamoto N, Rui A, et al. CGRA-ME: a unified framework for CGRA modelling and exploration. In: International conference on application-specific systems, architectures and processors, 2017. p. 184–9.
28. Kum KI, Kang J, Sung W. AUTOSCALER for C: an optimizing floating-point to integer C program converter for fixed-point digital signal processors. IEEE Trans Circuits Syst II Analog Digital Signal Process. 2000;47(9):840–8.
29. Ong SW, Kerkiz N, Srijanto B, et al. Automatic mapping of multiple applications to multiple adaptive computing systems. In: IEEE symposium on field-programmable custom computing machines, 2001. p. 10–20.
30. Levi G. A note on the derivation of maximal common subgraphs of two directed or undirected graphs. Calcolo. 1973;9(4):341.
31. Chen L, Mitra T. Graph minor approach for application mapping on CGRAs. ACM Trans Reconfigurable Technol Syst. 2014;7(3):1–25.
32. Mehta G, Patel KK, Parde N, et al. Data-driven mapping using local patterns. IEEE Trans Comput Aided Des Integr Circuits Syst. 2013;32(11):1668–81.
33. Mei B, Vernalde S, Verkest D, et al. DRESC: a retargetable compiler for coarse-grained reconfigurable architectures. In: International conference on field-programmable technology, 2002. p. 166–73.
34. Friedman S, Carroll A, van Essen B, et al. SPR: an architecture-adaptive CGRA mapping tool. In: International symposium on field programmable gate arrays, 2009. p. 191–200.
35. Bouwens F, Berekovic M, Kanstein A, et al. Architectural exploration of the ADRES coarse-grained reconfigurable array. Berlin: Springer; 2007.
36. Chenxi Z, Zhiying W, Li S, et al. Computer architecture. Beijing: Tsinghua University Press; 2009.
37. Stallings W. Computer organization and architecture. New York: Macmillan; 1990.

38. Hennessy JL, Patterson DA. Computer architecture: a quantitative approach. New York: Elsevier; 2011.
39. Trickey H. Flamel: a high-level hardware compiler. IEEE Trans Comput Aided Des Integr Circuits Syst. 1987;6(2):259–69.
40. Davidson S, Landskov D, Shriver BD, et al. Some experiments in local microcode compaction for horizontal machines. IEEE Trans Comput. 1981;7:460–77.
41. Pangrle BM, Gajski DD. State synthesis and connectivity binding for microarchitecture compilation. In: International conference on computer aided design, 1986. p. 210–3.
42. Paulin PG, Knight JP. Force-directed scheduling for the behavioral synthesis of ASICs. IEEE Trans Comput Aided Des Integr Circuits Syst. 1989;8(6):661–79.
43. Rau BR, Glaeser CD. Some scheduling techniques and an easily schedulable horizontal architecture for high performance scientific computing. ACM SIGMICRO Newslett. 1981;12(4):183–98.
44. Park H, Fan K, Kudlur M, et al. Modulo graph embedding: mapping applications onto coarse-grained reconfigurable architectures. In: International conference on compilers, architecture and synthesis for embedded systems, 2006. p. 136–46.
45. Park H, Fan K, Mahlke SA, et al. Edge-centric modulo scheduling for coarse-grained reconfigurable architectures. In: International conference on parallel architectures and compilation techniques, 2008. p. 166–76.
46. Kou M, Gu J, Wei S, et al. TAEM: fast transfer-aware effective loop mapping for heterogeneous resources on CGRA. In: Design automation conference, 2020. p. 1–6.
47. Yin S, Gu J, Liu D, et al. Joint modulo scheduling and VDD assignment for loop mapping on dual-VDD CGRAs. IEEE Trans Comput Aided Des Integr Circuits Syst. 2015;35(9):1475–88.
48. Yin S, Yao X, Liu D, et al. Memory-aware loop mapping on coarse-grained reconfigurable architectures. IEEE Trans Very Large Scale Integr (VLSI) Syst 2015;24(5):1895–908.
49. San Segundo P, Rodr I, Guez-Losada D, et al. An exact bit-parallel algorithm for the maximum clique problem. Comput Oper Res. 2011;38(2):571–81.
50. Yin S, Yao X, Lu T, et al. Conflict-free loop mapping for coarse-grained reconfigurable architecture with multi-bank memory. IEEE Trans Parallel Distrib Syst. 2017;28(9):2471–85.
51. Nowatzki T, Sartin-Tarm M, de Carli L, et al. A general constraint-centric scheduling framework for spatial architectures. In: ACM SIGPLAN conference on programming language design and implementation, 2013. p. 495–506.
52. Gurobi. Gurobi—The fast solver[EB/OL]. [2020-12-25]. https://www.gurobi.com
53. Achterberg T. SCIP: solving constraint integer programs. Math Program Comput. 2009;1(1):1–41.
54. Singh H, Lee MH, Lu G, et al. MorphoSys: an integrated reconfigurable system for data-parallel and computation-intensive applications. IEEE Trans Comput. 2000;49(5):465–81.
55. Lee D, Jo M, Han K, et al. FloRA: coarse-grained reconfigurable architecture with floating-point operation capability. In: International conference on field-programmable technology, 2009. p. 376–9.
56. Liu L, Zhu J, Li Z, et al. A survey of coarse-grained reconfigurable architecture and design: taxonomy, challenges, and applications. ACM Comput Surv. 2019;52(6):1–39.
57. Han K, Ahn J, Choi K. Power-efficient predication techniques for acceleration of control flow execution on CGRA. ACM Trans Archit Code Optim. 2013;10(2):1–25.
58. Han K, Park S, Choi K. State-based full predication for low power coarse-grained reconfigurable architecture. In: Design, automation test in Europe. EDA Consortium; 2012. p. 1367–72.
59. Hamzeh M, Shrivastava A, Vrudhula S. Branch-aware loop mapping on CGRAs. In: Design automation conference, 2014. p. 1–6.
60. Han K, Choi K, Lee J. Compiling control-intensive loops for CGRAs with state-based full predication. In: Design, automation, and test in Europe. EDA Consortium; 2013. p. 1579–82.
61. Anido M L, Paar A, Bagherzadeh N. Improving the operation autonomy of SIMD processing elements by using guarded instructions and pseudo branches. In: Euromicro conference on digital system design, 2002. p. 148–55.

62. Sha J, Song W, Gong Y, et al. Accelerating nested conditionals on CGRA with tag-based full predication method. IEEE Access. 2020;8:109401–10.
63. Hamzeh M, Shrivastava A, Vrudhula S. Branch-aware loop mapping on CGRAs. In: ACM design automation conference, 2014. p. 1–6.
64. Yin S, Zhou P, Liu L, et al. Trigger-centric loop mapping on CGRAs. IEEE Trans Very Large Scale Integr (VLSI) Syst 2016;24(5):1998–2002.
65. Yin S, Lin X, Liu L, et al. Exploiting parallelism of imperfect nested loops on coarse-grained reconfigurable architectures. IEEE Trans Parallel Distrib Syst. 2016;27(11):3199–213.
66. Liu D, Yin S, Liu L, et al. Polyhedral model based mapping optimization of loop nests for CGRAs. In: Design automation conference, 2013. p. 1–8.
67. Yin S, Liu D, Peng Y, et al. Improving nested loop pipelining on coarse-grained reconfigurable architectures. IEEE Trans Very Large Scale Integr (VLSI) Syst 2016; 24(2):507–20.
68. Xue J. Unimodular transformations of non-perfectly nested loops. Parallel Comput. 1997;22(12):1621–45.
69. Hartono A, Baskaran M, Bastoul C, et al. Parametric multi-level tiling of imperfectly nested loops. In: International conference on supercomputing, 2009. p. 147–157.
70. Lee J, Seo S, Lee H, et al. Flattening-based mapping of imperfect loop nests for CGRAs. In: International conference on hardware/software codesign and system synthesis, 2014. p. 1–10.
71. Rong H, Tang Z, Govindarajan R, et al. Single-dimension software pipelining for multidimensional loops. ACM Trans Archit Code Optim. 2007;4(1):7.
72. NVIDIA. NVIDIA Virtual GPU Software [EB/OL]. [2020-12-25]. https://www.nvidia.cn/data-center/virtual-solutions
73. Docker. Why Docker? [EB/OL]. [2020-12-25]. https://www.docker.com/why-docker#/developers
74. Hong C, Spence I, Nikolopoulos DS. GPU virtualization and scheduling methods: a comprehensive survey. ACM Comput Surv (CSUR). 2017;50(3):1–37.
75. Dowty M, Sugerman J. GPU virtualization on VMware's hosted I/O architecture. ACM SIGOPS Oper Syst Rev. 2009;43(3):73–82.
76. Jain AK, Maskell DL, Fahmy SA. Are coarse-grained overlays ready for general purpose application acceleration on FPGAS? In: International conference on dependable, autonomic and secure computing, 2016. p. 586–93.
77. Liu C, Ng H, So HK. QuickDough: a rapid FPGA loop accelerator design framework using soft CGRA overlay. In: International conference on field programmable technology, 2015. p. 56–63.
78. Chiou D. The microsoft catapult project. In: International symposium on workload characterization, 2017. p. 124.
79. Adler M, Fleming KE, Parashar A, et al. Leap scratchpads: automatic memory and cache management for reconfigurable logic. In: ACM/SIGDA International symposium on field programmable gate arrays, 2011. p. 25–8.
80. Kelm JH, Lumetta SS. HybridOS: runtime support for reconfigurable accelerators. In: International ACM/SIGDA symposium on field programmable gate arrays, 2008. p. 212–21
81. So HK, Brodersen RW. Borph: an operating system for fpga-based reconfigurable computers. Berkeley: University of California; 2007.
82. Redaelli F, Santambrogio MD, Memik SO. An ILP formulation for the task graph scheduling problem tailored to bi-dimensional reconfigurable architectures. In: International conference on reconfigurable computing and FPGAs, 2008. p. 97–102.
83. CCIX CONSORTIUM I. CCIX [EB/OL]. [2020-12-25]. https://www.ccixconsortium.com
84. HSA. HSA Foundation ARM, AMD, Imagination, MediaTek, Qualcomm, Samsung, TI[EB/OL]. [2020-12-25]. https://www.hsafoundation.com
85. Goldstein SC, Schmit H, Budiu M, et al. PipeRench: a reconfigurable architecture and compiler. Computer. 2000;33(4):70–7.
86. Eckert M, Meyer D, Haase J, et al. Operating system concepts for reconfigurable computing: review and survey. Int J Reconfigurable Comput 2016:1–11.

87. Brebner G. A virtual hardware operating system for the Xilinx XC6200. In: International workshop on field programmable logic and applications, 1996. p. 327–36.
88. Kelm J, Gelado I, Hwang K, et al. Operating system interfaces: bridging the gap between CPU and FPGA accelerators. Report No. UILU-ENG-06-2219, CRHC-06-13. Washington: Coordinated Science Laboratory; 2006.
89. Ismail A, Shannon L. FUSE: front-end user framework for O/S abstraction of hardware accelerators. In: International symposium on field-programmable custom computing machines, 2011. p. 170–7.
90. Wang Y, Zhou X, Wang L, et al. Spread: a streaming-based partially reconfigurable architecture and programming model. IEEE Trans Very Large Scale Integr (VLSI) Syst 2013;21(12):2179–92.
91. Charitopoulos G, Koidis I, Papadimitriou K, et al. Hardware task scheduling for partially reconfigurable FPGAs. In: International symposium on applied reconfigurable computing, 2015. p. 487–98.
92. Dehon AE, Adams J, DeLorimier M, et al. Design patterns for reconfigurable computing. In: Annual IEEE symposium on field-programmable custom computing machines, 2004. p. 13–23.
93. Govindaraju V, Ho C, Nowatzki T, et al. Dyser: unifying functionality and parallelism specialization for energy-efficient computing. IEEE Micro. 2012;32(5):38–51.
94. Mishra M, Callahan TJ, Chelcea T, et al. Tartan: evaluating spatial computation for whole program execution. ACM SIGARCH Comput Archit News. 2006;34(5):163–74.
95. Kim C, Sethumadhavan S, Govindan MS, et al. Composable lightweight processors. In: International symposium on microarchitecture, 2007. p. 381–94.
96. Pager J, Jeyapaul R, Shrivastava A. A software scheme for multithreading on CGRAs. ACM Trans Embed Comput Syst. 2015;14(1):1–26.
97. Park H, Park Y, Mahlke S. Polymorphic pipeline array: a flexible multicore accelerator with virtualized execution for mobile multimedia applications. In: International symposium on microarchitecture, 2009. p. 370–80.
98. Gschwind M, Altman ER, Sathaye S, et al. Dynamic and transparent binary translation. Computer. 2000;33(3):54–9.
99. Markoff J, Flynn LJ. Apple's next test: get developers to write programs for Intel chips. The New York Times, 2005:C1.
100. Anton C, Mark H, Ray H, et al. FX! 32: a profile-directed binary translator. IEEE Micro. 1998;18(2):56–64.
101. Dehnert JC, Grant BK, Banning JP, et al. The transmeta code morphing software: using speculation, recovery, and adaptive retranslation to address real-life challenges. In: ACM International symposium on code generation and optimization. San Francisco: 2003.
102. Ebcio UG, Lu K, Altman ER. DAISY: dynamic compilation for 100% architectural compatibility. In: Annual International symposium on computer architecture, 1997. p. 26–37.
103. Nair R, Hopkins ME. Exploiting instruction level parallelism in processors by caching scheduled groups. ACM SIGARCH Comput Archit News. 1997;25(2):13–25.
104. Klaiber A. The technology behind crusoe processors. Transmeta Technical Brief, 2010.
105. Clark N, Kudlur M, Park H, et al. Application-specific processing on a general-purpose core via transparent instruction set customization. In: International symposium on microarchitecture, 2004. p. 30–40.
106. Lysecky R, Stitt G, Vahid F. Warp processors. ACM Trans Des Autom Electron Syst (TODAES). 2004;11(3):659–81.
107. Beck ACS, Rutzig MB, Gaydadjiev G, et al. Transparent reconfigurable acceleration for heterogeneous embedded applications. In: Design, automation and test in Europe, 2008. p. 1208–13.
108. Uhrig S, Shehan B, Jahr R, et al. The two-dimensional superscalar GAP processor architecture. Int J Adv Syst Meas. 2010;3(1–2):71–81.
109. Ferreira R, Denver W, Pereira M, et al. A run-time modulo scheduling by using a binary translation mechanism. In: International conference on embedded computer systems: architectures, modeling, and simulation, 2014. p. 75–82.